The AMERICAN FOUNDING

brief edition

second edition

FRANK W. FOX
History Department
Brigham Young University

Pearson Custom Publishing

Cover art: "The Spirit of Our Forefathers" by Jessica Hewklit-Ridd.

Printed in the United States of America

10 9 8 7 6 5 4 3 2 1

Please visit our web site at www.pearsoncustom.com

ISBN 0–536–63269–3

BA 993181

PEARSON CUSTOM PUBLISHING
75 Arlington Street, Suite 300, Boston, MA 02116
A Pearson Education Company

This book is dedicated
to the memory of my grandfather,
Frank M. Openshaw,
who cared deeply about
the American Founding.

CONTENTS

3

American Republicanism 58

4

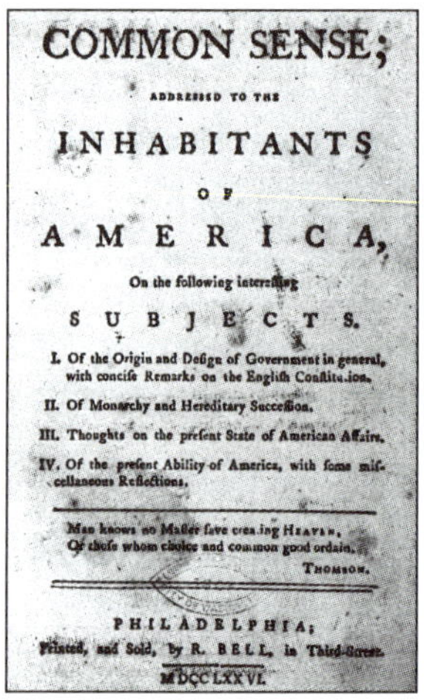

COMMON SENSE;
ADDRESSED TO THE
INHABITANTS
OF
AMERICA,
On the following interesting
SUBJECTS.
I. Of the Origin and Design of Government in general, with concise Remarks on the English Constitution.
II. Of Monarchy and Hereditary Succession.
III. Thoughts on the present State of American Affairs.
IV. Of the present Ability of America, with some miscellaneous Reflections.
Man knows no Master save creating Heaven,
Or those whom choice and common good ordain.
THOMSON.
PHILADELPHIA;
Printed, and Sold, by R. BELL, in Third-Street.
MDCCLXXVI.

The American Revolution 88

5

The Confederation and the Rule of Law 120

6

The Constitutional Convention I: The Framers **152**

7

The Constitutional Convention II: Compromise and Creativity **178**

8

The Constitutional Convention III: Structure and Dynamics **206**

9

The Meaning of the Constitution 240

10

The Ratification Campaign 270

11

The Bill of Rights 294

12

From the Constitution to the Republic 320

Conclusion: The Unfinished Founding 354

Appendices 373

Message from a former student and T.A.

One of the meanings of the word "heritage" is a gift passed down from one generation to another. This course is about a gift most people who have lived on the earth have never received—the gift of living in a stable, orderly, free society. We have been given that heritage from our nation's Founders. This class will help you better appreciate and understand America's heritage. But more importantly, if this class "works," you will be motivated to preserve it.

You will be rewarded for your hard work in this course. As you study America's story you will sense the reality that the future of America is whatever the American people want it to be. If America is to have a bright future we must make it happen. The material you study here will help you see how.

As a former student and teaching assistant I can tell you that no matter what your major or year in college, you can do very well in this class. You will have an excellent instructor. I predict that you will enjoy the lectures in this class as much as those in any other. While the class size is large, you will have a teaching assistant—a fellow student who has done well in the course—to review the material with you.

Now, a fair number of students do not earn a good grade in this class. You cannot simply attend most of the lectures, take casual notes, go to half of your review labs, do a little of the reading and get an "A." I have never seen it happen. This is not a fluff class. The students I have seen succeed have taken careful notes in lectures, read the assigned material, and come to their review labs ready to discuss and examine the material. Anyone can do it. The students who don't "get" the material probably haven't paid the price to understand it. Paying the price now will be a very rewarding experience.

Good luck!

ACKNOWLEDGMENTS

This book evolved over a period of some twenty years in the teaching of a course titled "American Heritage" at Brigham Young University. The course seeks to combine insights and methodologies from history, political science and economics in order to give students a way of understanding their own past and, more especially, their own institutions. Because the course is interdisciplinary, it has drawn on an extended group of faculty in its design and pedagogy. The following constitute only the primary contributions.

In the history department, James B. Allen and Thomas G. Alexander pioneered course development. They were joined later on by Mark Grandstaff, Gary Daynes and Susan Rugh. In political science, Martin B. Hickman, Noel B. Reynolds, and Louis Midgley led the way, to be followed by Ralph Hancock, Richard Vetterli, Brett Latimer and Matthew Holland. In economics, Clayne L. Pope and Larry Wimmer took part in the early discussions and helped establish American Heritage at the core of the university's general education curriculum.

Several teaching assistants have also borne a hand, as the exercises found at the end of each chapter will indicate. A team consisting of Rebecca Johnston, Bryan Norton and Christopher Karpowitz developed a host of teaching materials early on, and their work was recently enhanced by another team consisting of Karen Reeder, Ben Ramsey and Tom Lufkin. Their efforts were ably directed by Joseph Darowski.

Virginia Rush conducted a competition among BYU art students which resulted in the cover art and several other visuals. Adam Darowski wrote the study guides found at the end of each chapter.

Joseph and Kay Darowski undertook meticulous copy editing and assisted in virtually every detail of preparing the manuscript for publication, and Neil L. York read through it to spot errors of fact and make valuable suggestions. Without the help of these three individuals, the project would never have come to fruition.

Introduction

One of the most curious aspects of the American Founding—a story filled with curious aspects—is the lack of interest it commonly merits from American college students. Even the most patriotic of them often fail to find the subject riveting. I suspect the reason for this is either that they accept the Founding as a given—like the Rocky Mountains or the Mississippi River—or else that they regard it as an impenetrable mystery. Either way, why try to figure it out?

Accordingly, the purpose of this book is twofold. First, we want to see if we can shed some light on the mystery. And second, we want to show that there was nothing "given" about it. Real people had to take chances, make tough decisions, and try to come to grips with enormously difficult issues in order to make the Founding work. It was a dicey business all around, and for that reason alone an interesting one.

For it is the Founding, at bottom, that makes *America* interesting. Everything American, from Hollywood movies to computer software, seems to merit fascination in the world of the twenty-first century, but a rather special place is reserved for American constitutional forms. They were, after all, the first of their kind in the modern age, and beyond that they have proven to be the most stable and longest lasting. With the fall of communism in Europe, the interest in American institutions has increased. Peoples struggling toward nationhood in various parts of the world are asking anew questions about "founding" and are not infrequently looking to the United States for answers.

The Human Predicament

The story does not begin in 1776 with the ringing of the Liberty Bell. Indeed, it is as old as the human story—another reason it is so interesting—for it addresses a problem we can trace back to the very beginning. The Greeks knew the problem. So did the Romans. The Turks, the Chinese, the ancient inhabitants of the Americas, all of them ran up against it in one way or another and could give us an eloquent introduction to it. So, let's have one of them do so.

When the historian Thucydides looked back upon his world, the world of the ancient Greek republics, the sight he saw was little short of horrifying:

> During the few days that Eurymedon, with his troops, continued at Corcyra, the people of that city extended the massacre to all whom they judged their enemies. The crime alleged was their attempt to overturn the democracy. Some perished merely through private enmity; some, for the money they had lent, by the hands of the borrower. Every kind of death, every dreadful act, was perpetrated. Fathers slew their children; some were dragged from altars, some were butchered at them; numbers, immersed in temples, were starved. The contagion spread through the whole extent of Greece. . . .

Thucydides' observations in this quotation are paraphrased by none other than John Adams, who was combing through the ancient record for lessons he could apply to his own time and place. As he read Thucydides and many others, Adams concluded that humankind has had to put up with a good deal of such pain and suffering. History strongly suggests, in fact, that the overwhelming majority of human beings have had to choose between two exceptionally unhappy alternatives—tyranny and anarchy:

> ***Tyranny***. We often think of tyrants as cruel despots, and many of them have certainly lived up to the reputation. But the truly essential feature of tyranny is neither cruelty nor despotism—it is simple human will. *Tyranny is the rule of will.* It is possible to imagine a "benevolent tyrant" such as Louis XIV, who does what he can to improve his people's lot, but even here it is the tyrant's will and that alone which determines life and death outcomes. With the rule of will, those who are ruled have to *hope* for benevolence on the part of the tyrant, for they literally have no other safeguards.
>
> ***Anarchy***. If tyranny is the rule of will, *anarchy is the rule of chaos.* Where the tyrant gathers all power unto himself, anarchy disperses it among the multitude, with the effect that every individual can exercise his *own* will. For a picture of the result, think of any good movie you have seen about the French Revolution, or the Russian, or the Chinese. Mobs roam the streets, burning and pillaging. Private armies crisscross the land with swaths of destruction. It is Thucydides' description of Corcyra over and over again.

Tyranny and anarchy often have a circular relationship, each leading around to the other. A tyrant, let us say, becomes so detested by the people he rules that they begin to plot his destruction. Sooner or later a revolt breaks out and, after a few thousand partisans have lost their lives, the tyrant is overthrown. But then, who will replace him? His opponents resolve themselves into various factions, each based on a separate creed or interest or ethnic identity, and the factions commence a bloody free-for-all. If none of the contenders proves capable of dominating its rivals, the anarchy goes on and on. And if one does manage to gain a whip hand, its leader becomes the new tyrant.

And so we come to what might be called the *human predicament.* It is not universal nor entirely without exceptions, but most human beings throughout most of history have essentially had to make a choice between tyranny, on the one hand, and anarchy, on the other. They have suffered under kings, lords, sultans, Caesars, pharaohs, moguls, satraps, chieftains, dictators, masters and a hundred other designations for tyrant, taking what comfort they could in the fact that they at least had order in their lives; or they have thrown in with rebels, insurrectos, renegades, rogues, adventurers, dissidents, freedom fighters, private armies, and plotters of palace coups, hoping against hope that they could somehow ride the tiger.

The Good Society

With the human predicament in mind, we might note with small wonder that the longing for a "Good Society" has an equally long history. *The* ***Good Society*** *is one that escapes the human predicament.* It has had many different names over the centuries. Plato referred to it as a republic. In John Locke's day it was known as a commonwealth. John Adams used the term free government. We would probably call it a liberal democracy today. And as the terms are all different, so too are the various conceptions of the Good Society that political philosophers have advanced. Still, most of them have the following three attributes in common. The Good Society:

- Extends the benefits of social interaction beyond a single favored group or class to society as a whole.
- Strives for a reasonable balance between liberty—defined as the right of individuals to seek their own happiness in their own way—and order.
- Offers its citizens a variety of "goods of the soul"—including such things as political participation, a sense of individual empowerment, and a sense of human dignity.

In sum, the Good Society gestures toward human good in meaningful ways. We acknowledge it whenever we talk about "truth" or "justice" in public affairs, or conversely whenever we point up "wrongs" and demand they be righted. In our own history, the Good Society was under discussion in Boston when the tea ships landed; and in Lincoln's Cabinet when he pondered freeing the slaves; and on Omaha Beach in Normandy when thousands laid down their lives for a just cause. On the other side, our recollections of Indian wars, or race riots, or the incarceration of Japanese Americans during World War II often give us a contrary feeling. Like the judge's famous definition of obscenity, we know the Good Society when we see it.

The ideal of the Good Society has been enshrined in Plato's *Republic*, in Locke's *Second Treatise on Government*, in Rousseau's *Social Contract*, in Montesquieu's *Spirit of the Laws*, and in many other works of political philosophy. Indeed, whenever mortals have faced the horrors of the human predicament, the thoughtful among them have supposed that somehow there must be a better way.

USA

It is not the purpose of this book to argue that the United States of America is the world's first, best or only Good Society. Patriots, of course, have maintained that it is all of the above, while critics have denied that it is a Good Society at all. Our purpose is to study the American Founding as a way of understanding America itself, and from that perspective I would like to submit that America—and I use that term throughout as a reference to the United States of America—is the world's most self-conscious Good Society. That is to say, Americans, probably more than any other people, think of their nation-state in relation to the rest of the world, to universal truth, and to the experience of mankind.

The American Founders themselves are a good illustration. If the center of their world was a point on the map somewhere between London and Paris, they were living on the far-flung periphery of that world, socially, politically, economically and culturally. They had every reason to regard themselves as the most provincial of provincials. Yet as we read through their correspondence, the records of their deliberations, their explanations to the world, and most of all the founding documents themselves, we don't see them behaving like provincials at all—on the contrary, they spoke and acted as if they were the children of the ages. They laced their discussions with references to classical antiquity. They discoursed on their relationship to Nature (with a capital "N") and first principles. They spoke of their land in terms that reflected Eden and their hopes for the future in terms that reflected Utopia. They discussed the European Enlightenment as though its real purpose had been to provide background for their own undertaking. Something seems to have convinced them of their own surpassing specialness—which may be one reason why we look back on them and see demigods.

There was, however, another side to their experience, and the two are often difficult to reconcile. If the Founders had one foot in the realm of theory, the other foot was in the realm of practice. Virtually every one of them were practical men of affairs, not cloud-bound philosophers, and throughout the events this book describes, they were harried, overworked, caught up in events which seemed anything but clear, and wondering what on earth to try next. The consequence of this double-sidedness couldn't be more striking. Yes, the American Founders really were thinking about the Good Society and their own providential role in bringing it forth, and, yes, at the same time they were thinking about the nuts and bolts of actually building one.

Accordingly, what we will look for in the story of the American Founding are two interwoven strands. In one of them we will catch glimpses of an ideal world, the sort that Plato wrote about in the *Republic*, a world of truth and justice and human dignity. In the other we will see something more like, for want of a better term, "Peoria, Illinois." Politicos used to express doubt by saying: "Fine, but how will it play with the little old lady in Peoria?"—meaning how will it work in the real world as we know it? In the real world as we know it, it may not be possible to bring forth perfect justice, sublime happiness, total freedom, and unbounded prosperity for everyone—the best to hope for is considerable justice, a chance for happiness, conditional freedom, and a growing prosperity for most. Keep an eye on Utopia, in other words, but keep an eye on Peoria too.

The American Founding, then, was at once a stroke of extraordinary brilliance, approaching the very sublime of human achievement, and at the same time what a group of politicians was willing to settle for under the circumstances. Probe it as we may, it will partly remain a mystery. No single individual, no single point of view can be credited with the final accomplishment. There was a broad variety of contributions, some reflecting one idea and some another, some representing bold flights of theory and some meek acknowledgments of experience, often joined together in surprising ways. The resulting congruities and incongruities make the study of this exceptional phenomenon worthy of our attention. And as we sort through the tangle, it may help us to grapple with the deepest mystery of all—why the American Founding worked.

CHAPTER 1

The Art and Science of Founding

Not every political society can look back to a "founding" as such. Most, in fact, came into being as a result of military conquest, historical happenstance, or likeliest of all, some process of gradual evolution. But most attempts to establish a Good Society have included a founding, for the very idea of a Good Society implies conscious human intention.

This is not to say that there must always be a dramatic "founding moment"—a point in time when a document is signed and the bells are rung. Americans in particular like to celebrate such moments. There was July 4, 1776, when the Declaration of Independence was signed and, yes, the bells did ring. Many scholars would prefer September 17, 1787, when the Constitution was signed, or possibly June 21, 1788, when the ninth state ratified it. Thomas Jefferson thought of his election as President in 1800 as the founding moment—the "Second American Revolution"—and a number of his supporters agreed. Still other Americans pointed out that the United States as we know it today was essentially a product of the Civil War. It could be argued, in other words, that the American Founding is an episodic *re*founding.

Even so, we can still step back and visualize a founding moment. At some point in time the crucial elements really did come together and the political society we know as the United States of America experienced its birth. To discover when that moment actually occurred, and what actually transpired, might be to discover what the United States actually is.

Elements of a Founding

Founding: See study aid.

Most foundings include several elements. First, there is usually a specific name of the polity in question and perhaps a rough description of its geography. Second, there is generally a system of fundamental laws together with a sketch (at the very least) of an essential political order. Of the laws we will have more to say further on. The political order sets up a governmental structure of some kind and may also describe its operation. We use the word ***constitution*** to refer to either or both of the above, sometimes rather loosely. A constitution can exist as a set of unspoken agreements—a broad understanding of the rules of the game—or it can be meticulously spelled out in writing like a legal document.

The third element of a founding concerns citizenship. *Who rules*? is the way the ancients used to put the question, and the answers were never simple. In the modern world we generally resolve *who rules* into two separate questions:

> Who participates in the political process and why?
>
> Who is excluded and why?

The answers can be quite revealing. For example, Aristotle argued that women, children and slaves could never take part in the political process because they were by nature the "ruled part" of humanity rather than the "ruling part"—made to take orders. In our own time, we have denied participation to African Americans for reasons less straightforward. Asking the citizenship question is a way of asking what a given society is all about.

Polity: A group of people who organize themselves into a political society.

The fourth element of a founding is what we might call a sense of "**polity**" or "nationhood." That is, citizens feel themselves to be part of a people, an organic whole, not just a political organization. As a people, they will probably have distinguishing manners and mores, a "cultural identity," which may affect their political life in subtle ways. They might also have some sort of moral consensus—a fundamental agreement about right and wrong—binding them together.

Finally, most foundings are rooted in a "**founding myth**." The myth may be as romantic as an ancient hero pulling a sword from a stone, or as prosaic as a group of founders praying over some burdensome issue before resolving it. In either case, the people need to feel that the founding is somehow sanctioned by the gods, by nature, by truth and justice, or by history. The founding myth imparts a sense of legitimacy.

One might ask how all of these elements are supposed to come together. How does a group of people suddenly begin to see themselves as *a* people, find that they share widespread agreement on fundamentals, work out a system of laws and a structure of government, resolve the question of citizenship, and come to believe that all of this is the result of divine purpose? Or in other words, why do foundings happen at all? To pose the question, and to think carefully about the answer, is to conclude that foundings, and especially foundings aiming at a Good Society, must be among the strangest and most improbable of human achievements.

Ancient Foundings

The foundings of ancient polities suggest that, in the old days at least, the mythical element was the key. For once the gods had made their determination, all else fell into place. Thus, Yahweh wrote Israel's law on stone tablets and entrusted them to Moses. Jupiter dictated the laws of Crete to his own sons, Minos and Rhadamanthus, who went on to found Minoan civilization. The laws of Lacedaemon were given to Lycurgus by Apollo. And the ancient Romans could thank Numa Pompilius and the nymph Egeria for their ceremonial laws and religious rites.

These cases also illustrate the crucial role of law in the founding process. The fact that the law was handed down by the gods themselves took care of the legitimacy problem. The structure and operation of government were implied in the law if not openly stated in it. And those who recognized and obeyed the law automatically became *the people*.

Equally crucial was the role of the "**Lawgiver**." When he was recognized as a divine or supernatural figure, that fact alone endowed the law with a special sanctity. In time, the Lawgiver became both mortalized and pluralized. He was ever less likely to be a mythical hero and ever more likely to be a statesman, a jurist, occasionally even a philosopher. And the edicts of a single individual were gradually replaced by the deliberations of a small body reasoning together. The "Lawgiver" became the "Founder."

Examples of a **Lawgiver:** Moses for the Israelites, Hammurabi for Babylon, Solon for ancient Athens.

As the founding process became humanized, founders tried to account for such ineffable things as moral truth, human dignity, even public happiness. They drew upon past experience, future possibility, custom and tradition, public values, religious perceptions and the unique genius of the people. With the ever greater importance of law, foundings became less a matter of fire and blood and more a matter of "justice." The idea of power associated with tyranny was gradually replaced by the idea of authority. In the very concept of a founding, we begin to see outlines of the Good Society.

Revolution and Founding

Foundings rarely just happen. They usually result from some sort of political upheaval, such as a revolution, and the consequent need to rethink fundamentals.

Revolutions occur because large numbers in a given society become fed up. It has happened to relatively old peoples—the French, the English, the Russians, the Chinese—and relatively new ones. Revolution can also bring a people into being and give them a sense of themselves *as* a people.

Not infrequently, revolution takes the form of a cathartic "rebirth." A society suffering under tyrannical oppression builds steadily in its discontent until eventually something blows. Then, in a burst of energy, the people come together and throw off their shackles, tasting freedom for the first time. In the heady atmosphere of success, the revolutionists often consider making extreme changes. They may seek:

Utopia: Book by Sir Thomas More published in 1516 describing an imaginary perfect society.

> *A return to "Eden."* That is, they remember (or perhaps imagine) some earlier condition of happiness to which they now wish to return. In England's Glorious Revolution, for instance, many of the revolutionists looked back to what they fondly recalled as the lost Saxon democracy.

> *An advance toward* ***Utopia****.* Perhaps the past was never that good—but the chance has come to right all wrongs, eradicate injustice and create the world as it ought to be. Thomas More's *Utopia* painted an alluring picture of such a world, and it became the theme of the Bolshevik Revolution in Russia, among others.

Thomas More/Utopia: www.orst.edu/instruct/phl302/philosophers/more.html

Just as there are alternative ways to see revolutions, there are alternatives ways to see foundings. Revolutionists who seek a return to Eden will regard the act of founding conservatively, drawing upon the inspiration of the past. This was the way the world used to be, they will say, and we liked it a lot better. They will value experience, tradition, established institutions, even though these may be flawed or may have been perverted of late. Conversely, revolutionists who look to an ultimate Utopia will regard their founding radically, seeking to establish a new and unprecedented order. They will tend to invoke reason rather than experience, will rely on the human imagination, and may advance a bold new revolutionary doctrine, such as "Marxism-Leninism."

There is a third possibility too. A founding could also be rooted in the "**laws of nature**." Belief in such laws goes way back. They are understood to be the moral equivalent of physical laws, such as the law of gravity. By nature, one might say, it is wrong for one person to steal property from another or to harm another without justifiable cause. Philosophers who have emphasized natural law, such as Cicero or Thomas Aquinas, argue that it is accessible to all human beings through common intuition, or "natural reason," and thus that it inspires many of the black-letter laws passed by legislatures. A political society founded on natural law principles might be neither radical nor conservative in the ways we have described—and yet it might also be both.

While revolutions typically precede foundings and make them possible, all revolutions do not lead to foundings, certainly not to successful ones. Indeed, most revolutions lead to the episodic chaos and tyranny described above. In his study of the ancient republics, John Adams listed massacre after massacre (and gave the death toll for each) resulting from revolutionary disturbances. Any reader of the newspapers today could more or less do the same.

Accordingly, one of the intriguing questions of human history is why some revolutions "succeed"—that is, lead to a successful founding—while others fail. Those committed to establishing a Good Society have been especially interested in this question. But they have found few hard answers. In the course of the 1600s, England experienced two revolutions: one of them resulted in a bloody civil war with an inconclusive outcome, and the other led to a successful founding. Here is another puzzler. The American and French Revolutions occurred within a few years of each other and were propelled by many of the same ideas. The French Revolution resulted in years of confusion, then the Terror, then Napoleon, then the restoration of the monarchy, then two more revolutions, another empire, several rocky attempts at establishing a stable democracy, and finally a world war. The American Revolution resulted in the USA.

The Durable Founding

While many lessons can be drawn from such comparisons, one stands out. For a revolution to result in a successful founding, the people themselves have to be generally satisfied with its outcome. If the satisfied include only a small group or a single class, everyone else will continue to perceive injustice, and sooner or later there will be more turmoil.

Skulls of Khmers from Cambodian Killing Fields, circa 1980. Not all revolutions lead to the founding of a good society.

Epistemology

With the broad mass of the people in mind, we can think of foundings in one of two ways:

Universal. As the term suggests, a ***universal* founding** is correct for all people at all times and in all situations—"what ought to be everywhere," in the words of Condorcet. Universalism implies that the human animal is of a common type, and if the right means are employed, we could take a random sampling of that type and work them into a successful political society. So thought the Russian revolutionists, in any case. After what appeared to be the triumphant founding of the Soviet Union in 1917, they confidently predicted that Europe, Asia, the Americas, and eventually the entire world must follow suit.

Particular. Here the assumptions are just opposite. A ***particular* founding** is correct only at a given time, in a given place and for a given people. It might not be right for any other. When we say, in common speech, that something "works for me," we are using the particularistic mode of

Universal/Particular: See study aid.

Condorcet: French mathematician, philosopher, and politician (1753–1794, corresponded with John Adams).

Condorcet: www.britannica.com/bcom/eb/article/8/0,25538=1,00.html

> thinking. A particularist would point out that Soviet Communism did not take the world by storm after all. It was adopted by others only at gunpoint and not very successfully at that.

Each of these alternatives requires—in the words of political scientist Bradley Thompson—its own "constitutional epistemology." This formidable phrase simply means that the kind of reasoning employed by the founders will be very different in the two cases. Universalism will emphasize abstract, deductive, *a priori* reasoning. The founders will say, in effect: *The human animal being thus and such will accordingly require this, that and the other to live in a successful political society*. We can see how the ancient lawgivers were regarded as prophets, for they spoke with enormous confidence and seemed to have all the right answers beforehand.

Particularism, on the other hand, will emphasize empirical, inductive, *a posteriori* reasoning. In other words, the founders will examine the established facts of a particular political situation, possibly of other similar situations, and will ask empirical rather than logical questions:

> What has worked best for us?
>
> What seems to work for those most like us?
>
> What specific conclusions can we draw from this information?

There will be little abstract reasoning here and very little self-certainty. Conclusions will be limited and partial, often based on past mistakes, and made only after a great deal of discussion. Particularism, moreover, will emphasize things like "the character of the people," and will assume that such character is at least as important to a founding as any formal political arrangements.

Both types participate in most foundings, including the American. But in the American case it is the particularists who prevail. True, some of the American Founders liked to make blank assertions about politics and humanity, and their *Novus Ordo Seclorum* (New Order of the Ages), which is still visible on the dollar bill, was sweepingly universal. In the main, though, the Americans combed back over their English constitutional heritage and asked each other how this or that feature had worked (or perhaps failed to work) in actual practice. They were adapting traditional verities to new circumstances.

Stability

A durable founding also requires a stable polity. Two societies founded on similar ideals may have very different political arrangements, and one of them might turn out to be inherently more stable than the other. For a simple analogy, think about one pretty girl and two good-looking guys on a date—an inherently *un*stable arrangement. Under the Articles of Confederation, the American states had much in common and much to bind them together, but they had not achieved a stable political order, and as a result they experienced a good deal of grief. They had a founding of sorts but not a durable one.

Aristotle was especially interested in the durability problem, which is a very old one. He undertook a comparative analysis of the constitutional arrangements common in ancient Greece, noting what factors seemed to promote stability or instability in each. He had essentially three models to choose from:

Ancient Greece: http://eawc.evansville.edu/grpage.html

See study aid.

> *Monarchy*. The powers of government are concentrated in the person of a hereditary ruler. Aristotle called it "the rule of one."
>
> *Aristocracy*. The powers of government are concentrated in the elite of society, the wisest, best educated and most virtuous. For Aristotle, "the rule of the few."
>
> *Democracy*. The powers of government are exercised by all citizens. "The rule of the many."

Aristotle noted the tendency of each general model toward its own kind of instability and its own form of corruption. Monarchy often degenerated into **despotism**. Aristocracy could turn into **oligarchy**. And democracy easily fell into plain chaos. Yet there were strategies for strengthening each of the three so as to promote stability and integrity. Any of them, suitably modified, could become the foundation of a *republic*—the ancient world's term for a Good Society.

Despotism: Oppressive rule by a single individual.

Oligarchy: Oppressive rule by an elite few.

In the early 1700s, Aristotle's study was updated by a French political scientist named Montesquieu. His compendium, titled *Spirit of the Laws*, examined the structure and operation of governments all over the world, comparing and contrasting a variety of features. Montesquieu claimed to have no special favorites, but it was clear enough from reading between the lines that he admired many features of the English constitution. England had always had *government*, perhaps, but after the Glorious Revolution of 1688 and the established supremacy of Parliament, it had something that went deeper than government—a true *founding*.

Montesquieu: www.fordham.edu/halsall/mod/monstesquieu-spirit.html

Glorious Revolution: www.bartleby.com/65/sl/Glorious.html

America's two principal lawgivers, John Adams and James Madison, made studies of their own. Adams began his during the American Revolution. He examined no fewer than fifty republics, ancient and modern, asking himself over and over what it was that made them tick. Why did a few of them—a *precious* few—prove to be relatively durable, while the rest more or less self-destructed? he wondered. James Madison conducted his search through history when it became apparent to him that the American Confederation was foundering. Why did some constitutions work so well, he asked, while others, outwardly similar, failed so badly? Madison had an additional spur to his curiosity. He found himself in a veritable laboratory of comparative government, observing thirteen different republics in action, and once again he could see marked differences in their performance: some of the state governments appeared to be sailing along serenely, while others were heading toward the shoals. Why?

All of these observers, ancient and modern, drew some general conclusions. The first was that the durable constitution needed to be "mixed" or "balanced" in some way, so that the various orders in a given society might each have a say in public discourse, might each defend itself against the others, and above all might each check the drift of power into dangerous places. For it was when power began to drift that the difficulties occurred.

Their second conclusion was that the durable constitution needed to pay careful attention to the question of *who rules*. Given that all groups had a certain voice in public affairs, whose voice ought to be loudest and heard most clearly? For Aristotle it was the middle classes. For they, he argued, were the most accountable of all the contenders and the least likely to go to extremes. As property owners and taxpayers, the middle classes had a stake in society. They also embraced cohesive institutions such as family and local community, and had a deeper regard for religion and morality. Both Adams and Madison reached similar conclusions. They distrusted the very poor for their lack of personal independence, and the very rich for their "aristocratical passions"—love of gold, love of praise, love of power—and the influence these must have on politics.

See study aid.

The third conclusion was that the "**rule of law**" was all-important. We will learn later on precisely what was meant by those three simple-sounding words. For now, it suffices to repeat a phrase often used by the American Founders—"government of laws and not men"—which meant that people must be governed by general rules, not particular directives. It also meant that citizens had to revere the law as they would revere something sacred.

In time there would emerge an entire science of creating durable constitutions. It would include some very subtle considerations:

> How much power can a government safely be given? How much is too much? How much is not enough?
>
> How to apportion that power so that its possessors will tend to respond appropriately? What conditions would prompt a public official to behave better than he otherwise might? Or worse?
>
> How flexible should a constitution be? Who should be able to change it and under what circumstances?
>
> How to create a sense of legitimacy in the constitution? How to endow it with the mysterium of those ancient mythical foundings?

All of this was important. However, the single *most* important factor in creating the durable constitution was to base it on a correct understanding of human nature. A government of angels, for angels, would be easy to contrive, Madison noted dryly—the trick was devising a government for men.

Foundings and Human Nature

For the ancients, there was a close tie between human nature and politics. The *polis*, which was the primary political unit of the ancient world, was thought to be much like an individual person, with strengths and weaknesses, likes and dislikes, an inclination toward certain kinds of endeavor and a disinclination toward others. In a manner of speaking, the soul of the *polis* was the individual soul writ large.

This belief had two implications. First, any given *polis* would presumably reflect the moral and spiritual qualities of its citizens. Thus Sparta, which was often aggressive and warlike, was inhabited by aggressive and warlike people, and its constitution bore the markings of a military dictatorship; while Athens, a lover of the muses, not only sponsored a wide variety of activities, it favored the inclusiveness of democracy.

Second, a good political order—as opposed to a bad one—might have the capacity not only to *reflect* human nature but to *improve* it. The argument goes something like this. The human animal is what it is by virtue of learned behavior. If children are exposed only to positive influences, they will develop accordingly. So in a society focused on truth, justice, honor, dignity and concern for one's fellowman, we are much more likely to see those qualities reproduced. Plato went so far as to suggest that the ideal republic be guided solely by philosophers like himself.

These assumptions led to a level of political involvement we would hardly recognize today. In some of the ancient republics, citizens found highest meaning in political activity, the sort of meaning we might look for in professional careers or family life or even religion. The active shaping of one's own world—and thus one's own character—by means of the political process, with open horizons of possibility, we would associate with the most utopian of revolutionaries. It was what led Aristotle to observe that man was an "animal made for political activity"—a ***zoon politikon***.

In what specific ways did the *zoon politikon* (political animal) link politics to human nature? The question is as relevant to our own time as it was to antiquity, and the American Founders pored over it as thoughtfully as the ancients did:

Tribalism

Human nature is "tribal" in the sense that *Homo sapiens* apparently want to live among their own kind. Within limits, we wish to act alike, dress alike, talk alike, and see whatever movie is drawing the biggest crowds. The same (except for the movie) was true for the ancients. Societies have always been knit together by culture.

Tribalism creates a sense of community and common purpose—a sense of "we." It is the taproot of what we call patriotism. But it divides as well as unites. Those who cannot be a part of *our* tribe must necessarily be of some other, and this fact points up one of the fundamental problems of political society. How is the "other," the outsider, to be regarded? There were essentially only two answers for the ancients. Outsiders must either be excluded altogether or else included only as servants or slaves.

The problem of otherness would not be made easier in the modern world, especially in America. The bringing of various peoples together would create language differences, religious differences, ethnic differences, and racial differences to deal with. No political society, apparently, could escape the homogenizing tendency of tribalism, nor could a political founding contemplate the inclusion of outsiders. Yet could a Good Society include only insiders?

Reason

All human beings had a capacity to reason, the ancients observed, but some had more than others. Accordingly, philosophers of antiquity often focused on "the rational element" in society. Who were they? How could they be identified? How could their numbers and quality be increased? How could their influence best be brought to bear on public issues?

Closely tied to reason was education. The assumption was that learning made a vast difference in the promotion of rationality. Give human beings the best possible education and you will make them into the best possible citizens. It was for this reason, broadly considered, that Plato established his Academy and Aristotle his Lyceum, and that Socrates posed his mind-bending interrogatives.

Also tied to reason was something we might call "creativity." Educated men, employing reason, could surely create their own political world, it was thought, not just inherit one from the past. Why remain content with the *polis* as it was when human creativity might be able to improve it, pare away its flaws and enhance its ability to perfect human nature? Platonic ideals such as truth, justice and the good often came into play when creativity was under discussion, and politics itself became a kind of spiritual undertaking.

Some of the ancients trusted reason more than others. Aristotle, who ranked among the doubters, pointed out that pure reason, disconnected from everyday realities, could lead political theory down some dangerous paths. In the modern world we have come to appreciate Aristotle's skepticism. Viable foundings have rarely been based solely on reason. Still, who would consider eliminating reason altogether?

Passions

The ancients concluded that "passions" such as greed, lust, jealousy, anger, fear and hatred arose from a different part of the mind than reason did. And just as some elements in society were more given to rational behavior, others were more given to emotionalism. The passions led people to do unpredictable things—things they might later regret. Moreover, passion was contagious. A rousing orator could stir up a crowd and soon have it ready to do something rash.

In politics, passion often led to turmoil. The peaceful life of the *polis* could be thrown into confusion by impassioned behavior, and war between friendly states could precipitate the same way. The lesson seemed to be that the passions must be held in check at all cost. But if they were truly a part of human nature, what, precisely, could be used to check them?

The most troubling of the passions, as the ancients saw it, was "desire." Human beings, it seemed, had a covetous affection for exotic foods, rare wines, fine cloth and glittering jewelry. They sought wealth not only for its own sake but for the prestige it often conferred. And they certainly preferred a life of ease to one of dull drudgery. The very leisure to engage in politics was often a product of fortune.

But desire had harmful effects. For one thing, it drew citizens away from public life and into private enjoyment. For another, it had a caustic influence on the political virtues. How could a person develop a true sense of justice, for example, if his own appetites got in the way? Desire often exacerbated the other passions too. Fear, envy, jealousy, suspicion were all made worse by the unequal distribution of wealth.

The most disturbing effect of man's incessant desiring was factional behavior. The "haves" of society often tended to form a political faction according to their common interest, and the "have-nots" would respond in kind. Since the latter always outnumbered the former, some form of chaos or tyranny was often not far to seek.

Honor

For the ancients, courage was among the most important virtues. One found it on the battlefield, of course, but one also found it in the political world in the form of moral courage—the courage of one's convictions. We acknowledge it today whenever we salute patriotism. The term used by the ancients to describe this quality was *honor*.

Deeds of honor have often been rewarded with political power—think of Julius Caesar or Napoleon. Plato coined the term ***timocracy*** to describe government by the glorious, and it was not a rare phenomenon. Public figures still seek reputations for heroism, and when no battlefield presents itself, they might gain their laurels through acts of political bravery. The powerful, it seems, are given to boldness.

Timocracy: Form of government led by those who have proven themselves heroic.

John Adams referred to this tendency as the *spectemur agendo*—the "desire to be seen"—and he came to regard it as the single greatest peril to the Good Society. Adams' argument went something like this. Virtually all political persons seek a reputation more glittering than that of their rivals. Each wishes to be better known and more admired among the people, and as a result, each will reach for ever more power and resort to ever more brazen demonstrations of it. Too much, alas, is never quite enough.

Adams puzzled long and hard over the *spectemur agendo*. If a Good Society were to survive, he reasoned, this dark propensity must be dealt with. Since it couldn't be banished from the human heart, it would have to be transformed in some way, so that instead of tearing republics apart, as it had done throughout history, it somehow held them together instead. Was it possible, Adams wondered, to design a constitution in such a way as to make human nature in general, and the *spectemur agendo* in particular, a cohesive rather than corrosive influence?

Theodore Roosevelt's well-advertised exploits in the Spanish-American War propelled his political career. After being elected governor of New York in 1898, he received the Republican nomination for vice president in 1900 and often seemed to overshadow President McKinley.

Rule-Mindedness

Experiments have demonstrated that human beings almost instinctively search out rules for ordering their behavior. For example, young children on the playground, when presented with the idea of a new game, will quickly devise a set of rules for playing it. And we often find ourselves asking what the "rule" is for dealing with some new situation. A sense of "**rule-mindedness**" seems to arise from within.

What we call law is simply rules backed by ultimate sanction. History suggests a variety of ways law can come into being. It can be passed by a legislative body. It can be pronounced by magisterial decree. It can arise from custom or convention. It can be "discovered" in the natural order of things. It can be devised by a judge for the settlement of a specific dispute and then applied to similar disputes later on. It can be handed down by the gods.

Some legal theorists have argued that the source of law, and even its moral content, are relatively unimportant. What counts, according to this point of view, is the law itself and its recognition as such. In other words, the human need for rules is so great that virtually *any* rule, even a harsh or mindless one, can serve the required purpose. As long as the rule in question is truly a *rule*—a standing precept generally applied—and not just an arbitrary edict against some individual, it has the almost magical property of ordering human expectations.

On the other hand, something else in human nature prompts us to desire that the laws also be "just." Justice is a difficult term. For what seems just to one person may not seem so to another. (Is it "just," for example, to use indigent patients as guinea pigs in medical research?) As a result, human beings often disagree about the moral content of the laws they must depend on.

Habits of the Heart

Every society is set apart from all others by its customs and traditions. These "habits of the heart," as they have been called, have a life of their own. They are not in anyone's particular custody, nor are they subject to anyone's manipulation. And very often they make little rational sense. In England, for example, the institution of monarchy became so beloved that its color and pageantry were preserved for centuries after the monarch was essentially dethroned.

Precisely because custom and tradition cannot rationally be accounted for, they are not subject to ordinary scrutiny. Forms can become empty and meaningless, like the symbols attending college commencement ceremonies. Corruptions can creep in unnoticed, so that tipping the waiter for a job well done gradually turns into a compulsory "gratuity" added onto the check. Even so, our customs and traditions tell a lot about who we really are and how we really deal with the world. American colonists remained Englishmen for all intents and purposes, even with a revolution coming on, because they still maintained English habits of the heart.

Any significant political change, and certainly any founding, will have custom and tradition to reckon with. There will be cherished institutions (like the English monarchy) that make little sense but have enormous staying power. There will be assumptions about the world which are regarded as unchallengeable, but which in fact are only traditional beliefs. And there will be instances where reason points to a given course of action but custom makes it impossible to follow. More than one ardent reformer has denounced "the dead hand of the past." "What is established already," said David Hume, "is as important as what ought to be established."

Spiritual Sense

In God We Trust

Even the most rational of the ancients acknowledged the gods' supervision of temporal affairs. If the ancient Athenians had not believed that Pallas Athena was smiling down on them, no constitution imaginable could have made their *polis* sound, for people have but scant faith in purely human contrivance. This helps us to understand why, in the American Founding, references to deity were insisted upon by some less-than-active churchgoers.

Man's spiritual sense is relevant to the Good Society in other ways. For one thing, the citizens of a Good Society need to agree about what is in fact good, and such consensus often reflects spiritual values. For another, political actions are often based on a sense of the sacred. Some religions take eternal rewards and punishments into account in making this-world decisions, while others acknowledge the spiritual realm more indirectly. Finally, there are those "goods of the soul" the Good Society promises, many of which are spiritual in nature. The whole point of the Good Society is to create a life worth living.

Most foundings have been sacred in character, as we have seen, for they have required supernatural legitimation. And most political societies, until fairly recently, have been accompanied by the equivalent of a state church, the assumption being that secular authority needs all the help it can get. Even so, the spiritual sense poses dangers. It has driven human beings toward brittle dogmatism, seething prejudice, and with sad regularity, holy war. Ironically, no aspect of human nature has done more to thwart the Good Society.

Two Approaches to the Problem of the Ages

See study aid.

Given these and other aspects of human nature, what strategies could be employed in the design of a Good Society? Here are the brief outlines of two, one of them ancient and one modern.

Areté

Areté: Greek form of virtue; the selfless pursuit of excellence.

Areté—a Greek term we usually translate as "virtue"—was an abiding preoccupation of the ancients. It referred not so much to moral virtue as to the quest for excellence in all things. Thus, there was the *areté* of the athlete, all about strength and timing, the *areté* of the soldier, all about battlefield courage, and so on. The sculptor, the poet, the artist, the musician—each had his own *areté*. Each called upon different resources of the body or mind. All had to do with perfection.

Areté proposed that the elements of human nature, once they were rightly understood, could be mobilized to resolve the human predicament and form the basis of a Good Society. To put the argument in simplest terms, the Good Society could be founded on virtue. In other words, if all elements of a political society could be persuaded to behave according to their highest human potential, the pieces of the age-old puzzle would finally fall into place.

Consider the role of the statesman, for example. In a regime inspired by *areté*, the statesman would constantly seek to improve in wisdom, in prudence, in courage, in personal integrity, in a sense of justice. He would also strive to perfect the rhetorical skill of the orator, the sophistication of the jurist and the shrewdness of the diplomat. Above all—and this was where *areté* became truly challenging—the statesman must come to hold the public good above any sort of private advantage.

The *areté* of the ordinary citizen, or "public virtue," was equally important. Like their leaders, citizens were expected to cultivate wisdom, prudence and the like, and they must also place the public good above their own self-interest. But the virtue of the citizen had its own dimensions too. One of these was to participate willingly in the political process regardless of any partisan concerns. Another was to abide the decision of the electorate no matter who won or lost. A third was to revere the law above all else—whether or not there was a cop on the corner.

Areté tended to draw people into public life and make political activity an end in itself. Many of the ancient regimes were aristocratic, for the *aristoi*—a Greek term literally meaning "the best"—were those most devoted to the pursuit of excellence. But democratic regimes emerged as well when the *demos* (ordinary citizens) caught the spirit of politics as self-expression.

Any *polis* that diligently practiced political virtue stood a good chance of becoming a Good Society. And thus *areté* was stressed by the ancients with an almost fanatical zeal. It was taught in the schools, preached in the temples, defended with eloquence in the symposia. If statesmen and citizens were but virtuous *enough*, it was argued, the destructive urges of humankind could be overcome and the constructive urges find fullest expression. Yet there was a troubling question below the surface, a question made all the more insistent by the actual experience (as opposed to the professed theory) of the Greek *polis*, and the American Founders would still be wrestling with it two thousand years later. Did human nature—*areté* or no *areté*—support a level of public-spiritedness *that* high?

Self-Interest

Let us fast-forward from the world of the ancient Greeks to the world of the European Enlightenment around the middle of the 1700s. Political thinkers continued to extol the importance of virtue—which they still referred to as

High Noon

On the very day that Marshall Will Kane turned in his badge, word arrived in Hadleyville that the notorious Frank Miller, whom the Marshall had sent to prison five years ago, had inexplicably been pardoned and was on his way back. "I'll kill you, Will Kane," he had vowed. "I swear I'll come back and kill you!" He was due to arrive in an hour and twenty minutes, on the high noon train. Three members of his old gang were already waiting at the station.

For a western movie, the plot of Stanley Kramer's 1952 classic, *High Noon*, started out ordinarily enough. Viewers expected to see plenty of action, a fisticuff or two, some fancy gunplay, and in the end, the unambiguous triumph of right. What they saw instead was a psychological drama that left many of them shaken. High Noon did not seem like a western at all. Rather, in a vague and indefinable way, it seemed to be about the viewers' own lives.

Hadleyville owed a lot to Marshall Kane. Before he came it had been a two-fisted frontier hell, ruled by the Miller bunch. Now, thanks to the law and order Kane had established, Hadleyville was peaceful and prosperous. "You cleaned up the town," says an admirer, "Made it fit for women and children to live in."

But had peace and prosperity been good for Hadleyville? The viewer began to wonder. As Marshall Kane pinned his star back on and attempted to recruit his old posse, it became clear that all was not well. Those who had helped to break the Miller gang five years earlier were not willing to break it again. Their reasons were a study in political psychology.

The deputy, Howard Pell, for instance, merely wants to play politics. He will strap on his gun and help the Marshall out if the latter will make him his successor. Otherwise, Kane can go to hell.

The Marshall's friend, Sam Fuller, cravenly hides in his bedroom and orders his wife to lie. "You want me to get killed?" he shrieks to her. "You want to be a widow?" The boys at the saloon are more subtle in their cowardice. "Things were different then, Kane," one of them explains of the earlier time. "You had six steady deputies to start off with, every one a top gun. You ain't got but two now."

Gradually it becomes clear that certain elements in the town are not upset about the return of Frank Miller. Business has been off at the local saloon; gambling is in a slump; prostitution has declined. Explains a resentful hotel clerk: "There's plenty of people around who think he's got a comeuppance coming."

The broader significance of this remark is established by the town judge, while packing his bags to clear out. "No time for a lesson in civics, my boy," he says but delivers one anyway. "In the fifth century B.C. the citizens of Athens, having suffered grievously under a tyrant, managed to depose and banish him. However, when he returned some years later with an army of mercenaries, those same citizens not only opened the gates for him but stood by while he executed members of the legal government."

So much for the corrupt elements of Hadleyville. What about the law-and-order people—those whose way of life Kane is defending? Their reasons for bailing out appropriately given at a church meeting are even more convoluted. Why should we get involved in a personal feud between Kane and Miller? Whose problem is this, anyway, but the politicians up north who turned Miller loose? Why should we fight Miller when we pay taxes for law enforcement?

Kane's staunchest supporter then delivers this speech: "People up north are thinking about this town. Thinking about sending money down here to build stores and factories. It'll mean a lot to this town, an awful lot. But if they're going to read about shooting and killing in the streets, what're they going to think then? If Will isn't here when Miller gets into town, my hunch is there won't be any trouble. Not one bit."

The last, and most devastating, defection is Will's best friend and fellow lawman, Mart. "Yeah," he says, "been a lawman all my life. It's a great life. You risk your skin catching killers, and the juries turn 'em loose so they can come back and shoot at you again. If you're honest, you're poor your whole life. And in the end you wind up dyin' all alone on some dirty street. For what? For nothin'. For a tin star."

Kane, however, refuses to buy any of this. He will stay and fight at whatever cost. "I've never run from anybody before," is the sum and substance of his explanation. As the townspeople, in their corruption and cowardice, make ever more rational sense, Will Kane makes ever less. All he seems to know is that he stands for the right—come what may. If need be he will face the Miller gang alone.

Which is, of course, precisely what he has to do. As the last deputy quits—"There's a limit how much you can ask of a man. I got a wife and kids"—the marshall loads his guns, makes out his will, and heads through the deserted streets to meet the high noon train.

What was it that made *High Noon* so relevant to its American viewers? Without using the terms themselves, the film was exploring two fundamental concepts—virtue and self-interest. And it was asking which of the two was more important to the survival of the Good Society. It was a question that Americans had asked before.

From America: A Study in Heritage, *Kendall/Hunt, 7th Edition, p. 24–25.*

Sheriff Will Kane (Gary Cooper) at a church meeting in 1952's High Noon. *(Bettmann Movie Still)*

areté—but their emphasis was beginning to shift toward a different approach to the Good Society. Human nature might not answer *areté*'s demand for selfless and unwavering public devotion, they reasoned, and even when it did so there were corollary difficulties, as we shall see further on. A more reliable understanding of mankind would place its stress not on the quest for perfection but on the tendency toward self-interest.

Someone noticed that there was a central theme running through almost all observations about human nature. Whether it was appetites and aversions under discussion or something exotic like the *spectemur agendo*, human beings displayed a marked tendency to consider themselves first. This appeared to be the case in spite of any spiritual sensitivity, any desire for excellence, or any inborn love of truth and justice they might have. And there was a cogent explanation for it. If human beings had a strong social sense bringing them together into communities, they could only feel pleasure or pain *as solitary individuals*. Accordingly, while I might sympathize with my neighbor whose house has burned down, I really can't share his sense of loss because I can't feel the chill of night as he can. While self-interest did not say much for the dignity of mankind in the way *areté* did, it had the redeeming value of being wholly reliable. People could be counted on to act in their own interest as surely as the winds blew and the tides surged. Individuals, on occasion, might behave generously, even altruistically, especially regarding close friends or family members, but behavior as a whole was almost mathematically predictable.

Enlightenment thinkers, no less than the ancients, understood the downside of self-interest. It could make people cold and uncaring. It could draw them away from community life. It could run to excess and lapse into corrupt, irresponsible, even criminal behavior. But several of the Enlightenment's best minds—Thomas Hobbes, John Locke, David Hume, Adam Smith—saw a side to self-interest that had escaped the ancients (Aristotle notably excepted). For one thing, it could lead to cooperative as well as destructive behavior, especially in economic life. For another, it was amenable to a certain type of control mechanism called ***counterpoise***. That is, it was possible to structure political situations so as to balance, or "counterpose," the self-interest of one individual or group against that of another, each holding the other under control. And if self-interest drew hearts away from politics, that wasn't always such a bad thing. People who had what we call "a life" also had something to lose. They were less apt to go for wild experiments or easy utopias—and they did not make good fanatics.

It was less obvious how self-interest might lay a steadier groundwork for the Good Society than *areté* had. A great deal of analysis and experimentation would be required before political thinkers could really place full confidence in it. But we can guess at some of their main ideas. They would emphasize the structure of government rather than any guiding spirit, and hope to achieve as much counterpoise as possible. They would emphasize the limitation of government too, so that government was less likely to intrude upon the private lives of its citizens. And they would emphasize participation in the political process, not for *areté*'s sake, but for acknowledging and serving as many interests as possible.

The modern 'Good Society?'

And the modern Good Society? It would be calmer than its classical predecessor, more private in outlook, more materialistic and comfort-driven, and much more interested in individual freedom. It would be less inclined to deeds of honor and thus less inclined to war. And it would constantly work against tribalism and toward an ever broader sense of inclusiveness. It would be a convoluted, multi-layered society and it would need a government to match. Where the governments of antiquity had required *arête* and not much else, a government in the modern world, capable of sustaining a Good Society, would have to be crafted like a fine watch.

This is why the story of the American Founding is a story, not simply an event. Among its characters would be some who looked back to the English past, some who looked forward to Utopia, several apostles of natural law, and quite a few of the practical sort focused on the here and now. The narrative would wind among every facet of human nature, from rule-mindedness to the *spectemur agendo*, and touch on every idea for building a stable foundation.

There would be as much discussion of *areté*—using the modern term *virtue*—as there had been in ancient Athens, together with every means the Enlightenment had come up with for cashing in on self-interest. The story would tell of founding myths, of a conclave of lawgivers, of a written constitution (virtually without precedent), of an accountable political system, and of an emerging national character. Monarchy would play a role in it—rather larger than most Americans appreciate—but so would aristocracy and democracy, the three joined together in a unique way. And from the ancient world, which would furnish much of the vocabulary and not a few of the plot ideas, there would come a description of the whole as a *republic*.

CHAPTER 1 STUDY AID

(These aids are not all-inclusive, nor are they intended to be a substitute for thorough study of the material presented in the textbook.)

Elements of a Founding:

- Identity (name and geographic boundary)
- Constitution (fundamental law and political order, source of authority and governmental structure)
- Citizenship (who rules–participation and exclusion, who is in and who is out)
- Polity (sense of nationhood, cultural identity)
- Founding Myth (source of legitimacy)

MAD (Monarchy, Aristocracy, and Democracy)

GENERAL MODEL	PRO	CON
Monarchy the King (rule of one)	• Strong • Decisive leadership • Figurehead symbol • Can act quickly	• Despotism • Tyranny • Abuse of power • One man, even a good man, makes mistakes
Aristocracy the Nobles (rule of the few)	• Admirable, can be looked up to, can be aspired to • Advances the arts and learning • Best educated • Most virtuous	• Oligarchy • Snobbery • Self-interest • Ignores the poor • Does not seek the general welfare
Democracy the People (rule of the many)	• Greater equality • Civic involvement • Welfare of the poor sought	• Chaos • Slow to make decisions • Much bickering • Low opinion of arts and learning • Susceptible to following a demagogue • Base desires

The negative tendencies of each estate can be remedied if the interests of all three estates are balanced in a mixed government under a rule of law.

Constitutional Epistemology (Founding Archetypes)

Universalist Founding

- General and Inclusive
- All people
- All times
- All places

A Priori

- Abstract
- Deductive
- Based on self-evident 'first' principles

What ought to be possible?

- Associated with Plato
- Theoretical
- Philosophical
- Idealistic
- Utopian

Revolutionary

- Virtue can be learned
- Self-interest can be suppressed
- Attitude dominates
- Human nature malleable, it can be made to change
- Virtuous elites should rule

Particularist Founding

- Limited or Exclusive
- Specific people
- Specific time
- Specific place

A Posteriori

- Empirical
- Inductive
- Based on 'facts and experience

What is possible?

- Associated with Aristotle
- Practical
- Scientific
- Pragmatic
- Temperate

Evolutionary

- Virtue not dependable
- Self-interest innate, but can be channeled into positive end
- Behavior dominates and must be controlled by institutions
- Human nature not malleable, but it can evolve

Human Nature, Politics, and the *Zoon Politikon*

(Anciently, the *polis* or 'political society' was thought to reflect the moral and spiritual qualities of a people. Moreover, it was thought that the political order did not just reflect human nature, but that it could actually alter human nature, for better or worse. Thus, political activity was seen as an imperative and man was seen as a *zoon politikon*, or 'political animal.')

Elements influencing political activity:

- *Tribalism*—Creates sense of community and common purpose while identifying outsiders to be excluded or discriminated against.

- *Reason*—Represents the 'rational element' in society. Fostered by education and creativity if they were focused on the pursuit of truth, justice, and the 'good.'
- *Passions*—Greed, lust, jealousy, anger, fear, and hatred. In politics this could lead to turmoil, factionalism, inequality, and apathy if 'desire' or covetousness dominated.
- *Honor*—Courage and boldness which could lead to seeking after fame and glory (*spectemur agendo*) and the creation of a timocracy (government by the glorious).
- *Rule-Mindedness*—An innate seeking out of order and justice as opposed to chaos and anarchy.
- *Habits of the Heart*—Cultural habits, traditions, and customs may become institutionalized and thus come to carry political implications.
- *Spiritual Sense*—Reflected in the invocation of deity, imputed sacredness, and identification of the 'good' in a political society.

Two Approaches

The Ancient Greeks (Areté/Virtue)

- Areté = Virtue = Excellence
- A 'Good Society' could be founded on virtue
- Statesmen seek public good
- Citizens willingly participate for the good of all
- Virtue could overcome excesses of human nature
- Upside—pursuit of excellence
- Downside—human nature unpredictable and a system built on the principle of areté susceptible to foibles of that nature.

The Enlightenment (Self-interest)

- Self-interest not innately negative, could produce a 'Good Society'
- Most people do not act nobly, but out of self-interest
- Self-interest is reliable, predictable
- Self-interest could be regulated through mechanisms which created structural counterpoise
- Upside—cooperation and improvement and caution
- Downside—exploitation

CHAPTER 2

Puritanism and the Founding

Aboard the Arbella ***A Modell of Christian Charity***

Wee shall finde that the God of Israell is among us, when tenn of us shall be able to resist a thousand of our enemies; when he shall make us a prayse and glory that men shall say of succeeding plantations, "The lord make like that of NEW ENGLAND." For wee must Consider that wee shall be as a Citty upon a hill. The eies of all people are upon Us, soe that if we shall deale falsely with our god in this worke we have undertaken, and soe cause him to withdrawe his present help from us, wee shall be made a story and a by-word through the world. . . . Therefore lett us choose life, that wee and our seede may live by obeying his voyce and cleaving to him, for he is our life and our prosperity.

—John Winthrop (1630)

"Realism and Idealism," still side by side in the American culture.

The notion of God's elect—or "saints" as they called themselves—supplies a basic theme of American history and some important implications for American politics. One implication might be called the assumption of guilt. In the godly community, Puritans reasoned, there would always be some, perhaps many, unregenerate persons. So there would always be jealousy, avarice, lust, contention, pride, and other marks of damnation. Corruption might literally be found anywhere.

But there was a second, contrary assumption, the assumption of innocence. For, in the same godly community there would also be those whom God had saved. These saints were not subject to the ordinary frailties of mankind—they were significantly stronger and wiser than their fellows and were clothed with an undauntable righteousness. They could be counted on to provide the community with spiritual guidance and moral backbone no matter what it might have to face.

This "schizoid" view of human nature made it possible to have coldly realistic assumptions about political conduct, and at the same time to be capable of a certain warm idealism.

Confidence

Armed with divine grace, the Puritans had an almost legendary sense of confidence. Nothing seemed too difficult for them to tackle. Their courage in facing the wilderness on a far Atlantic shore was only one of the many benefits of that confidence. There were several others.

To begin with, the Puritans were so convinced of the soundness of their orthodoxy that they wasted little time debating or defending it—as they had had to do in England—and turned their main attention to practical tasks. Their "errand into the wilderness," as they put it, was to show by example what God could really do when it came to community building. This purpose gave rise to a metaphor that would come to define not only Puritan America but America as a whole. "We shall be as a city upon a hill," **John Winthrop** prophesied, "the eyes of all people are upon us."

John Winthrop: A founder and governor of Massachusetts Bay Colony (1588–1649).

Their confidence made it possible for the Puritans to do whatever they deemed necessary in order to accomplish this mission. They rarely had qualms about double-dealing with Native Americans, about launching wars of extermination, or later on, about participating in the slave trade. But they also rarely had qualms about clearing the wilderness, laying out towns and cities, founding colleges, setting up libraries, or doing the thousand other things necessary to build a society from the ground up.

City upon a hill: www.mtholyoke.edu/acad/intrel/winthrop.htm

Although the Puritans' initial perceptions of America were fraught with images of the "howling wilderness," that would eventually change. They came to see America, graced by their presence, as nothing less than the Promised Land and their city upon a hill as the New Jerusalem. The imagery of blessedness and bounty began to replace that of the wasteland.

Several traits of the American character can be traced to Puritan self-confidence. Our reputed narcissism, smug superiority and evangelical enthusiasm (complaints lodged by early visitors) would appear to arise from Puritan beginnings. But so would the assurance necessary to face down the British Empire and found a successful republic.

Newness

Although the Puritans' response to the New World was backward looking in some ways—in the design of houses, for example, or the layout of villages—it was often forward looking in others. In building their city upon a hill, the Puritans proposed to make all things *new* again, re-creating the world by dint of first principles.

New lands are seedbeds for new ideas. With something like that fact in mind, John Adams would call the United States "the nation of futurity." By the time of the Revolution, Yankee tinkerers were already becoming known for their inventive skills, and in the years following they would create an astonishing array of new mechanical contrivances. As for new political ideas, the American Founding would either create or make use of dozens of them, and some would be directly traceable to the Puritan experience, as we shall see.

Newness also made it possible to "invent" oneself. Just as the Puritans came to America and began calling themselves "Americans," the re-creation of self became an American theme. Fiction writer Horatio Alger Jr. would one day tell engaging tales of immigrant paupers who declared themselves to be tradesmen, accountants and entrepreneurs—and went on to become precisely what they declared. "Rags to riches" would become one of America's founding myths.

Finally, newness laid the groundwork for an attitude historians have called "**American exceptionalism**." Americans would come to believe that they were quite unlike any people, any political society or any national state in the world's experience. America, they would say, breaks the rules.

Learning

Scholasticism: A philosophical and theological movement from the Middle Ages which argued that reason and faith are compatible.

The founders of English Puritanism had great respect for learning. Many of them were academics themselves, mostly affiliated with the university at Cambridge, and they were well versed in **Scholasticism**, a highly rationalistic mode of inquiry taught at the time. The scholastic investigator began from certain premises, usually taken from the Scriptures, and worked his way through a maze of careful deductions to final conclusions. It required a sharp mind and rigorous training.

One result of this heritage was a high regard for human reason. The Puritan ministers in New England were among the most learned men of their time, and their sermons were models of careful thought and wide reading. Having worked through the Scriptures by the light of Scholasticism, they saw little or no conflict between religious faith and secular knowledge, and therefore they saw nothing to fear from scientific investigation. American Puritans were among the first to

Religion and science, is there a conflict? During the Scopes trial in 1925, Christian fundamentalism squared off against Darwinian evolutionism. Would the Puritans have understood such a schism?

embrace the new psychology of John Locke and the new physics of Isaac Newton. In Massachusetts Bay, they tried out inoculation for smallpox, based on the available data supporting it—the very cutting edge of applied science. Later on, the children and grandchildren of the Puritans would find it easy to modernize their faith with transfusions from the European Enlightenment.

Another result was that the American Puritans promoted learning at all levels. The alternative, as they saw it, to perishing in ignorance on the Atlantic frontier was to pursue knowledge with heroic single-mindedness. Thus, at a time when education in most colonies was virtually nonexistent, Puritan New England was an educational pioneer. Town and village schools taught the classics along with the three Rs, and turned out generations of exceptionally well-schooled children. And in order to maintain high standards in the ministry, they founded both Harvard and Yale.

Many of the American Founders would be heirs of Puritan learning. Their wide reading in the classics, their interest in science, their ready acceptance of Enlightenment ideas, and on a deeper level their assumption that knowledge served the moral and spiritual purposes of mankind were all outgrowths of the Puritan heritage.

Paranoia

There was a dark side to the city upon a hill. For if the Puritans were squarely on the Lord's side, they were squarely in opposition to Satan and his minions. As a result, they tended to see forces of darkness all around them.

Satan, they believed, had a special advantage in his ability to exploit man's fallen nature. Almost anyone might be a target. In the most pious congregation, one might find those who were under the devil's influence. This fact posed a delicate problem. The Puritans believed God's judgments to be swift and sure. If the godly community tolerated sin in its midst, it might expect fire, flood, pestilence, even earthquake, not in the remote future but here and now. It was up to the elect to take stern measures when such were called for.

Occasionally, this mindset led to tragedy, as in the infamous Salem witchcraft trials. More often it simply created wariness. The enemy was at the gates in Puritan consciousness—but the enemy might also be within. Awareness of this possibility gave the Puritans a peculiar way of understanding change, especially social change. What a sociologist might see as the stresses and strains of a developing society a colonial Puritan would likely attribute to plain wickedness. And wickedness was never to be tolerated.

No wonder Puritanism erupted in periodic bursts of religious revivalism such as the Great Awakening. Spiritual healing, taught the ministers, could only be brought about by recognizing sin, crying repentance, purging the soul of its accumulated corruptions—chief among which were luxury and vice—and taking a stand for the Lord. This habit of mind gave the Puritans a high potential for reform. They had an urge to root out every iniquity, right every wrong, and when the occasion presented itself, they rarely hung back. It also gave them a high potential for revolution.

Calling and Community

John Calvin had little patience with the attempts of many Christians to rise above the human condition. Man, born in sin, should be content to live a mortal life, he argued. Rather than seeking to emulate God through self-denial or spiritual exertion, man ought to concentrate on worshipping God and glorifying His name.

One did not glorify God by singing hosannas, Calvin said, but by showing forth God's work among men—well-tended farms, busy shops, merchantmen plying the seas. If this was a European vision, it was all the more an American one, for America offered a blank slate on which the godly could write to best effect. We begin to see what Winthrop's city upon a hill was really about.

Calling: A Puritan belief that by pursuing worldly careers virtuously, a person was in effect glorifying God.

The Puritans' means for building that city was the Christian "**calling**." The calling was a worldly pursuit—banking, say, or shipbuilding—but it was also a religious pursuit. It was the manner in which a finite human being could glorify God. President Calvin Coolidge, a descendant of the Puritans, stated the theory of the calling in a nutshell. "One who builds a factory builds a temple," he said, "and one who works there worships there."

The calling mandated righteous behavior in the workplace. The godly merchant could not be a swindler, nor the godly wheelwright a shoddy craftsman. And all were expected to perform their labors with honesty and diligence. They must rise early, work soberly, save frugally, spend wisely, live modestly, and impart a portion of their substance to the poor.

In this picture of things, prosperity had a special meaning too: that God was pleased with the community's worship. Prosperous individuals were very likely to be counted among the elect (we still believe that today), and there was nothing wrong with a little tasteful display of their wealth. There was a line, however, and one crossed it at one's peril, for real extravagance paved the way to hell. The point was that the entire *community* prospered, not just a few individuals. Businesses made good profits, artisans made good wages, and the poor were decently cared for.

This vision of the American community would cast a long shadow. The small town, celebrated in hundreds of novels and plays, would become a keystone of Americana. When things seemed well on Main Street, they would seem well with the world. And when the community was beset by crime, vice, grinding poverty, politicians on the take, it would be a problem for everyone's concern.

Later on, we will see how Calvin's ideas helped get the American Founders around a difficult problem. For the time being, we need only note that for the Puritans, there was no necessary choice to be made between virtue and self-interest in the godly community. The two could fit together seamlessly.

Calling and community: virtuous self-interest benefits all.

Puritan Constitutionalism

In much of the foregoing we see the dim outlines of things American. In their baggage, the Puritans brought myths and images, concepts and attitudes we would recognize today. But did they also bring the building blocks of a founding? Probably so. We might pay special attention to the following three:

Puritan Law

As educated Englishmen, Puritan magistrates were versed in the English common law, and they often made appropriate use of it. For most Puritans, however, the "real" law was God's law, as revealed in the Holy Scriptures. They paid particular attention to the commandments they found in the Old Testament, those given to Moses most of all. God's law—as opposed to man's law—must necessarily be coherent, consistent and in keeping with the truths of nature. A society could not go wrong by obeying it.

We have seen that all foundings require fundamental law of some kind. And if the law in question is divine—that is, given directly by the gods—then men do

A Quirk of the American Mind

This experiment, mind you, was not performed at a secular university but a religious and specifically Christian one. It could have been Notre Dame, or Southern Methodist, or Loma Linda, or Baylor. (And at any of them, I suspect, the outcome would be the same.)

A professor, addressing a large class, asked this question: "How many of you consider yourselves to be deeply and truly religious?" A majority of the hands went up. He then asked: "How many of you find it consistent with your religion to be rich—not just comfortable or well off, rich?" Again, a majority of the hands went up. Indeed, they were the same hands.

The professor then singled out individuals who had raised their hands and asked them specific questions:

What about the wealthy young man who wanted to be Jesus' disciple, and was told to sell all he had and give it to the poor?

What about the scripture that says: "It is easier for a camel to go through the eye of a needle, than for a rich man to enter the kingdom of God?"

What about those early Christians who gave up private ownership of property and shared their goods with each other? What about those vows of poverty that carried tens of thousands into monasteries and convents? What about priests in the Third World today who decry the evils of wealth?

The students had answers for all these questions and more. When the discussion ended, the professor asked if any had changed their mind. None had.

It may not have gone this way anywhere else in the world. For what the professor was dealing with was not an abstract question of morals so much as a quirk of the American mind. As a matter of national character, Americans believe that wealth and spirituality easily and comfortably fit together—despite considerable evidence to the contrary. That unique synthesis of virtue and self-interest, in a nutshell, is John Calvin's legacy to modern America.

From America: A Study in Heritage, *Kendall/Hunt, 7th Edition, p. 58–59.*

Christ and the rich young ruler.

not have to spend time and energy debating its merits. The law forms a reference that is beyond the wit of man.

In many societies of the time, the laws were uncertain. Kings could claim divine sanction for their edicts but so could legislatures, and in seventeenth-century England those two authorities were battling it out. The American Puritans were troubled by no such ambiguity. In their political affairs, which were at times very lively, they could scrap over particular applications or interpretations of divine law but the law itself remained inviolable. It became a kind of constitution.

Puritan Politics

The Puritans were not republicans, liberals, or democrats, as we understand these terms today. (The terms themselves will be explored in greater detail in the next chapter.) Yet there were elements of all three in the Puritan makeup.

***Republicanism*.** The English Puritans steadfastly opposed the Stuart monarchs, as we have seen, and raised their voices against the so-called "divine right of kings." The king, they insisted, did not rule by God's pleasure and did not stand above the law. This was only one of their republican ideas; they embraced several others. And when civil war finally broke out in 1642 and the king was eventually executed, the Puritans pioneered England's first experiment in republican government known as the Commonwealth and Protectorate. Their dark view of human nature and their distrust of arbitrary power were engines that drove them on.

Driven by those same engines, the Puritan colonies in New England essentially became republics in their own right. In theory, of course, they were under the sovereignty of the English crown. But in fact they were based on corporate charters that granted them powers of self-government, and as a practical matter the crown could supervise them precious little. They were autonomous political societies in effect, and they behaved like it.

***Liberalism*.** It was the belief of English liberals that government belonged to the substantial elements of society, the owning classes, and that its main purpose was to protect the rights of property. Most Puritans agreed. They believed that God intended the earth for the earnest and the toiling, and whatever property a man could accumulate, it was a sign of God's pleasure.

Liberals also favored religious toleration. Freedom, to them, meant freedom from established churches quite as much as from heavy-handed magistrates. The Puritans certainly agreed with part of that—being out from under the Church of England. However, when *they* became the established church, as they did in the colonies, they were no more tolerant of dissenters than the London episcopate had been. But in America the dissenters could do precisely as the Puritans had done—go somewhere else. Soon there were congregations of schismatics here and there, feuding and bickering with one another and causing the authorities to ponder the costs of enforced religious orthodoxy. Almost in spite of themselves, the Puritans were led toward toleration.

Other seeds of liberalism seemed to sprout among the American Puritans. Take the idea of "natural liberty." Liberals held that human beings were born free, not captives of the social order, and accepted government in their lives because they freely chose to. Puritans found themselves making exactly the same argument against the Church of England—that is, God required men to choose the correct path for themselves, not have it thrust upon them. Or take "natural equality." While the Puritans were not exactly levelers, they acknowledged that all souls were equal in the sight of God, and hence that all members of a given congregation deserved to participate. These ideas would someday spur on the American Revolution—and they would come straight out of Puritan New England.

Covenant: A formal agreement among the members of a Puritan congregation which functioned as a founding document.

Democracy. Just as the Puritans did not think of themselves as republicans or liberals, they did not think of themselves as democrats. Theirs was not a world in which ordinary people held the reins of power. Still, there were democratic strands in Puritan life, and in the American colonies they became more visible. Consider, for example, the role of the "**covenant**" in Puritan society. It was laden with democratic implications.

The covenant was a formal agreement worked out among every Puritan congregation. As a given congregation came together, its members would draw up and sign a compact among themselves making their congregational status official. This covenant, as it was called, was itself a kind of founding. (It foreshadowed both the Declaration of Independence—"we mutually pledge to each other our lives, our fortunes, and our sacred honor"—and the Constitution—"We the people of the United States . . . do ordain and establish this Constitution. . . .") In its text, each male head of household covenanted with God to live the commandments and walk uprightly. Each then covenanted with all the others to effect the same end. Everyone was in it together, for better or worse—their brother's keeper indeed.

Once established, the congregation proceeded to hire its own minister and arrange its own affairs. It answered to no external authority imposing uniformity of worship or orthodoxy of belief. For all intents and purposes, the congregants had formed themselves by covenant into an autonomous political society—a striking instance of people creating their own government. And "government" it truly was, for upon migration to America, each congregation became an independent town or village in one of the New England colonies. The members generally lived in close proximity to one another, with adjacent town lots and a village commons—this in order that they might "bear one another's burdens," as the scriptures commanded—and thus protected the integrity of the community. Outsiders were not welcome.

Town or village selectmen—those who actually ran things—were chosen in annual elections. Virtually all male heads of household were allowed to vote. Beyond that, there was a good deal of grass-roots participation in the political process, and town meetings became famous for blunt outspokenness. Colony leaders may have found the mince-no-words style a little troubling. One of them, John Winthrop, in his well-remembered "little speech on liberty," had to remind the people that Massachusetts wasn't a complete democracy. The magistrates might be chosen by the people, he admitted, but once they were chosen, their authority came from God.

All the same, "chosen by the people" took on real meaning in Puritan New England. If all church members had an equal right to participate in community affairs, the machinery of democracy—if not precisely the theory of it—was in place and operating.

Governance of the Self

In Winthrop's "little speech on liberty," he argued that there were two kinds of liberty, which he termed "natural" and "civil." **Natural liberty**, he explained, was the liberty to do whatever came into a person's mind, and this was the kind often favored by the unregenerate of mankind. **Civil liberty**, by contrast, was the liberty to do only that which was right—the unerring choice of God's elect. Puritan society had but scant means to enforce Winthrop's civil liberty. The institutions of police power were few and rudimentary. People could be hauled before the magistrates, but beyond expulsion or public humiliation, there was little of practical consequence to coerce them.

Little speech on liberty: www.law.du.edu/russell/lh/alh/Winthrop.html

Natural Liberty: Freedom to do whatever one wants.

Civil Liberty: Also known as moral self-governance; the freedom to do what is right.

Instead, individuals were expected to govern their own lives. They might get encouragement from their neighbors, and even from the congregation assembled, but the most important source of rectitude was the self. To "walk uprightly before the Lord," a person was required to be honest, hard working, sober, thrifty, temperate, and visibly respectable. Much stress was laid on appearances, to be sure, but appearances were assumed to reveal the actual condition of the soul.

Such a belief system assumed universal standards of right and wrong. The Puritans, in other words, looked to an unwritten law of moral behavior, and the individual conscience was expected to hold to that law unswervingly. There were important implications here. Aristotle had pointed out that in order for a Good Society to exist, there must be general agreement about what was "good." The Puritans had such agreement and they took it seriously.

Puritan freedom, then, might be called *freedom as self-governance*. Individuals governed their own actions according to the moral law. Families and communities did the same. The principle was easily extended, right up to the colony itself, as long as discipline remained strong and the moral law could be assumed. Moreover, self-governance was what made it possible for politics to become so democratic. Those who could govern their own behavior could be trusted to participate politically, and the decisions they made were more likely to win general approval. There was a final benefit. Those who had a hand in making their own laws were more likely to respect and obey them.

All of this was according to theory, of course. In the real world of colonial life, there were any number of gaps between theory and practice. All the same, Puritanism performed well enough in the real world to create a tough and resilient society, able to hold its own in the wilderness.

The First Founding of America

Puritan New England provided a sort of "first founding," a rough draft of what the United States was to become. And the first founding would prove to be an important precursor to the second.

As a religious experience, Puritanism was simply too intense to last. Less than fifty years after the first ships had dropped anchor off the Massachusetts coast, the signs of its decline were everywhere. Yet burnout did not bring an end to the Puritan influence. On the contrary, that influence transformed itself in ways that would amplify its long-term effect. The intellectual side of Puritanism became even more intellectual, soaking up the European Enlightenment like a sponge and providing the generation of John Adams (one of its great-grandsons) with much of its curiosity and learnedness. The emotional side of Puritanism gave rise to periodic bursts of religious revivalism, still evident in the American experience, and imparted a certain intensity to the American character.

Puritan political ideas, such as their deep distrust of power, and Puritan economic ideas, such as the calling, seeped into the American consciousness. So did their preoccupation with community, with self-governance, and with that hands-on style of political behavior. And so did their belief in moral law. The notion of a fundamental right and wrong permeating the universe would give coherence to American family life and shape the content of American education.

Our perception of selfishness (not to say sinfulness) as a basic human trait is undoubtedly of Puritan origin, and so is the restless, guilt-ridden, and sometimes self-tortured American conscience. The events leading up to the American Revolution were but one example of how the political world might take on a startling new appearance when viewed through Puritan spectacles. Was Eng-

America: The Promised Land?

lish society hopelessly corrupt? Were the lords of empire plotting to destroy American liberty and sully American virtue? Were the Americans themselves in peril of losing their souls by succumbing to temptation? Credit John Calvin.

The Puritans' unique view of the world would also provide the basis of an American political consensus. Americans could agree on essentials—and thus disagree only on *non*essentials—because Puritanism gave them a more or less common way of seeing life. Another kind of social cement was provided by the Puritans' assumption that religion mattered. For them, all sorts of ideas and institutions that we would think of as secular were permeated by spiritual significance. It has been observed of our own time that "the screws holding the world together often seem dangerously loose." In Puritan America the screws were all tight.

Finally, the Puritan worldview provided an elementary sense of nationhood. Winthrop's vision of a city upon a hill—prosperous, righteous, envied—would in large measure become America's vision of itself. And Puritan metaphors like the Promised Land and the New Jerusalem would eventually be transformed into America's founding myths. For there was something thrilling, indeed mythic, about the elect of God sorting themselves out of a depraved world and migrating westward, something to fire the mind and make the spirit soar. Nothing is more central to the American psyche than that feeling of chosenness.

CHAPTER 2 STUDY AID

(These aids are not all-inclusive, nor are they intended to be a substitute for thorough study of the material presented in the textbook.)

John Calvin

- Born Jean Chauvin or Chaulvin, a 16th-century French theologian and religious reformer.
- Author of *Institutes of Christian Religion* (1536).
- Calvin argued that God is inscrutable and unknowable, and that man is degenerate and depraved due to the fall of Adam.
- Furthermore, he contended that all men are justifiably damned, but that God in his inscrutable wisdom chooses to save some portion of mankind at His own pleasure—the elect.
- The power of the atonement is limited to the elect.
- Only the elect can perform works which please God.
- Those who are among the elect can not fall from grace.
- One might suspect one's election but never know with certainty.
- Man should not seek to become like God; that is blasphemy.
- Regardless of whether one is among the elect or not, all should worship and glorify God.
- All need to accept rather than evade the human condition by marrying, raising families, being educated, and working at one's calling in life.
- One glorifies God by doing the best one can in any endeavor.
- If blessed with success and prosperity, it is all right since it is but a byproduct of glorifying and worshiping God.
- One should, in effect, pursue what is in one's self-interest in life, but do it worshipfully and for the glory of God. In doing so one can live a virtuous life.

From the preceding arguments some deduced the following:

- Though one cannot know with absolute certainty, it seems reasonable to assume that those who prosper the most are among the elect.
- Only the elect can perform work that is truly pleasing to God.
- Why would God prosper someone who was not among the elect?
- Thus, if one did lead a virtuous life, seeking to worship and glorify God, and in doing so prospered, was it not probable that one was among the elect?
- Therefore, it is in one's self-interest to be virtuous, industrious, and prosperous, since this would imply that one is most likely among the elect.

Puritan Ideas and Their Impact:

- *Human Nature:* All men are cut off from God's influence. A select few, the elect, are to be saved. Thus, corruption is found everywhere, as are the elect of God. One must recognize the weaknesses of human nature but have hope in the potential of the elect to create a better world.

- *Confidence:* Divine grace gives the elect confidence. Nothing is impossible for the Lord. The elect need to lead the way. The wilderness would be tamed and New Jerusalem built. From this the Puritans derived incredible courage and an insufferable sense of superiority.
- *Newness:* The building of the "city upon a hill" would be a new thing. All things would be renewed or made new. Because of this newness one could, in effect, reinvent oneself. One might even be said to be exceptional, beyond the customary rules and expectations.
- *Learning:* Learning is to be esteemed along with human reason. One could not properly worship and glorify God in ignorance. Knowledge is not to be feared. Knowledge serves moral and spiritual purposes.
- *Paranoia:* The forces of Satan are everywhere and must be guarded against. God's judgments are swift. If opposition or adversity arise, they must be the consequence of Satan gaining the upper hand. Sin must be rooted out at all costs. The community must constantly be reformed.
- *Calling and Community:* One glorifies God by excelling in one's calling. The "city upon a hill" is to be built through the efforts of the members of the community working in a virtuous manner. The community's needs and problems are everyone's problems. Virtue and self-interest sustain each other.
- *Puritan Constitution:*
  - Puritan Law:

    'Real' law is God's law. There is nothing ambiguous about this. It provides a constitution for the community. One may disagree about how precisely to apply the law, but not about the law itself.
  - Puritan Politics:

    Republican element—The king is not above the law. Communities can be self-governing.

    Liberal element—Property is a sign of God's pleasure. The propertied class should lead out. Dissent is not to be tolerated, but in America, dissenters could leave and set up their own communities. Men enjoy a 'natural equality' and must choose the correct path themselves, not have it thrust upon them.

    Democratic element—Communities are governed by covenant. The community makes decisions for itself. All the elect can participate in regulating the community.
- *Governance of the Self:* Natural liberty is the liberty to do whatever one wishes; civil liberty is the liberty to do what is right. Individuals must govern themselves for the community to succeed. They must act with virtuous self-interest to worship and glorify God. Everyone knows what is right and wrong. God's law is clear.

CHAPTER 3

American Republicanism

The American Revolution was of far more than local significance. Indeed, from the perspective of the present, it was a seismic, world-changing event, nothing less than a turning point in human history. Yet it was also among history's great surprises. Why, historians ask, did it occur when and where it did, not at the center of Western culture but out on the Atlantic frontier? Why did a colonial rebellion metamorphose into a true revolution and begin delving into new conceptions of the human experience? And why did its leaders insist that they were not revolutionizing anything, only restoring some prior condition of happiness? It was all very puzzling.

One clue must be that the Revolution began at a time when new ideas were in the air and, like the genie out of the bottle, they took on a life of their own. But why did the new ideas flourish in America? What made the English mainland colonies fertile ground for classical republicanism, or Whiggish liberalism, or political democracy? How did fishermen and tradesmen and tobacco farmers become interested in cutting-edge philosophical issues? Those who have spent their lives studying the American Revolution still feel a sense of wonder about it. On the one hand it seems vaguely inevitable, and on the other it seems like the most improbable thing that could have happened.

Ideas and Experience

The ideas in question came from the European Enlightenment. It was called that because it broke with the past in significant ways and prompted a feeling that the "light" had dawned at last. After centuries of Christian orthodoxy and the turmoil of the Reformation, there were new ideas about religion. After centuries of monarchy and feudalism, there were new ideas about society and government. After centuries of knowledge based on authority, there were new ideas about learning and discovering. And after centuries of looking to the medieval world for inspiration, there was a breathtaking rediscovery of antiquity.

The latter was especially important. For the texts of classical antiquity, many of which had reposed in musty vaults for two millennia, opened a world of new significance for the West. There was Plato's *Republic* and Aristotle's *Politics* to reckon with once again. There was Cicero on the law of nature, Polybius on politics and Livy on jurisprudence. There was Herodotus and Thucydides with their searing tales of tyranny versus liberty. Not least there was Cato, a moral hero of the Roman Republic who would inspire, among others, the American Founders.

New ideas may take root more easily in new lands. America certainly was that. Yet there were other new lands in which Enlightenment ideas had scarcely any impact at all. To be sure, these were not colonized by England—which gives us one important clue. England was admired by intellectuals across Europe not only for the quality of its thought but also for the character of its practice. It became a kind of working model of Enlightenment political philosophy. Still, the American colonies didn't just replicate the English experience either; they went far beyond it and in the end broke away from it entirely.

So the puzzle comes down to this. Some remarkable, dynamic process appears to have taken place in the North American colonies of Great Britain. Some rare and potent mix of environment, circumstance, and English culture seems to have made the soil of the New World especially fertile to Enlightenment ideas. We have seen the process hinted at in the previous chapter, where the American Puritans responded to New World conditions in highly significant ways, but the process existed beyond New England and with other groups besides the Puritans. Accordingly, what we want to focus on in this chapter is the relationship between ideas and experience in the American colonial setting. How did each affect the other? How did each cause the other to be modified, developing new ideas and altering experience? And how did ideas and experience together create the American Republic?

The term ***republic*** is elusive. Like other powerful words—*freedom, democracy, equality*—it means different things to different people. Thus, the American Republic isn't remotely similar to ancient Athens or Renaissance Florence, even while it does share a few of their characteristics. In order to understand what the American Republic is, we must explore three different strands of republican thought—three somewhat similar, yet also somewhat different conceptions of what the ideal republic should be.

For each of the three we must make the central concept as clear as possible. We must then see how each became manifest in the American colonial experience. Each would have a shaping influence on the United States, as we will see later on, yet the three reflected different political ideas, worked from different assumptions about human nature, and proffered different visions of the Good Society. Small wonder the American Founding is complex.

Classical Republicanism

See study aid.

The classical republics originated in Greece and Rome, as we have seen. Their history was generally troubled and their reach toward the good often short. The republics of Greece typically flared into tyranny or anarchy, while that of Rome swelled into an empire. But the ideal of the Good Society continued

Classical Republicanism looked upon antiquity as a superior source of inspiration for architecture, the arts, literature, politics, and government.

to glow within the ashes of experience. The rediscovery of classical texts in the Renaissance and a growing awareness of the ancients' wisdom and sophistication combined to keep those embers burning.

Renaissance political thought saw itself as taking up where the ancients had left off. And by the time of the Enlightenment in the latter 1600s, conceptions of the Good Society took Plato's *Republic* and Aristotle's *Politics* as their starting point. This was particularly true in England, where an education in the classics was *de rigueur* for a gentleman. In the great public schools (which were actually *private* schools), wellborn young Englishmen were required to memorize long passages of Herodotus, Thucydides, Plutarch, Cicero, Polybius, Livy—all of them committed republicans. Perhaps they absorbed the ancient politics unawares, along with the ancient literature.

In Stuart England, moreover, republican ideas had a uniquely romantic appeal. English writers often imagined their own society, in a mythic past before the Norman Conquest, to have been a republic in its own right. The "olde commonwealth," as they called it, had supposedly existed without feudal land tenure, without a standing army, without an established church, and best of all, without a swaggering and disdainful nobility. Its monarch, so the belief went, had been elective, not hereditary, and his power had been shared with a democratic parliament. There had even been an ancient "constitution" to guarantee the rule of law.

Civil War and Revolution

The lost "Saxon democracy" was made all the more poignant by the bitter controversy between king and Parliament that ensued in the 1600s. Beginning with James I, who assumed the throne in 1603, the Stuart monarchs advanced a claim to rule Great Britain by a theory known as "**divine right of kings**." As the theory was expounded, it held that the king inherited the authority given by God to Adam—an authority without limitations. Those who resisted that authority were in violation of the Fifth Commandment. They were also liable to be drawn and quartered.

It is important to note—with the American experience in view—that the controversy centered on taxation. King James claimed the power to levy a variety of imposts—ship money, tunnage and poundage—more or less at his pleasure, along with a prerogative right to dun wealthy subjects for forced loans whenever it suited his pleasure. Parliament counterclaimed that as the voice of the people, it alone had the right to assess taxes, and that such right was well settled in the English constitutional experience.

Many of the king's parliamentary opponents gave voice to classical republican ideas. They pointed to the rule of law and argued that the king must obey the laws like anyone else. They pointed to the public good and argued that society didn't exist for the benefit of kings and nobles alone. They pointed to classical virtue and argued that the king's "evil" ministers—Buckingham, Strafford, Laud—were sources of moral decay.

Commonwealth to Glorious Revolution:
http://www.britannia.com/history/narrefhist5.html
http://www.britannia.com/history/naremphist.html

Oliver Cromwell:
http://www.britannia.com/history/monarchs/mon48.html

Civil war broke out in 1642 and for the remainder of that decade England was torn with conflict. In the end, Charles I was beheaded by the victorious parliamentary forces and England was proclaimed a "Commonwealth and Protectorate." But the Commonwealth, under Oliver Cromwell, seemed more like Caesar's Rome than Arthur's Camelot—a bitter disappointment to the classical republicans. Cromwell ruled in despotic fashion throughout the 1650s, and there were enough abuses growing out of his dictatorship to unsettle the stoutest supporter. Upon his death in 1658, the English reluctantly restored the Stuarts to the throne. For England, as for so many others, revolution failed to bring about the Good Society.

Even so, republican ideas continued to flourish after the Restoration. They took on new relevance when Charles II returned to some of the same practices as his father. In 1679, opponents of the restored monarchy formed England's first political party. The Whigs, as they were called, would become both an inspiration and a model for American republicanism later on. For in their railings against the king, the Whigs made repeated reference to England's "ancient constitution," to the traditional rights and liberties of the people, and to a dozen other republican concepts. And these gained an ever wider influence.

The Whigs' greatest achievement—and one of the great achievements of all time—was the Glorious Revolution of 1688. James II, yet another Stuart despot, was driven from the throne in a bloodless uprising and this time there was real change. Parliament made it clear—indeed, spelled it out in writing—that when the next heir was invited to assume the throne, the monarchy was to be constitutionally limited. That is, the king of England was no longer "sovereign" in and of himself. For all intents and purposes, Parliament became the real sovereign power—which meant that England was now a kind of republic. It would take many years to work through the implications, both in England and America, but to the extent that Parliament represented the people as a whole, a true revolution had taken place.

The Commonwealth Ideology

Constitutional Monarchy: Form of government in which the monarch is held in check by a written constitution.

Now the Whigs, who dominated Parliament, had to figure out how to make a constitutional monarchy work. For answers, they drew upon Aristotle's political science, exploring such ideas as the rule of law, the fragmentation of power, and Aristotle's notion of intermixing monarchy, aristocracy and democracy. But their main focus, indeed their preoccupation, was that of the classical republicans—virtue. What is it, they asked, that maintains a high moral tone and sense of integrity in the body politic? What keeps a Good Society from going bad?

In response, Whig political thinkers began to develop a modern version of *areté*, which we recall from chapter 1 as the foundation of the classical republics. Historians refer to this as the "commonwealth" or "country" ideology. It was manifest most clearly in a series of newspaper essays by John Trenchard and Thomas Gordon, published in book form as *Cato's Letters*. By 1730 the commonwealth ideology had diffused to all parts of the British Empire. It was especially popular in the North American colonies, where it gave rise to dozens of pamphlets, broadsides, and newspaper articles. If there was a single intellectual taproot of the American Revolution some forty years hence, this was it.

See study aid.

Cato's Letters: http://www.constitution.org/cl/cato_000.htm

As a coherent view of the political world, the commonwealth ideology went something like this. Man is inherently evil, blind, and foolish. Government must arrest his fallen tendencies by checking and balancing itself against the insidious wiles of the few. Yet in spite of the best efforts, there is a tendency for those close to the center of power (the "court") to usurp authority and corrupt the constitutional system. The rot can be impeded by only one thing—the voice of the truly virtuous in Parliament, the "country."

"Court" and "country" were highly charged terms in this dramatic interplay. The court party was the affiliation of those evil counselors of the king, of whom history had told much, but it was also the affiliation of a few Whigs themselves, stout republicans, indeed *reformers*, who nevertheless proved Lord Acton's dictum that "power corrupts." For those who got too close to the center of things, the "court," could not resist its beguiling influence, and eventually they were mired in depravity. The country party possessed all the opposite qualities—at least the way the story was told. It consisted of gentlemen farmers living out in the country, close to the soil, far from power's temptations, still in touch with the verities of life and above all graced with independence. They and they alone, it was held, could stem the tide of corruption.

See study aid.

The commonwealth ideology led to a fierce preoccupation with independence. Everything depended on it. The true patriot must think his own thoughts, chart his own course, and be beholden to no one. Furthermore, he must be a man of property, for property alone could guarantee his personal autonomy. This explained why the commonwealthmen were so distrustful of taxes. The power to tax was the power to take property away—the power to destroy independence. Parliament alone should exercise that power because Parliament alone represented those who would have to pay. Finally, every stalwart of the country party must be constantly at vigil lest he fall prey to the court's machinations and forfeit his precious integrity.

Just as they had their own ideology, the commonwealthmen had their own style. They were all landed gentlemen, well off for the most part but not given to extravagant display. They valued sober conduct. Their schooling covered both classical and modern learning, and often included an enthusiasm for science. Since they didn't "work" for a living—relying on the income of their estates—they had no "interests" as such and thus possessed the detachment thought necessary for public service. They also possessed a strong sense of honor and were capable of great personal sacrifice.

This romantic ideal, which traced all the way back to Cicero, made an important contribution to the American Republic. For it provided the setting in which the American elite, including the Founders themselves, came of age.

Lord of the Rings: The Temptation of Power

The psychology of tyranny has often been explored in literature. One problem, certainly, has been to explain why it is that ordinary people can become tyrants. Lord Acton offered a clue when he observed that "power tends to corrupt and absolute power corrupts absolutely." For most of us the "something more" required by human beings is unquestionably good. We want love, companionship, beauty, and grace in our lives. But what happens when chance presents us with power—a lot of power? Does the "something more" remain good? Or is it soon corrupted?

In J.R.R. Tolkien's *The Lord of the Rings* (1954), familiar throughout the western world, power exerts precisely the effect that Lord Acton attributed to it. For in the book the Ring of Power—which is something like possessing your very own nuclear arsenal—sorely tempts anyone who comes in contact with it. Staunch and faithful Boromir turns into a madman when he thinks the ring might be his; the once-fair Gollum degenerates into a loathsome reptile; and even the book's hero, Frodo, experiences moments of dark temptation. But the Ring's most unusual transformation is wrought upon Samwise Gamgee, Frodo's honest, humble, self-effacing manservant, who unexpectedly finds the ring in his custody:

> As Sam stood there, even though the Ring was not on him but hanging by its chain around his neck, he felt himself enlarged, as if he were robed in a huge distorted shadow of himself. . . . Already the ring tempted him, gnawing at his will and reason. Wild fantasies arose in his mind; and he saw Samwise the Strong, Hero of the Age, striding with a flaming sword across the darkened land, and armies flocking to his call as he marched to the overthrow of Baraddor. And then all the clouds rolled away, and the white sun shone, and at his command the vale of Gorgoroth became a garden of flowers and trees and brought forth fruit. He had only to put on the Ring and claim it for his own, and this could be.

At this rate, it is remarkable that anyone has courage enough to cast the ring to its destruction. And, indeed, no one does. At the critical moment, Frodo can't bring himself to pull the thing off his finger, and only the greed of Gollum, making a last desperate grab for it, plummets it into the fiery Crack of Doom.

From America: A Study in Heritage *Kendall/Hunt 7th edition, p. 15*

Even more to the point, it provided the conception of patriotism that would lie at the heart of American republicanism. The patriot is the person whose virtue operates as a bulwark against political corruption and moral decline. He is the backbone of the constitution.

Classical Republicanism in America

There is a legend to the effect that many of the gentry described above settled in the American colonies and shaped their political development. In fact, most early colonists were of the more humble yeoman class. Owning their own land and running their own lives, they were not the poorest of the poor, to be sure, but they were far from the rank of gentleman. In any event, political matters were not uppermost on their minds. Survival was their immediate goal, and then simply making a go of it.

But America did work a change in many of them. Their lives had been modest in England. In the colonies they beheld a breathtaking expanse of available land and began to contemplate the opportunity that must surely go along with it. Depending on how the cards fell, a small stake in Virginia or the Carolinas could be sizably increased in a generation or two. And after the introduction of slavery in the 1620s and the development of tobacco, rice, and indigo farming in the South, truly substantial gains became possible. Gov. William Berkeley of Virginia made a concerted effort to induce gentlemen to immigrate to his colony and take up large grants. In his eyes, and in those of other colonial overlords, America might well come to boast of a dazzling Old World gentry.

That didn't happen—exactly. But in the latter half of the 1600s, thousands of yeomen farmers increased the size of their holdings so dramatically that they could begin to put on gentlemanly airs, as it were, with imposing country houses, imported furnishings, fancy equipage and much else. These self-made squires were joined by another group similarly on the rise. Successful merchants in New England and the Middle Colonies often secured their own rural estates and adopted the same general lifestyle—such was the power of the gentlemanly ideal.

The rising colonial elite has been criticized for its artificiality and want of refinement. However, as historian Wesley Frank Craven points out, it was no less "real" for all that. Indeed, an up-by-the-bootstraps gentleman, while he may lack the essentials of etiquette, might rather outdo his European counterpart in some ways. In the eagerness with which the American upstarts assumed the country squire's political leadership role, for instance, they were marked by a positive zeal. In Virginia alone there were some 1,600 justices of the peace exercising both executive and judicial functions in local government—and most of them were regarded as gentlemen.

There was an even stranger sort of mirroring. The political institutions evolving in the colonies came to bear a marked resemblance to those of the mother country. The governor, for example, became not only the king's representative in a given colony but often the king's stand-in as well. And the governor's council, where sat the most distinguished members of the local elite, occasionally behaved like the House of Lords, demanding an outsize voice in the conduct of affairs.

The most striking parallel development was that of the colonial assemblies, which came to regard themselves as local "parliaments." Probably no one intended this result or even foresaw it clearly. Assemblies began as convocations of landowners discussing common problems. But consider how the participants might come to see themselves. Here is a governor and his "court," most of them powerful and well connected, sitting at Williamsburg or some other colonial seat. What keeps them from becoming a power unto themselves, and running roughshod over the people? Why, those independent men of property and accountability, of course, who live out in the country and raise their voices in "parliament." Perhaps it was not a good description of reality but, as Bernard Bailyn has pointed out, it certainly provided Americans with a potent self-understanding.

No wonder they were so taken with the commonwealth ideology. And taken they were. Pamphlets extolling the "country" and warning against the "court" saturated the colonies, as did more ambitious works such as *Cato's Letters*. By its very nature, the colonial situation framed the commonwealth drama in such a way as to flatter American sensibilities. After all, those farmers-turned-gentlemen were undeniably the good guys.

The commonwealth ideology imparted a slightly spooky cast to colonial politics. Governors and other crown-appointed officials could not conduct affairs the way they did back in England, for their authority, so far from home, was much weaker. Accordingly, they were often forced to govern by means of under-the-table bargains. Besides, graft, bribery and other forms of chicanery were only too common among the kind of individuals often appointed to such posts. Imagine how all this looked to American Puritans, who were convinced of their own rectitude, skeptical of human nature—and always on the lookout for conspiracy. Small wonder that the men of the "country" had to be on their guard.

Here in embryo, then, was the essence of American republicanism. It was based on what Jefferson would call "the aristocracy of virtue and talent"—an aristocracy quite unlike any other. For its places of honor were not reserved for the highborn few, as in the Old World, but were open to all, or at least to most, on the basis of intelligence, drive, fortitude, and other bootstrap qualities. The style of this squirarchy was one of relative simplicity, which Americans equated with purity, and in some way its members felt sanctified by the land they lived on, by the communities of which they were a part, and by Nature itself—a word they often capitalized—which in America seemed close at hand. They saw themselves as a bit more righteous than their English cousins—and a lot more righteous than their English administrators—and for that reason they tended to view politics in terms of good and evil. Aristocracy means "rule by the best." In America, it could be argued, the best really was the best.

Classical republicanism explains many of the issues that arose between England and the colonies. It explains a great deal about the American Revolution. It explains what many Americans had in mind when they declared independence and reconstituted their provinces as "republics," and what the delegates to the Constitutional Convention had in mind when they explored republican concepts and devised republican mechanisms. Most importantly, it explains why the conception of America as a "republic of virtue" would become a central founding myth.

Liberal Republicanism

Growing alongside classical republicanism was a second set of ideas which we will call *liberal republicanism*. Many of the same people who embraced the one also embraced the other, yet not without certain difficulties. Liberalism was not necessarily "republican" in all respects and republicanism was not necessarily "liberal." For example, classical republicanism preached restraint of trade, the "just price" doctrine, and a variety of sumptuary laws—and it held commercial prosperity to be the high road to perdition. Liberal republicanism embraced all the opposite points. Yet somehow the Founding had to straddle them both.

See study aid.

Just Price Doctrine: The "just" price at which every good should be sold.

Sumptuary Laws: Laws regulating personal behavior on moral or religious grounds; everyone must "toe the mark."

The term *liberal* is used here to designate personal freedom—the freedom of the individual against society. We take it to mean the freedom to go one's own way, think one's own thoughts, and live one's own life. Where classical republicanism offers the citizen freedom *in* government—that is, freedom to participate in the political process—liberal republicanism offers him freedom

Liberal Republicanism embraced principles that were particularly conductive to commercial activity.

from government. The two are rather different propositions and yet they are connected by an important tie. What sort of government would respect the citizen enough to keep out of his personal affairs *except* a government in which the citizen himself was powerful?

Personal freedom played virtually no part in the ancient republics. There the citizen found meaning in politics not private life. The very idea of such freedom would have seemed odd to the ancients. How could a person be virtuous, they would ask, if virtue's great projects were of no interest to him? (Imagine, say, the citizens of Athens convened to discuss an urgent political matter, when someone notices that a key figure is missing—he had stayed home to paint his house.) However, personal freedom did become important in seventeenth-century England, and even more important in the American colonies.

John Locke and English Liberalism

John Locke: http://www.orst.edu/instruct/phl302/philosophers/locke.html

The importance began with the rise of English commerce in the 1500s. England's had been a sleepy subsistence economy. Now, rather suddenly, common land was fenced off for the grazing of sheep, and the woolen industry opened the way to expansive commercial development. Foreign trade turned the inward-looking English outward and gave rise to a new class of merchants, artisans, and functionaries, whose way of life would be considerably different from that of the countryside.

A new outlook followed. Where Englishmen had once thought of their society as an organic whole, some now came to see it as diverse and divergent groups, each with different requirements. While the older groups preferred the status quo, the newer ones stood to gain by innovations such as religious toleration and the free expression of ideas. They would also benefit from the abatement of aristocratic influence, of course, and from enhancements in the rule of law. Theirs was a world of property, not privilege, and property required them to rethink almost everything.

In one way or another, the new middle class worked for these ends through political activity in the House of Commons, and they played into the great controversy between king and Parliament described above. Whiggism, too, embraced the new politics, for many Whigs were merchants or manufacturers themselves. It remained only for someone to put liberalism all together in a single coherent package. That person was John Locke.

As secretary to the Earl of Shaftsbury, the founder of the Whigs, Locke saw liberalism close at hand. Indeed, he was in exile for political activism when his most influential work was written. What Locke set about to do in his *Two Treatises on Civil Government* (1690) was take issue with the Stuart monarchs' claims that English subjects owed their king an absolute and unqualified obedience because of his "divine right" to govern. Such claims, Locke argued, missed the whole purpose of government as originally constituted. Government was not about *ruling* the governed so much as *protecting* them.

Locke's Five Principles: See study aid.

In the *Second Treatise*, Locke carefully reasoned his way through five principles, which together spelled out the liberal case. They are important to us, all five, because they were to be echoed so clearly in the American Founding.

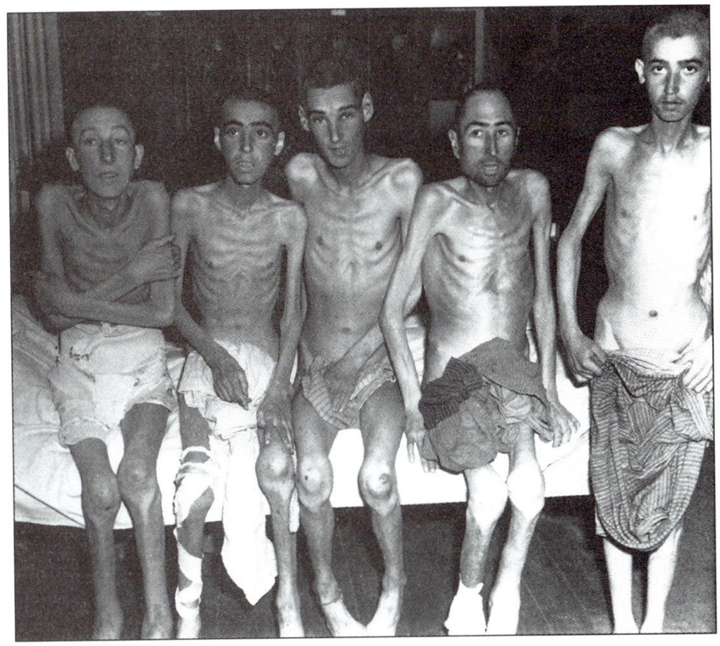

Nazi concentration camp, 1945. What are natural rights? What happens when a society systematically violates such rights?

To begin with, said Locke, in the original "state of nature"—a hypothetical condition assumed to exist in the absence of government—human beings must have lived in perfect freedom and a rough equality. In such a world, moreover, all had the same rights, for "rights" were granted not by man but by Nature. All had the right to live their lives, enjoy their liberty, and make the most of their property. As long as they did not disturb the rights of others, things promised to be well with them.

Some few, however, did disturb the rights of others, and beyond that there inevitably arose disputes. So the need was manifest for law of some kind and for a common judge. Accordingly, groups of people came together and agreed to form political society and establish government. There was no divine mandate for this "social compact," Locke asserted, only a simple and straightforward need. Government was a human invention and made to serve a human purpose.

The nature of that purpose was Locke's third principle. Government could have but a single end in view, he argued, and that was to protect the rights of its citizens. (After all, why else had they created it?) Those rights could never be surrendered, as the Stuart monarchs maintained, for they had been granted by Nature, but they could be disregarded by persons ill disposed. Government's job—its *only* job—was to make sure that didn't happen.

Government, then, existed by the *consent* of the governed and that alone—Locke's fourth principle. It meant that government was beholden to the people for its legitimacy. It could not presume to govern in God's name and the people be damned. There had to be elections, not democratic elections necessarily but "popular" elections nonetheless, where those who paid the bills had a chance to have their say.

Finally, if government violated the terms of consent, if it lost track of what it was supposed to do and whom it was supposed to serve, the people had the right, indeed the duty, to alter or abolish it, even if that meant revolution. The Glorious Revolution took place after the *Second Treatise* was written but before it was published. Nothing could have dramatized its argument more effectively. For here was the king, James II, accused of violating the trust of his people, and here was the result for all the world to see.

Of all the implications of Locke's philosophy, the most important was personal freedom. For if government stuck to securing the rights of its citizens, the

Natural Rights and the Uniqueness of Man

The appointment of [Peter Singer] to a tenured professorship of bioethics was unanimously recommended by a Princeton search committee and was approved by President Harold Shapiro, who chairs the National Bioethics Advisory Commission.

Peter Singer comes to campus to teach that truly ethical behavior will not flourish until humanity abandons the fallacy . . . of "the sanctity of life." He rejects "the particular moral order" which supposes that human beings are extraordinarily precious because God made them so. He also rejects secular philosophies that depict human beings as possessing a unique and exalted dignity that sharply distinguishes them from, and justifies their "tyranny" over, other species of animals.

Singer subscribes to utilitarianism, which holds that there is a single goal for human conduct— satisfaction of preferences and avoidance of suffering. Singer believes that . . . one pleasure is as good as another. Regarding humans, he says that assigning intrinsic moral significance to birth is arbitrary and logically indefensible. Birth is morally insignificant because a newborn, like a fetus, is incapable of regarding itself as "a distinct entity with a life of its own to lead." Because there are, he says, degrees of personhood, the intrinsic moral significance of the taking of the life of an individual gradually increases, like the physical being of the individual, from near nothingness in infancy.

He does not deny that killing a fetus or a baby involves killing a human being. Singer says infanticide is not necessarily more morally important than abortion, which is morally negligible. In fact, some infanticide is not even as important as, say, killing a happy cat. Killing an infant is never killing a person and is morally permissible in at least two kinds of situations. One is when a handicapped baby faces a life in which suffering will predominate. Another situation when infanticide is justified is when parents with a handicapped baby . . . will, if relieved of the burden of the baby, have another baby which will be happier than the handicapped baby would be, and will bring the parents more happiness. The logic of his position is that until a baby is capable of self-awareness, there is no controlling reason not to kill it to serve any preference of the parents.

From "Life and Death at Princeton" *by George F. Will,* Newsweek, *September 12, 1999, pp. 80 & 82.*

citizens were bound to enjoy such freedom in wide measure. They could do what they wanted with their lives. They could enjoy their liberty to the fullest. And most of all, they could buy and own and improve and sell their property. They had, as humorist P.J. O'Rourke insightfully put it, "Life, Liberty, and Ka-ching!, Ka-ching!, Ka-ching!"

One other implication is worth noting. Like the classical republicans, Locke often bowed to the importance of virtue. But for Locke, virtue was significant only in its milder bourgeois forms, such as obeying the law and conducting oneself civilly; it was not like the two-fisted *areté* of old. The people Locke was writing about were middle-class Englishmen. They weren't driven by a quest for honor on the battlefield or distinction in the halls of government so much as garden-variety comfort and prosperity. In other words, Locke's assumptions about human nature were different from those of the classical republicans, and his approach to the Good Society was not ancient at all but modern.

American Liberalism

Just as classical republicanism worked itself out in the American colonies, so too did liberal republicanism. Indeed, the American experience of building political societies seemed to reflect the very world Locke had described.

To begin with, Americans were far more equal than their European counterparts. They were not equal by present-day standards, of course, for there were property-owning requirements for the right to vote, and some groups—women, blacks, Native Americans—could not vote at all. But there was no formal aristocracy—it couldn't take root in a land made for hard work—and that fact was immensely important. It meant no titles of nobility, no special legal privileges, no feudal rents or dues. Consequently, where legal equality was an ideal to be striven for in Locke's England, it was a working reality across the Atlantic.

Legal Privileges: For example, some had the legal privilege of wearing powdered wigs and some did not.

Feudal Rents or Dues: Yearly rents and other obligations owed to titled noblemen or the crown by landholders and tenants.

With available land, high wages and burgeoning opportunity, American society was also much more mobile than in England. Fortunes were made and lost. Wealth circulated. One never knew from year to year who one's betters really were, for the social register changed so rapidly. People gradually stopped thinking in the deferential terms typical of aristocratic societies. Instead of wondering where their place was in the social hierarchy, they dreamed of scrambling to the top themselves. A few went into politics, too, for in most of the colonies yeomen could easily meet the property requirements for political participation. Here was yet another striking contrast to the English case.

Second, if Americans were more equal than their European counterparts, they were also more free. Colonial governments were small and understaffed by English standards, and they often lacked the capacity to be meddlesome. Besides, settlers were dispersed over a vast terrain—Virginia alone was larger than England—and those in the backcountry were effectively out of range. Where a household in England would be under the constant watch of the parish authorities, for example, their cousins on the American frontier might never see a state-supported clergyman in their whole lives. Americans carried firearms too and knew how to use them. The local militia, out drilling on leafy afternoons, added to the sense of rugged self-sufficiency.

This fact pointed to still another Lockean feature of the American scene. The real government for most colonials, that of the township or county, was a government in which the local citizens were likely to have a good deal of influence. Indeed, it was a government they might well have created themselves. John Locke would have been impressed. He never supposed that the "state of nature" he described had actually existed, or that its inhabitants had actually executed a "social compact"—these were more in the way of theoretical constructs. In America, however, there actually were places where people lived without government, and there were any number of instances in which settlers had devised their own.

Proprietary Colonies: Royal grants of land to individuals or groups who then retained discretionary authority over the designated region, i.e. Maryland and Pennsylvania.

Entire colonies had been created that way. Several of them had begun as corporate enterprises, and the officials of the company in question had drawn up the charter of government for royal approval. Other colonies had begun as proprietary grants to an individual or a group, and here again the ground rules had been worked out by those who would have to live under them. In a few of the colonies the settlers themselves had literally sat around a table and worked up their own charter agreements. They took the idea of a church covenant, as described in the preceding chapter, and applied it to civil society, each signatory pledging to obey the laws and abide by the majority will.

In none of these cases was there anything mysterious about the origin of the government or the source of its sovereignty. These governments, at least, were clearly creatures of practical utility, not divine mandate, and they clearly drew their authority from the consent of the governed, just as Locke said. Their charters, moreover, were drafted by men who knew their Aristotle, and they included provisions for mixing and balancing the various social orders. Americans were not only creating their own governments, they were attempting to create them thoughtfully.

American colonials also experienced "rights" in a different way. Since legal equality was the rule, everyone (excepting the aforementioned women and minorities) had more or less the same rights as everyone else. It is true that symbolic activities such as horse racing or wearing a powdered wig were frowned upon if one was not a bona fide gentleman. But honest-to-goodness privilege, which affected the English in so many ways, did not really exist in America. In a world that favored concrete ability and practical results, it was difficult to maintain, for instance, that any individual ought to have the hereditary privilege of bringing his neighbors to trial and passing judgment on them. Locke would be impressed once again. Where his "natural rights" were largely theoretical in England, in the colonies they were a working reality. And when the colonists boasted of their traditional "rights of Englishmen," they had more to boast of than most Englishmen did.

American liberalism existed not only in the conditions of society but in its institutions as well. We have seen how colonial assemblies came to play a role similar to Parliament's. These bodies vigorously represented the interests of the propertied—a fundamental aspect of the liberal worldview. ("Ka-ching!, Ka-ching!, Ka-ching!") And as colonial leaders gained ever more experience, they were better and better able to hold their own against imperial administrations. The royal governor found himself confronting precisely the same go-easy-on-the-taxes opposition as the king was facing in Parliament.

The assemblies operated on the basis of an important principle, also borrowed from England, called *representation*. As practiced in the ancient republics, representation played an aristocratic role in government, according to Aristotle's idea of the "mixed regime." Since the people would likely choose their representatives from the wisest and most virtuous elements, an assembly of representatives would consciously reflect *areté*. But there was a liberal aspect to representation as well, and it was especially evident in the colonies. Representatives were regarded as spokesmen for specific, definable interests—Boston shipping, Philadelphia banking, tobacco planting on the lower James. It was only when such interests had an audible voice in government that the liberal society worked.

In England, liberalism and republicanism jostled each other uneasily. After all, one of them was all about virtue and honor and the public good, and the other was all about personal freedom and material self-interest. In America they got along pretty well together. The secret of their friendship lay in the relaxed, apolitical attitude of most Americans. As long as they felt free, the colonial middle classes were content to leave the governing process to their betters. They did participate in politics, but only in limited and deferential ways. There were plenty of other things of more interest to them—expanding the farm, adding onto the house, investing in a voyage to the Gold Coast. Let the truly virtuous worry about governance, they said.

The governing classes, in turn, knew their countrymen. They knew that the liberal society was an accomplished fact, and that most Americans had a vested interest in it. They were willing to live in a world that was fundamentally ambiguous—part classical, part liberal—and broad enough for both.

The quarrel with England in the 1760s imperiled liberal interest as well as republican virtue. New rules and regulations for the empire raised the specter of busybody government. New taxes laid by a "foreign" Parliament undermined the security of property. And new judicial procedures, such as trial without a jury, threatened the rule of law. In a liberal society, jeopardizing the "natural rights" of citizens was the unpardonable sin, according to John Locke. Indeed, it was grounds for revolution.

Democratic Republicanism

In the ancient world, there was a good deal of suspicion against rule by the *demos*. The many, as Aristotle called them, had their own behavioral style and it was not considered good. They could be shortsighted and self serving. They were easily divided into competing interests. They were often willing to buy the flatteries of demagogues and fall in with their ambitious designs. All too frequently, it seemed, government by the many led to some sort of factional anarchy, and that in turn led to tyranny.

Yet Aristotle also noted an inherent drift in republican politics toward that very thing, the rule of the demos, owing to the fact that the many, tugging and pulling against the few, would always find a way to win in the end. This explained why Aristotelian political science sought to refine democracy with checks and balances. The hope was that a proper constitution might at least slow the inevitable decay into chaos.

Democracy was nothing to be feared in Georgian England. For the glittering titled "few" of that world not only monopolized political and social power, they monopolized the English mind itself with their ancient names, sprawling estates, and ever so graceful manners. Liberalism was hardly more democratic itself. In theory, liberals held that all men were created equal. In practice, they represented a propertied elite almost as small as the aristocracy. In England, in other words, there simply weren't enough franchised individuals to pose a credible democratic threat.

In the colonies, however, things were different.

American Democracy

To begin with, owing to the absence of a formal aristocracy, the democratic element was necessarily stronger in the American mix. Moreover, with the exception of a few urban areas, most Americans were of at least modest means, so there was no large desperately poor class, as in Europe. Equally important was the feeling of equality that most Americans shared, blurring class lines and making success seem more achievable.

Politics adapted to these New World realities. We have seen that most freehold farmers were able to meet the property requirements for voting, and even in the cities many were well enough off to participate. There were steady pressures to expand the franchise even further and to represent an ever greater variety of interests, for freedom included the desire to use public power for one's own ends. For example, small farmers, seeking to expand their holdings, behaved like any other interest by seeking loans on favorable terms and supporting laws favorable to debtors.

There were democratic implications in American Puritanism, as we have seen, and even more pronounced ones in Puritanism's successors. In the Baptist movement, for example, the Puritan doctrine that souls were equal in the sight of God came to occupy a central place in worship. Even more central was the Baptist idea that souls must be free to choose the right.

Classical republicanism did not mix well with democracy at first, for many commonwealthmen, following Aristotle, doubted the virtue of ordinary people. Here again, however, the American experience began to suggest otherwise. Ordinary people had to bear a great deal in the travail of the Revolution. The embargo movement, for example, called upon them to forgo imported luxuries in order to send the British authorities a strong message. Overwhelming numbers complied. And their valor on the battlefield was tested time and again, to say nothing of their willingness to endure hardship in the unspeakable British prisons. There was nothing soft about democratic virtue.

Even so, democratic virtue was appreciably different from classical *areté*. Ordinary farmers lacked the passionate identification with politics that typified landed gentlemen. Urban tradesmen lacked the "stake" in society that property conferred, not to mention any mystical connection to the land. And the have-nots of the world, wherever found, were the needful elements, those who by definition lacked independence.

For these and other reasons, democracy was held in deep suspicion by most of the American Founders. Indeed, when they saw mobs in the streets at the time of the Stamp Act, they thought of the Athenian democrats crying for the blood of their rivals. But the Founders had no difficulty with a second cousin they referred to as "**popular government**," a mixed and balanced form in which some democratic elements were allowed. With popular government the common people would have their own role to play even if it was limited. Their lot would be to uphold the law, yield to proper authority, chose wise and virtuous leaders, and lead quiet, sober lives. The democratic genie must be kept in its bottle.

And yet there were signs that the genie escaped anyway, in the turmoil of the Revolution. For if the Revolution was partly a republican movement and partly a liberal one, it was partly a democratic movement too.

We have seen above how both classical and liberal republican developments in America appeared to be thwarted by British imperial policy. The same was true of democratic republicanism. For several historians of the period have concluded that there were widespread fears that Britain's contemplated administrative reforms would threaten the democratic gains already made. Driven by such fears, artisans and laborers in the colonies became intensely active politically, and in key cities created the roughhouse that brought the Revolution to a flash point.

There were other evidences of democracy at work. In several states the electorate widened dramatically after the war was over. And lower houses of assembly, reflecting the democratic influence most directly, were given sweeping powers in some of the new state constitutions. The tone of politics became democratic too. In fact, much of the clamoring of special interests that James Madison found so objectionable in Virginia—and that led him to think about a new federal constitution—was simply what we would call politics as usual.

Threading its way among these changes was an outline theory of democratic government. Radicals like Samuel Adams and Thomas Paine searched for it in the writings of Jean Jacques Rousseau, a figure of the French Enlightenment who had stressed the role of the common people in political morality. Others looked to influential Scotsmen of the day such as Lord Kames and Thomas Ferguson. In Kames they found the concept of society as an organic whole in which all classes have legitimate roles to play. In Ferguson they found the concept of "common sense"—holding that truth was accessible not just to philosophers but to everyone.

In time, Americans would expand on these ideas. They would borrow from Puritan notions of grace and their own feeling of chosenness, and argue that the common people of the New World, unlike those of the old, were endowed with a practical wisdom they called "horse sense," with a form of virtue more relevant than *areté*, and with an uncorrupted sense of right and wrong. They would believe in regenerate human nature—that is, Calvinism flip-flopped—which political empowerment might further improve. They would have a liberal's dislike of government together with a romantic's enthusiasm for volunteer service and humanitarian causes. They would often be detained by questions of justice, and once in a while preoccupied with reform. They would disparage privilege, frown at pretentiousness, and eventually conceive a violent dislike of slavery.

Democratic republicanism would create its own hero figure, too. This character would begin as the "minuteman" of the Revolution and wind up as the

Jean Jacques Rousseau:
http://www.utm.edu/research/iep/r/rousseau.htm
http://www.wsu.edu:8000/~dee/ENLIGHT/ROUSSEAU.HTM

Jean Jacques Rousseau: (1772–1778) Swiss-French philosopher and political theorist. Celebrated for his theories of the "natural man" and "social contract."

Sam Adams:
http://www.dell.homestead.com/revwar/files/adams2.htm

Thomas Paine:
http://www.bartleby.com/133/index.html

Democratic Republicanism created a series of hero figures that included the mythical American cowboy.

Hollywood "cowboy." He would lack the burnish of aristocracy, perhaps, but everything else about him would call ancient Greek virtue to mind. He would be inflexibly moral and unflinchingly brave. He would live close to Nature out on the wide prairie and play hunches to catch the bad guys. And he would save the widow's ranch from the shady railroad lawyer for no better reason than it was the right thing to do.

Thomas Jefferson was to become the new democracy's patron saint. In his supple mind he connected the idea of opportunity with the idea of the common people—and both of those with the idea of a regenerate America. Jefferson was a classical republican who truly believed in rule by the best. But his aristocracy of talent and virtue would eventually be open to anyone, not just a privileged elite. He would, as he said, comb through each new generation for the best and the brightest, and provide them with a good education. If their parents were hardscrabble farmers, so much the better.

Jefferson's vision of America was not unlike John Winthrop's city upon a hill. In such a place, the Virginian thought, human nature might shed its acquired dysfunctions and find its way back to first principles. Wealth would lose its odor of decadence, liberalism would become generous and civic minded, and the common people would cease to behave like an Athenian mob. For Jefferson, the American Founding might be a triumph for the whole human race.

Everyone's Good Society

We can see all three conceptions of republicanism in the lives of the American Founders. Some of them were rising merchants. Some were entrepreneurs. Some were members of that budding American gentry. A few were lowborn and self-made.

A small but significant number of the Founders truly were figures out of ancient Greece. These "marble men," as Christopher and James Lincoln Collier called them in their study of the Founding, took classical *areté* as their guiding principle. They made patient and systematic efforts in the pursuit of excellence, achieving a kind of secular sainthood in the eyes of their contemporaries. George Washington certainly belongs in that category. But there were others as well.

John Wayne: American

(Excerpts from *Reagan Eulogy for John Wayne*)

In 1979 Ronald Reagan was gearing up for his successful 1980 campaign for the presidency. He took the occasion of John Wayne's death to sound some of the patriotic themes that would serve him so well in 1980 and 1984.

We called him Duke, and he was every bit the giant off screen he was on. Everything about him—his stature, his style, his convictions conveyed enduring strength. In this country and around the world, he was the most popular box-office star of all time. For an incredible 25 years he was rated at or around the top in box-office appeal. Yet John Wayne was more than an actor; he was a force around which films were made. As Elizabeth Taylor stated last May when testifying in favor of the special gold medal Congress struck for him: "He gave the whole world the image of what an American should be."

He was born Marion Michael Morrison in Winterset, Iowa. [After] his family moved to California . . . he picked up the nickname Duke. He rose at 4 A.M. to deliver newspapers, and after school and football practice he made deliveries for local stores. He was an A student, president of the Latin Society, head of his senior class and an all-state guard on a championship football team.

He accepted a full scholarship to play football at the University of Southern California. There coach Howard Jones, who often found summer jobs in the movie industry for his players, got Duke work in the summer of 1926 as an assistant prop man on the set of a movie directed by John Ford.

From his job in props, Duke worked his way into roles on the screen. During the Depression he played in grade-B westerns until John Ford finally convinced United Artists to give him the role of the Ringo Kid in his classic film *Stagecoach*. John Wayne was on the road to stardom. He quickly established his versatility in a variety of major roles.

When war broke out, Duke tried to enlist but was rejected because of an old football injury, his age (34), and his status as a married father. He flew to Washington to plead that he be allowed to join the Navy but was turned down. So he poured himself into the war effort by making inspirational war films. To those back home and others around the world he became a symbol of the determined American fighting man.

Duke could not be kept from the front lines. In 1944 he spent three months touring forward positions in the Pacific theater. Appropriately, it was a wartime film, *Sands of Iwo Jima*, which turned him into a superstar.

As one of the true innovators of the film industry, Duke tossed aside the model of the white-suited cowboy/good guy, creating instead a tougher, deeper dimensioned western hero. He discovered Monument Valley, the film setting in the Arizona-Utah desert where a host of movie classics were filmed. He perfected the choreographic techniques and stuntman tricks which brought realism to screen fighting. At the same time he decried pornography, and blood and gore in films. "That's not sex and violence," he would say. "It's filth and bad taste."

Duke went to Vietnam in the early days of the war. He scorned VIP treatment, insisting that he visit the troops in the field. Once he even had his helicopter land in the midst of a battle. When he returned, he vowed to make a film about the heroism of Special Forces soldiers. The public jammed theaters to see the resulting film, *The Green Berets*.

I never once saw Duke display hatred toward those who scorned him. His work habits were legendary in Hollywood—he was virtually always the first to arrive on the set and the last to leave. To him, a handshake was a binding contract. Duke's generosity and loyalty stood out in a city rarely known for either. When a friend needed work, that person went on his payroll. When a friend needed help, Duke's wallet was open. He was also loyal to his fans.

Like any good John Wayne film, Duke's career had a gratifying ending.

In the 1970s a new era of critics began to recognize the unique quality of his acting. The turning point had been the film *True Grit*. When the Academy gave him an Oscar for best actor in 1969, many said it was based on the accomplishments of his entire career. Others said it was Hollywood's way of admitting that it had been wrong to deny him Academy Awards for a host of previous films. There is truth, I think, to both these views.

Yet who can forget the climax of the film? The grizzled old marshal confronts the four outlaws and calls out: "I mean to kill you or see you hanged at Judge Parker's convenience. Which will it be?"

"Bold talk for a one-eyed fat man," their leader sneers.

Then Duke cries, "Fill your hand, you sonofabitch!" and, reins in his teeth, charges at them firing with both guns. Four villains did not live to menace another day.

"Foolishness?" wrote *Chicago Sun-Times* columnist Mike Royko, describing the thrill this scene gave him. "Maybe. But I hope we never become so programmed that nobody has the damn-the-risk spirit."

Earlier this year, when doctors told Duke there was no hope, he urged them to use his body for experimental medical research, to further the search for a cure. He refused painkillers so he could be alert as he spent his last days with his children. When he died on June 11, a Tokyo newspaper ran the headline, "Mr. America passes on."

"There's right and there's wrong," Duke said in the *The Alamo*. "You gotta do one or the other. You do the one and you're living. You do the other and you may be walking around but in reality you're dead."

Duke Wayne symbolized just this, the force of the American will to do what is right in the world. He could have left no greater legacy.

From Fulfilling the Founding,
Pearson Custom Publishing,
1999, pp. 247–252.

Virtually all of the Founders resonated with some version of the republican ideal and frequently with more than one. There were gentlemen with an entrepreneurial flair. There were members of the elite who inclined toward democracy. There were town-bred merchants saving up for a rural estate and the life of a country squire. There were scholars proficient in practical politics, and vice versa.

For all of them, republicanism had given rise to some uniquely American conception of the Good Society. Its main elements included:

See study aid.

Government for the benefit of the people, not the rulers.

Real and practical freedom for the individual.

Expanding economic opportunity.

A virtuous citizenry.

The rule of law.

Not all of these elements fit together comfortably, as we have seen. Moreover, each of them grew out of a different understanding of human nature. Before the Founding was completed, a great deal would have to be sorted out, and even then ambiguities and inconsistencies would permeate the final result. Was the US Constitution addressed to classical republicanism, liberal republicanism or democratic republicanism? critics asked. The answer: all three.

Perhaps this was a sign of strength rather than weakness. If there was an intellectual fuzziness to the American Republic, there was also a plastic, protean quality. America really could be different things to different people—which certainly helps to make it interesting.

Unifying all three conceptions was the assumption that an Everyone's Good Society was truly possible. Some idea of America as a promised land, a place apart, a historical exception, a *Novus Ordo Seclorum*—some founding myth, in other words—worked to fuse the disparate parts into a whole.

CHAPTER 3 STUDY AID

1. Enlightenment:

- Calvinism, as filtered through the American Puritans and their "New England Way," represented a theological solution to the human predicament.
- This was not the only influence on early American thought. In the secular world of understanding, the European Enlightenment generated profound changes in how western society interpreted the human predicament and the nature of political society.
- Enlightenment philosophers tended to discount virtue as a motive force in human affairs. They placed their emphasis on the role of self-interest in human interactions. They argued that reason, represented by 'science,' offered a better prospect of providing solutions to the human predicament than any set of ambiguous precepts might.
- Rather than hope for the universal adoption of lofty sentiments, they concerned themselves with the practical and pragmatic—what could be proved or demonstrated (*a posteriori*) rather than what dwelt in the realm of the metaphysical outside objective reality (*a priori*).
- Self-interest was not seen as being inherently evil since it did lead to significant advances and cooperative behavior. It was simply predictable and reliable.
- Because it was predictable, it was assumed to be governable through the counterbalancing of competing interests by the judicious application of some form of structure (counterpoise).
- In the British North American colonies, this 'Enlightenment' was first filtered (refracted) through English experience and then the colonial experience. Out of this background emerged several strains of republican reasoning that suggested how a good society could be established and maintained.

2. Classical Republicanism:

- Espoused the rule of law, the common good, virtue and honor, and constitutionally protected rights and liberties.
- Property (land) was the source of individual independence. Being possessed of 'independent means' meant one could remain free of specific 'interests' and act in the best interest of society as a whole.
- After the Glorious Revolution of 1688, England became a constitutional monarchy.
- In practice, power became centralized at the Royal Court through the use and abuse of patronage and other forms of royal political influence.
- This opposition perspective became identified as the 'commonwealth' or 'country' position.
- Taxes were particularly obnoxious to those of the country persuasion.
- A variation of this situation occurred in America where yeoman farmers

who became landed gentry imbibed this philosophy.

- Gradually, the American variant of classical republicanism came to place its emphasis on the creation of an aristocracy of talent, merit, and virtue as well as on simplicity.

3. Liberal Republicanism:

- Classical republicanism represented freedom within government through virtuous participation. Another approach espoused freedom from government to the fullest extent possible.
- This position tended to be adopted by the middle class in colonial society, the merchants and entrepreneurs. They too endorsed the rule of law, but also favored toleration, protection of all forms of property, and the abatement of privileges and restrictions—in effect, anything which might encourage commerce.
- John Locke had provided a conceptual framework for this position by arguing that government obtained its legitimacy and authority directly from the governed:
  a. Initially, mankind lived in a state of nature
  b. By entering into a social contract, people created a political society—both society and government were inventions.
  c. Government was instituted to protect 'natural rights' such as liberty, equality, and property.
  d. Government's legitimate authority was derived from the consent of the people.
  e. The people were therefore sovereign and could revoke a government if it no longer protected their rights.
- Though this approach focused on self-interest, it did assign a modest role to virtue in the form of 'civic virtue'—obey laws, act civilly, etc.
- Some elements of classical and liberal republicanism contradicted each other. However, they did not come into direct conflict in colonial America. If liberal republicans felt free then they did not feel threatened if a virtuous minority of classical republicans sought to govern.

4. Democratic Republicanism:

- In colonial America there was a large middle class, social and economic mobility, relative equality, prosperity, easily met voter requirements for males, and no formal aristocracy.
- Democratic precedents and practices had inadvertently been introduced through the Puritan experience and ready access to property ownership and thus to the vote.
- Popular government through representation was viewed as acceptable, though literal democracy was not.
- The working class was not directly repressed and came to have its own

aspirations.

- A belief developed that the common people of America were more pure and wise than elsewhere and that their common sense could generally be trusted. In addition, they possessed what could be called a heroic form of virtue.
- It was assumed that the common people were too bogged down in the mundane aspects of existence to govern other than through representation. However, they could rise in a mobile society, through talent and virtue, to positions of leadership.
- Jefferson believed that talent and virtue were to be found in all classes. Through equality of opportunity, a natural aristocracy of virtue, talent, and merit would arise and lead.
- The rise of an aristocracy of talent, virtue, and merit would enable America to achieve its destiny as a 'city upon a hill,' epitomizing the principles of virtue, wisdom, and prosperity.

5. American Synthesis:

- In America all three republican styles could be blended without coming into direct conflict. In a relatively open society, the differences among the three styles were not forced into a confrontation with each other.
- Classical republicanism and its propertied class contributed virtuous representatives.
- Liberal republicanism and its merchant and entrepreneurial class contributed civic virtue and enlightened self-interest.
- Democratic republicanism through the working class contributed a seemingly inexhaustible pool of talent, heroic virtue, and commons sense when called upon.
- The 'good society' in colonial America was a composite drawn from the following prescriptions:

  a. Republicanism—for the people, by the people.
  b. Liberty—individual freedom.
  c. Opportunity—equality, mobility, and prosperity.
  d. Order—founded on virtue (classical, civic, and heroic) and a rule of law premised on natural rights.

- Such a constellation of possibilities has seemingly never existed before. Perhaps America was truly exceptional, after all.

CHAPTER 4

The American Revolution

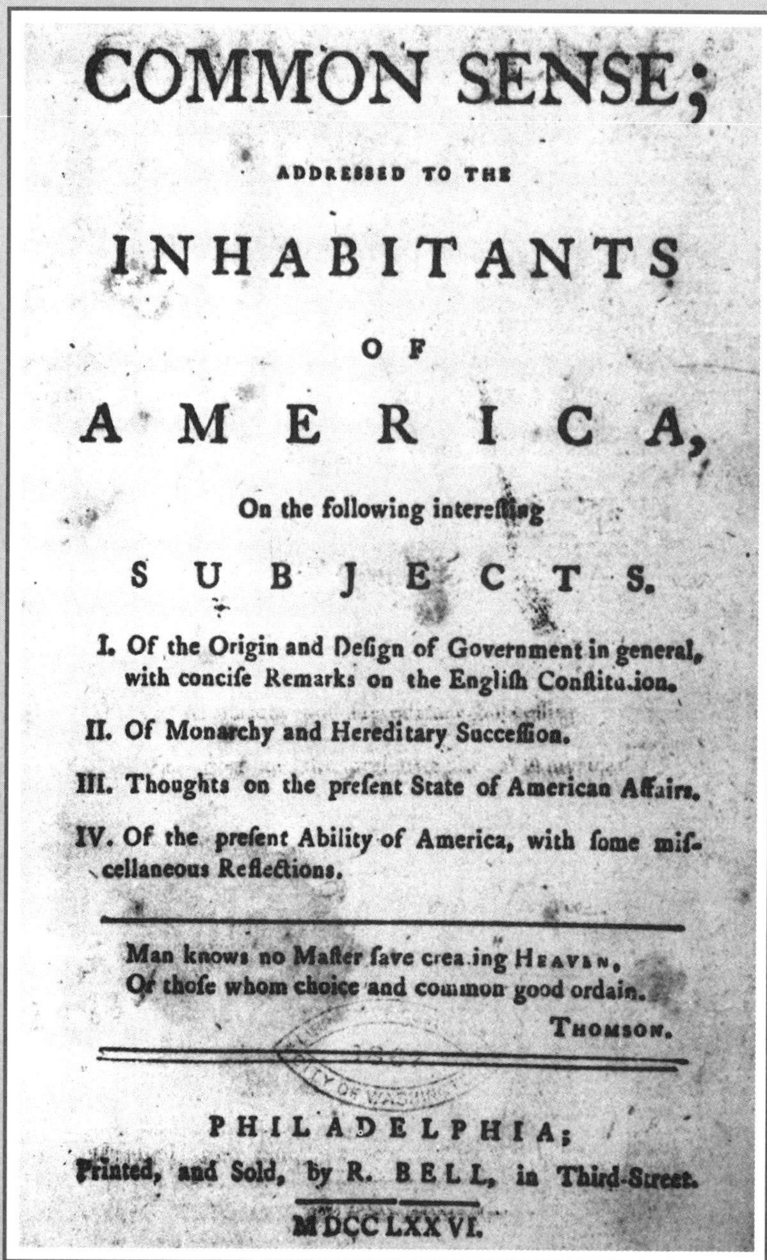

COMMON SENSE;

ADDRESSED TO THE

INHABITANTS

OF

AMERICA,

On the following interesting

SUBJECTS.

I. Of the Origin and Design of Government in general, with concise Remarks on the English Constitution.

II. Of Monarchy and Hereditary Succession.

III. Thoughts on the present State of American Affairs.

IV. Of the present Ability of America, with some miscellaneous Reflections.

Man knows no Master save creating HEAVEN,
Or those whom choice and common good ordain.

THOMSON.

PHILADELPHIA;

Printed, and Sold, by R. BELL, in Third-Street.

MDCCLXXVI.

The American Revolution was more than just a glorified colonial rebellion. It was a true revolution—the first of several in the modern world. It inspired a wave of republican upheavals throughout Europe and the Americas, and made people rethink politics from the ground up. Arguably, in fact, it effected a turning point in human history.

The Revolution was also the premier event of the American Founding. At its beginning there were thirteen British colonies with little in common and few bonds of sympathy. At its end there was a United States of America. It is easy in the light of subsequent history to miss the epoch-marking significance of that transformation. But consider it. An entirely original polity, of continental size and global influence, emerges from a world of monarchy and aristocracy and religious establishment, proclaiming new doctrines, advancing new principles, relying on new mechanisms, and describing the whole as "self-evident truth." Strangest of all, it works.

The Coming of the Revolution

As schoolchildren, Americans learn all about the events leading to the outbreak of hostilities in April of 1775. These events—like Paul Revere's ride—have long since taken on the hyperrealism of myth.

Yet they were by no means inevitable. It is true that American and British institutions had been on divergent courses of development from the beginning. It is also true that political maturity in the American colonies pointed toward eventual home rule. But American institutions were as English as England's, growing out of the same historical experience and the same seedbed of ideas. And if the two societies were on divergent courses in some ways, they were on parallel courses in others. We could easily imagine American nationhood coming about by *e*volution rather than *re*volution, as it did in Canada or Australia. Virtually on the eve of hostilities, Americans were still toasting the king's health and celebrating his birthday, picturing themselves the most loyal of subjects. The first thing we must explain, then, is what happened to this particular corner of the British Empire—what made it mysteriously self-destruct?

In the beginning there was no empire as such, only colonies. These had been established in a variety of ways and for a variety of purposes. The colonies

Mercantilism: http://www.westga.edu/~cscott/history/mercan.html

sometimes doubled as religious havens, as we have seen, but they also came to play a broader role described by the term ***mercantilism***. That role was to strengthen the British economy by providing crucial products—such as tobacco, rice or naval stores—and by extending the British market for manufactured goods. The governing idea of the mercantilists was to achieve complete economic self-sufficiency.

Accordingly, toward the close of the 1600s, Britain organized the colonies into an empire. The subordinate relation of "chicks" to "hen," as the authorities sometimes put it, was codified into a series of laws known as the Navigation Acts, designed to insure that the colonies continued to advance English commerce generally and certain political interests in particular. All trade was to be carried in English ships. Certain colonial exports must go directly to London before being transshipped elsewhere. And virtually all colonial imports must come through London as well. In theory, the empire was supposed to benefit the colonies as well as the mother country. Some historians argue that it essentially did so, at least in the beginning, while others maintain that the benefits went mostly to London merchants. Either way, the colonies enjoyed a cartel monopoly in supplying important products, along with assorted franchises, bounties and other incentives. They were by no means cynically exploited.

The colonies also had the international situation to consider. Other empires were established in North America, and when relations among them grew tense in Europe—or erupted into actual war—their colonies played the role of proxies and pawns. Between 1689 and 1763 a series of four major conflicts was fought between Great Britain and France, and in all of them colonial militias saw action, colonial cargoes were confiscated on the high seas, and colonial settlements became targets of attack. As long as the world remained so perilous, the colonies had no choice but to seek protection from the British.

This situation changed dramatically in 1763. The last of the Anglo-French Wars came to an end, and a defeated France handed Canada over to its British foe. All over North America, the scales of power shifted. Suddenly the colonies had much less need of British protection. Conversely, the British had much less need of imperial solidarity in the face of a common enemy. Indeed, the lords of empire considered that the time had come to undertake some long overdue reforms.

Imperial Reform

The chief problem requiring reform was the growing political maturity of the colonies. By 1763 Virginia had been established for more than a century and a half—there were residents of the Old Dominion whose great-great-grandparents had been born on American soil—and the other colonies weren't much younger. Between 1670 and 1730 colonial leadership had quietly passed from English to American hands. In the process, "colonies" had become "provinces."

At the same time, American institutions of government had also come of age. We saw in the previous chapter how assemblies had come to view themselves not as local administrative units but as Westminster-like parliaments. The leaders of these bodies had gained in sophistication accordingly. They had learned to organize committees in order to get things done, to circumvent obstacles with parliamentary maneuvers, to use structure and procedure to their advantage. Most importantly, they had learned to manage strong-willed royal governors by such expedients as funding the governor's salary.

The lords of empire were also uneasy with American prosperity. Colonial merchants were among the most successful in the world, not excepting their London counterparts, and growers of tobacco, rice and other subsidized crops were doing equally well. American fishing and furring operations, American shipbuilding, American ironmongering and liquor distilling were all thriving industries, while ordinary farmers were selling their surplus foodstuff around the hemisphere. The famous "triangular trade" in sugar, molasses, rum, horses, ivory, gold dust, and African slaves was reaping fabulous profits and personal fortunes were on the rise.

Americans, moreover, paid few taxes. They were taxed by their own assemblies for the support of local government, of course, and their trade was subjected to regulatory imposts. But when it came to, say, supporting the British armies that had come to *their* defense in the last French war, they had laid out very little. Their British cousins, meanwhile, had been taxed to the limit.

Though the Stamp Act was deemed reasonable by the British Parliament, it represented repression to the American colonists. One publication parodied the official revenue stamps, depicting them as viewed by some colonists.

At the war's end in 1763, all of this would abruptly change. In what seemed like a series of lightning strokes, Parliament laid new duties on sugar and other commodities, tightened up restrictions on colonial trade, and blocked off further settlement in the West. Then, before the Americans could catch their breath, it followed with the notorious Stamp Act. This was not an impost or trade restriction but an internal tax—the sort of thing colonial assemblies had always been in charge of—and it was clearly meant to be enforced. By its provisions, revenue stamps were required for just about everything from legal documents to college diplomas. The colonists could well afford to pay such taxes, Parliament reasoned, and it was high time they did so.

Nothing could have better aroused American ire. For it was the old issue of taxation—the one that had driven the Stuarts from the throne—in new guise. When James I had demanded tunnage and poundage without his subjects' consent, Englishmen had wondered what protection was afforded *any* of their property. Americans now wondered the same thing. How could Parliament tax *them*? they asked. *They* had no voice in Parliament.

Ordeal of the Empire

The British Constitution was unwritten, consisting of traditional understandings about the origin of government and the distribution of its power.

There ensued a decade of tumult. Colonial spokesmen penned angry remonstrances to Parliament, explaining why the Stamp Act violated the British constitution. Then mobs gathered in the streets and harassed the would-be stamp vendors. Shouting "no taxation without representation," the rowdies hurled both stones and epithets. They were not above resorting to tar and feathers either, and occasionally even worse. And pamphleteers had a field day. They cast British intentions in the darkest hues possible and harked back to the commonwealth

In the Declaratory Act (1766) Parliament reasserted its right to govern the colonies "in all cases whatsoever," even though the Stamp Act was repealed for political reasons.

ideology. This wasn't about funding imperial services, they accused; it was about corruption and tyranny.

Eventually the Stamp Act was withdrawn, grudgingly, only to be replaced by a new series of imposts called the Townshend Duties. Colonial resistance to these was equally intransigent. Such "duties" were not trade regulations at all, the Americans cried, but Parliamentary taxes in disguise—clear evidence of deception.

Trans-Atlantic communication made things worse. Policy makers in England had to rely on American envoys, such as Benjamin Franklin and James Duane, to explain to them why the colonials found the measures so odious, and some explanations only confused matters the more. To many Britons, it appeared that the Americans were either bad citizens or free riders, and that their arguments were mere hypocrisy.

At the same time, Parliament overhauled the customs service and appointed a Board of American Commissioners to supervise rigorous enforcement of the Navigation Acts. (A number of colonial merchants were free traders even then—the usual term was smugglers—and they evaded the commerce laws when they could.) Unfortunately, the new officialdom was no more aboveboard than the old, which many knew to be corrupt. Officers took bribes as a matter of course, and some even practiced a sort of customs racketeering, condemning entire cargoes through legal technicalities and gaining handsome bounties for themselves.

The imperial reformers were on a roll nonetheless. In order to curb smuggling further, they gave the customs inspectors broader legal options, some of which circumvented the notoriously lenient local juries, and made use of high-powered search warrants. These writs of assistance, as they were called, gave the inspectors a much wider latitude of discretion in their investigations. In the hands of corrupt officials, such warrants could be used to raid colonial warehouses on the flimsiest of pretexts—and create further opportunities for racketeering.

Americans complained that these measures abridged their historic English rights. Jury trial was sanctioned by centuries of practice. And the invasion of property by snoopy government agents violated an equally long-standing tradition of privacy. Indeed, in arguing against writs of assistance, Boston attorney James Otis held them to violate the very law of nature. Such objections honestly puzzled the imperial authorities. If Americans were the law-abiding subjects they claimed to be, then why would they protect smugglers, and why all this fuss about rights and procedures?

Matters continued to escalate on both sides. Boston became so disorderly that the British government brought in troops from Nova Scotia and stationed them on call. This raised yet another specter of Stuart tyranny—standing armies. It was inevitable that soldiers and citizenry would come to blows sooner or later. They did so on the evening of March 5, 1770 in the infamous Boston Massacre.

Revolution of the Mind and Heart

In the afteryears, John Adams remarked that the real revolution had taken place long before the outbreak of hostilities in 1775—"in the minds and hearts of the people." It was this quiet revolution, he said, that made separation from the British Empire inevitable.

While the American colonists thought of themselves as loyal Englishmen, they had also begun to think of themselves as Englishmen with a difference—and to recognize the things they shared in common. Their lack of a collective identity was proverbial back in the days when Benjamin Franklin had attempted to unite them in a defensive alliance against the French. Once the Stamp Act was passed, however, they commenced a vigorous dialogue with one another. Committees of correspondence shared analysis and strategy with their counterparts in the nearby towns—and eventually the nearby colonies. The appearance of a "tyrant" accomplished more than a foreign enemy could in drawing them together.

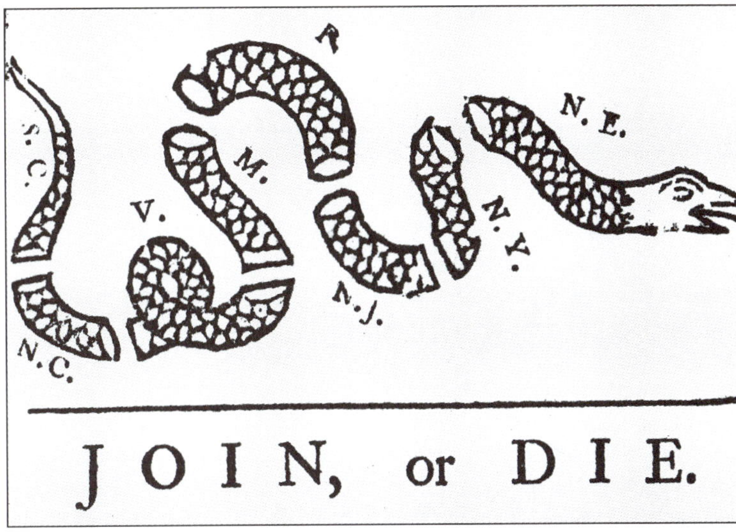

As early as the 1750s some colonial leaders sensed that a union of the North American colonies would be essential to their political survival. Benjamin Franklin created this political cartoon in 1754, appealing for colonial unity.

It might be argued that the idiom used by Americans to explain the imperial controversy was the language of Puritanism. For even though Puritanism had long since cooled as an active force, its perspective and assumptions were still much a part of the American psyche. Adams himself may have been a case in point.

Consider, for example, how the lingering Calvinist mind-set reshaped the commonwealth ideology and gave it new meaning. For the commonwealthmen, there were essentially two character types: the corrupt official (the court party) and the virtuous patriot (the country party). These happened to correspond to two Calvinist counterparts: unregenerate humanity and God's elect. We begin to see how British and American spokesmen could talk past one another with such regularity.

But the parallels went further. The virtuous patriot of commonwealth lore possessed two essential qualities: personal independence and a sense of civic duty. There were Puritan analogues for both. Personal independence could also mean independence from the Tempter's snares—from the vice and debauchery he used to entrap the unwary. And civic duty could also mean responsibility for the community's moral well-being, as we saw in chapter 2.

To understand why these analogies were dangerous, we must examine a style of human relations prevalent in Great Britain at the time, a style generally known as ***patronage***. In a patronage society—which was always markedly hierarchical—each person was obliged to reckon with a "patron," someone higher up on the totem pole, for the day-to-day advantages of life. In everything from getting a job to finding a spouse to locating a suitable school for the children, you had to cultivate your patron assiduously: pay him little visits, write letters of appreciation, bestow small gifts—and sometimes large ones. Patronage, with its never-ending "favors," oiled the world's machinery—the classic case of who you know, not what you know, that counts.

Life Under Patronage

What life could be like in a patronage society was illustrated by colonial New Hampshire under Gov. Benning Wentworth. Assuming the governorship in 1741, Wentworth immediately packed the Council with friends and relatives. He then subverted the elected Assembly by blackmailing new communities into electing pro-Wentworth representatives. Next, the governor began to make free with civil commissions, appointing so many of his followers justice of the peace that the office became a standing joke. As for military commissions, the governor's faithful supporters instantly became full colonels in the militia, while his two adolescent sons and a ten-year-old nephew had to be satisfied as captains.

Having passed around the favors, Wentworth now exacted his price. Every new township in the province had to reserve eight hundred acres as the governor's personal property. The New Hampshire mast industry—the mainstay of its economy—gave him a cut on every mast shipped to England. Smuggling operations also fell under Wentworth's control, and he skimmed his percentage off of every keg of molasses and every barrel of rum sneaked past the royal customs. The list went on. New Hampshire was compared to a "field of battle" and the governor's minions to "the vultures and ravens glutting on the carnage."

In vain did the Assembly try to fight back. It withheld Wentworth's salary—but he seemed to need none. It hauled his friends into court—but the governor controlled the courts too. It refused to seat his handpicked representatives—but Wentworth executed a deft maneuver and neatly tied its proceedings into knots. Finally, in desperation, the Assembly appealed over the governor's head to the British lords of trade. Unfortunately, the man they appealed to, John Tomlinson, happened to be a Wentworth crony himself, and he politely told them to take a hike.

On and on the saga went, a study in the politics of patronage. In his old age, Benning Wentworth could sit back on the veranda of his fifty-two-room mansion and look out happily on the bustling panorama of Portsmouth Harbor below him—master of all he surveyed.

From America: A Study in Heritage, *Kendall/Hunt 7th edition, p. 49*

In the commonwealth ideology, patronage spelled dependence, as we have seen. In the Puritan mind, it spelled something even worse. All the flattery and kowtowing that patronage made necessary seemed to Calvinists but a step from outright bribery. And in the administration of the colonies, it had often *become* outright bribery, for if a person wanted to get along with the powers that be, he had to perform his share of "favors." In other words, what seemed like politics as usual to the British authorities appeared far more ominous to Americans tinged with Puritanism. To them, the issue was nothing less than their own moral integrity. Were they to be corrupted by the landwaiters and tidewaiters of the customs office, by the royal governor and his minions, by the Board of Trade and its multitude of lackeys? Or were they, before God, to be their own men?

There were Puritan implications in the rhetoric of taxation as well. A person whose taxes supported a corrupt officialdom was helping to accomplish the Devil's work, or so it could be argued. And such a work it was. British officials were waxing proud and engrossing upon the fat of the land—all good biblical metaphors. They lived in sumptuous houses, rode the streets in gleaming carriages, and fairly wallowed in moral delinquency. And, worse, Americans were following suit. Honest toil was giving way to idleness. Simplicity in dress and manners was being eclipsed by frippery. A tea-drinking, hand-kissing, Old World effeteness was seeping into the American soul.

If this picture seems overdrawn, we must remember that a streak of paranoia lay embedded in the Puritan mind. Just as Winthrop's generation had believed in the Devil's war against them, so too did descendants see their own dark shadows, and these were not irrelevant to the present situation. If they refused to drink the tea of iniquity freely, some supposed, Satan might employ any means at hand to force them—including the might of the British Empire.

In the five years between the Stamp Act and the Boston Massacre, then, American colonists reacted to Britain's reform measures in a way that defied normal logic. They flooded the streets with an accusatory pamphlet literature. They turned out in mobs to prevent the execution of parliamentary laws. They boycotted British goods, preferring homespun to silk and West Indian coffee to English tea. They argued confusedly that Parliament had no right to tax them "internally," then no right to tax them "externally," and finally no right even to regulate their trade. And all the while they alleged a demonic conspiracy to enslave them.

The Crisis

Two years of comparative calm followed the Boston Massacre, the result of steady hands on the tiller and forbearance on both sides. But nothing had really changed. British authorities were still determined to consolidate their power and reform their ramshackle empire. Another opportunity to do so arrived in 1773 with the ill fortunes of the British East India Company. In order to bail out the nearly bankrupt—and politically influential—corporation, the Board of

Boston Tea Party:
http://odur.let.rug.nl/~usa/E/teaparty/bostonxx.htm

Boston Tea Party: (December 16, 1773) Destruction of large quantity of British tea in response to the Tea Act resulted in the Coercive Acts, stationing of British troops, closing of Boston Harbor, and thus indirectly to Lexington and Concord (1775).

Trade hatched a bold plan. The entire inventory of tea in company warehouses would be shipped to the colonies and sold at a bargain price, which ought to please the Americans, and the resulting revenues would boost company profits. But there was a pill within the sweetener. When the Townshend Duties had been repealed five years earlier, the tax on tea had been left standing—a reminder that Parliament truly did have the power to tax the colonies if it desired. The Americans had responded by virtuously forgoing tea. The new cut-rate price surely ought to tempt them into submission.

It didn't. It only confirmed their suspicions that conspiracy was afoot. Mobs prevented the landing of the tea in New York and Philadelphia. In Boston there was a standoff. The tea ships stood at anchor in the harbor while British authorities played a game of cat and mouse with colonial representatives. Then, when the former attempted an eleventh-hour landing, the Boston Tea Party ensued.

After that, as far as the British were concerned, all games were over. When Bostonians refused to pay for the tea they had dumped overboard, Lord North, who was then in charge of imperial policy, opted for a showdown. He asked Parliament to close the port to all shipping—assuming that rivals would eagerly snatch at Boston's commerce—and followed up with a series of kill-or-cure measures to end the disturbances. It was either this, reasoned North, or else it was to see the empire dissolve in tatters.

Harsh treatment for the worst offender while all others went free was a policy calculated to divide the Americans. It had the opposite effect. What happened to Boston today could happen to New York tomorrow and Philadelphia the day after, the colonists reasoned. Tyranny was attempting to conquer them piecemeal.

Both sides jockeyed for time and tactical advantage. Colonial militias began rounding up arms and ammunition, readying themselves for some sort of face-off. On the other side, General Gage, in command of the troops stationed in Boston, sent for reinforcements. The British now assumed that blood would have to be spilled, perhaps in quantity, before the rebellion fever subsided.

The question Americans now faced was whether or not to swerve from their course. There was still time to do so. Many of them, indeed most, regarded themselves as law-abiding British subjects—no more rebels than pirates or bandits. But the crisis had unlocked several doors, each of them opening upon some aspect of republicanism. Slowly, haltingly, Americans began to peer inside.

The first was the door of democracy. While issues flew back and forth in the pamphlet literature and were loudly debated on street corners, many normally apolitical elements—mechanics, tradesmen, country farmers—had taken an ever more active part. And for good reason. Corruption, subversion, conspiracy, tyranny—these were issues that concerned not just the political classes but everyone.

Equally clearly, the *demos* had its own stake in the outcome. Most institutions of local government were already strongly democratic by English standards, as we have seen, and by threatening these, the British government was in effect threatening democracy itself. After shutting down the port of Boston, for example, the North ministry went on to reorganize the Massachusetts government so as to minimize the democratic influence. The result was not to get the people out of the streets, as hoped, but to commit them all the more to revolution.

The second was the door of liberty. The cry was on everyone's lips. There were liberty poles and liberty trees and liberty bonfires. Orators like Patrick

Henry knew they could always score by inserting the word into their speeches. And parliamentary friends of the American cause, such as Edmund Burke, urged colleagues not to misapprehend the colonists' sincerity. When Americans said they loved freedom, they meant it.

But freedom from what? Americans were arguably the freest people on earth. They had the most power over their own lives, faced the least officious government, and paid the lowest taxes in the western world. So why did they sound like so many prisoners clanging on the bars? Because they had truly come to believe, reasonably or otherwise, that their freedom was about to be snuffed out. It might be argued that by its very nature, liberty tends to beget more liberty and becomes ever more suspicious of restraint. Like democracy, in other words, liberty was a kind of genie out of the bottle.

The third was the door of virtue. For the same events that appeared to endanger democracy and liberty also placed republican *areté* at risk. If the world was truly divided between a "court party" given to vice and corruption and a "country party" cherishing virtue and independence, what could the presence of the king's troops in Boston mean other than a showdown between the two? And could Americans really back away from that?

Virtually all colonial leaders belonged to that untitled aristocracy for whom the commonwealth ideology made greatest sense. They were steeped in a sense of personal *areté* and confirmed in political self-righteousness. America *was* the "country," as far as they were concerned, and it *must* hold out against the "court" no matter what. And if the court manipulators could not achieve their ends by the usual means—such as that come-hither offer of low-priced tea—of course they would use force.

Democracy, liberty, and virtue—the three strains of republicanism we examined in chapter 3—and all were under challenge by Lord North. All three, moreover, tapped into the colonies' latent Puritanism. Weren't God's elect the people of extraordinary virtue? Weren't all souls equal in the sight of God? Wasn't there a spiritual liberty that accompanied political liberty, where the individual soul needed to freely choose the path of righteousness?

Old-time predestinarians questioned that last point, of course, but they were rapidly losing ground to believers in free will—another of those Puritan flip-flops. Indeed, the Protestant clergy in New England came to lead the crusade for American liberty, so important did they see the tie between political and spiritual freedom. While some colonial ministers were politically indifferent, those of the Congregational and Presbyterian churches—whom Loyalist Peter Oliver called "the Black Regiment"—were in the fray from the beginning. They inserted political messages into their special sermons on election days, fast days or thanksgiving days, many of them remarkably direct.

Even more direct was a sermon called the *jeremiad*, a formal call to repentance that generally preached industriousness, frugality, and simplicity of manners—all good republican themes. As the trouble with Britain worsened, jeremiads began to thunder against passive obedience to wicked rulers, too, reminding worshippers that the profligacy of a people might provoke God's wrathful judgment. Repentance and revolution, as historian Perry Miller points out, were coming to be thought of as one.

Jeremiad: Term derived from Old Testament prophet Jeremiah, noted for prophecies lamenting the present and the future.

Given the Puritan conscience of New England, it was little wonder that on that April night in 1775 when Paul Revere made his historic ride, there were more than just a few who came in answer.

Forging the Bonds of Nationhood

It has been said that the United States of America was "invented" in a single day—July 2, 1776, when Richard Henry Lee's resolution for independence passed the Continental Congress. The actual genesis took much longer, of course, but a lot of it was accomplished between the night of Paul Revere's ride and the ringing of the Liberty Bell some fifteen months later.

With the outbreak of open warfare, the die of independence was essentially cast. There had been a flurry of backstage negotiations, peace proposals and the like, but when that memorable British expedition to Lexington and Concord limped back with 273 casualties, the war was clearly on and reconciliation clearly off.

Shared experiences, sufferings, and vision on and off the battlefield helped knit the American colonists into a fragile new nation.

American nationhood was another matter. Thirteen separate colonies had entered the war, and most Americans assumed that thirteen separate states would emerge from it. What these observers missed was the tendency of war to forge bonds of nationhood. What were those bonds, exactly?

Sovereign Government

What were loosely termed "continental congresses" had been called in the past in time of crisis. The one that came together in May of 1775, after the bloodshed outside Boston, became the de facto government of America. Its charter lay in public opinion, not the stroke of a royal pen, and not all Americans acknowledged its authority. Even so, it managed to raise an army, commission officers, plot strategy, supply and arm military forces, conduct diplomacy, coordinate the efforts of thirteen quarreling sovereignties, fund its own operations, and by hook or crook fight a war.

For an extralegal body, Congress was remarkably orderly. Its delegates, chosen by self-called provincial conventions, quickly organized themselves, elected their own officers and established parliamentary procedures. Its politics was as bumpy as one might suppose under the circumstances, yet somehow it carried on with its job. It was a striking instance of people coming together in a "state of nature" and creating their own government—straight out of Locke's *Second Treatise*.

A National Military Force

After Lexington and Concord, General Gage's troops were essentially under siege in Boston. The besiegers were a motley consortium of local and regional militias, indifferently organized and poorly led. Congress lost no time in appointing George Washington to general command. The selection of a Virginian to take charge of New England forces showed a good deal of political acumen: this was not to be New England's fight alone.

In time and with the approach of independence, a national army would grow out of this unlikely beginning. While most of the officers serving in it were not without some military experience, the enlisted men were for the most part only armed civilians, and they lacked the vaguest idea of discipline. When Washington told one man to fetch him a shovel, he was answered with a crisp "Git it yerself. I'm busy." Before the war was over, the army would profit from the services of European drillmasters, spit-and-polish sergeants, artillery experts, battlefield tacticians and fortification engineers. And it would learn to hold its own.

Service in a national army imparted a national point of view. While the activities of Congress required only a few dozen to reach beyond colonial borders, the activities of the military required thousands. Later on, a lopsided number of Continental Army veterans would back the Federal Constitution.

Demystified Monarchy

Throughout their quarrel with Parliament, the colonists had continued to hold the king in high regard. As the sovereign, he was supposed to stand above the fray of politics and in some mystical way embody the idea of **patria**. George III, however, could never quite keep above the fray.

MPs: Members of British Parliament.

The cause of the problem was patronage. While in theory king and Parliament were supposed to be independent authorities, in fact the king, through the bestowal of titles, honors, appointments, pensions, and assorted other munificences, had brought a considerable number of MPs under his control, making them in a sense members of his "party." Americans had turned a blind eye to this fact and blamed their troubles on the king's "evil counselors" instead. Bute, Mansfield, Grenville, Townshend, North—there was a long and growing list.

But the king's actions after the shooting started brought blindness to an end. Indeed, his employment of Hessian mercenaries—famous for their ruthlessness—was both a shock and rude awakening to the Americans. The king, they gasped, was as evil as his counselors! We must remember that the mystique of monarchy had provided a good deal of the glue holding the empire together. When George III "betrayed" his own subjects, it not only made them fighting mad, it demystified royal governance.

The demystification process was speeded up considerably by an elegant piece of rhetoric titled *Common Sense*. In this pamphlet to end all pamphlets, Thomas Paine said nothing new, perhaps, but he cast it into such mellifluous rolling cadences that readers were swept along. Monarchy, Paine argued, was washed up in the modern world. Kings were ridiculous figures—part sponger and part brigand—who had no claim to anyone's allegiance, least of all that of virtuous Americans. The indictment was so devastating that it amounted to a kind of symbolic regicide, according to social scientist Max Lerner, a killing of the European father that cleared the way for "natural-born republicans" to address the world on their own.

Bunker Hill

The fourth bond of nationhood was an old-fashioned military victory, at the Battle of Bunker Hill. On June 17, 1775 a patriot force of 1,200 under William Prescott was sent to fortify the high ground on Boston's Charlestown peninsula, overlooking the British encampment. Desperately needing to break the siege, the British opened fire on the American works, and later in the afternoon some 3,000 infantry prepared to attack. Two ferocious assaults were repulsed by withering fire from the defenders. The third British charge carried the hill, but only because the Americans lacked ammunition. There was stubborn hand-to-hand combat at the summit. And when the smoke cleared, the British had suffered 1,150 casualties, more than one-third of their total number.

While technically a defeat, the battle and its outcome caused American spirits to soar, for soldiers dismissed as "rag, tag and bobtail" had stood their ground against the best the British could muster.

Common Suffering

While it is hard to calculate the effect of common suffering, it may well have been substantial. Americans—*as* Americans—were being shot and killed, and given the realities of insurrectionary warfare, few could doubt what lay ahead. Cannon balls would soon rip through their towns. Men-of-war would cruise their harbors. Raiding parties would descend on their frontier settlements. Lives and fortunes would be torn asunder, and no group or class would be above risk. The common people would be in the thick of battle, facing professional soldiers armed with Brown Bess muskets and twenty-inch bayonets. The gentry would see their country homes plundered, their plantations burned, their families persecuted, and their own names placed on a wanted list. Furthermore, because there were plenty of loyalists, it would be a civil war, with all the bitterness and atrocity such wars commonly bring.

Working in and among these bonds of nationhood was the republican idea, whose effect was to remind Americans of the things that set them apart. What Rousseau referred to as a sense of "we," that mysterious togetherness which defines a people, grew not only out of the events of the Revolution but out of the feelings produced by those events when Americans stopped to think who they were, what they stood for, and what they stood to lose.

The Declaration of Independence

In the spring of 1776, the military situation took a turn. The British had given up the idea of holding Boston and had withdrawn to Nova Scotia to regroup. With the passing of winter's chill, they now launched a major thrust at New York. The Middle Colonies were appreciably calmer than New England: establishing an overwhelming presence there would divide the Americans geographically and isolate the New England firebrands—or so the British hoped.

The force they landed under Gen. William Howe truly was overwhelming. Howe defeated the Americans at the Battle of Brooklyn Heights, then again at the Battle of White Plains, and finally drove them deep into New Jersey. By the end of the year, Washington's original command of 20,000 had been reduced to a bedraggled 4,000. Nevertheless, they retreated without panic, for the most part, and even displayed moments of real soldiering. It was not the rout the British had expected.

This fact foreshadowed the faintest possibility of an eventual American victory. For time and circumstance were on their side. The territory of the colonies was very large, thinly populated, and surrounded by wilderness. The British had to pin their hopes on a European-style "great battle" whose outcome might prompt the rebels to throw in the towel. Yet, clearly, a different scenario was possible. The American army might fight smaller engagements, skirmish a bit, and then slip away as it had in New York. This could go on almost indefinitely.

Moreover, if the patriot army was not crushed soon, it stood to gain increasing moral ascendancy. People might come to believe that it couldn't be crushed at all. Accordingly, every day that Washington held out would tilt the balance a little more in his favor.

And finally, France had an interest in the outcome. After all, here was a chance to break the power of its English archrival once and for all, and at a bargain price. Who knew, the French might even be induced to enter the conflict themselves—*if* they were assured of a favorable outcome.

While these facts were only dimly glimpsed that spring, Americans did begin to glimpse them. And they pointed convincingly toward declaring independence. But how to do the job effectively? The powers of Europe, who were looking on with great interest, must be persuaded that American intentions were genuine and that this was not just a passing family quarrel. The American people were looking on too. Most of them were still lukewarm to the patriot cause and many were openly hostile. They wondered what sort of cause it was, and where it might ultimately lead.

A committee of Congress was put in charge of drafting a formal declaration. Thomas Jefferson became the designated draftsman. His task was to present the case for American independence in terms which all could understand and which the "right thinking" would be inclined to approve. Although still a youthful thirty-three, Jefferson was intelligent and sophisticated, and his reputation for a facile pen marched ahead of him. He knew the words of the document in question must be simple and lucid, their tone dignified, their logic unstrained, and their effect utterly convincing. This couldn't be a disquisition in political philosophy or a rambling insurrectionary tirade. And it must communicate on several different levels at once.

On the surface level, the declaration had to be a legal brief, arguing an ironclad case against the British. It was they, not the Americans, who were the revolutionaries, Jefferson contended, for they had taken it upon themselves to abridge historic rights. Jefferson had to aim his barbs at the king, not Parliament, for the colonists had argued that Parliament had no place at all in the colonial scheme.

On a deeper level, the Declaration was bound to play a role in the American Founding—though few guessed how important that role would be. Jefferson, accordingly, set forth his idea of what the colonies were now to become and what their fundamental principles were to be. The prose here soared toward poetry and packed broad swatches of meaning into simple household words. In the crucial second paragraph, it offered a concise but eloquent theory of American nationhood:

Unalienable: Inalienable, incapable of being separated or transferred from an individual or a polity.

> We hold these truths to be self-evident, That all men are created equal, that they are endowed by their Creator with certain unalienable Rights, that among these are life, liberty, and the pursuit of Happiness. That to secure these rights, Governments are instituted among Men, deriving their just powers from the consent of the governed, That whenever any form of Government becomes destructive of these ends, it is the Right of the people to alter or to abolish it, and to institute new Government, laying its foundation on such principles and organizing its powers in such form, as to them shall seem most likely to effect their Safety and Happiness.

These words are often taken to be a digest of Locke's *Second Treatise*, and indeed, they did encapsulate Locke rather nicely. It has been pointed out, however, that they might also reflect Jefferson's acquaintance with the Scottish Enlightenment, whose philosophy differed with Locke on crucial issues. The point is, Jefferson may not have "voted" for a single conception of republicanism—he may have sought to encompass all three.

Whatever the intent, Jefferson's prose poem articulated a vision of nationhood that was enormously appealing to ordinary Americans and capable of being read open-endedly. Historian Joseph J. Ellis has underscored the seductiveness and inclusiveness of that vision:

> Two monumental claims are being made here. The explicit claim is that the individual is the sovereign unit in society; his natural state is freedom from and equality with all other individuals; this is the natural order of things. The implicit claim is that all restrictions on this natural order are immoral transgressions, violations of what God intended; individuals liberated from such restrictions will interact with their fellows in a harmonious scheme requiring no external discipline and producing maximum human happiness.

Matters of equal significance lie buried in Jefferson's first paragraph, where he appealed to "the Laws of Nature and of Nature's God." On the surface level, once again, he was arguing his case for American independence (as opposed to mere resistance) on grounds that Britain had violated the very laws of the universe. Deeper down, it was a way of recognizing an authority that lay beyond the participants themselves. What they were doing was not simply in accordance with their own desires but in accordance with **Nature** and Providence. Jefferson was acknowledging universal moral truth, in other words, of which the Puritans had often spoken. It was this "self-evident" truth, more than anything else in the Declaration, that established the parameters of American nationhood.

For, whatever else the American nation must do, Jefferson was saying, it must recognize the legal equality of its citizens, protect their natural rights, and govern by consent. There was the shadowy form of a constitution here. The Declaration of Independence ruled out both enlightened despotism, which some Americans might have favored, as well as popular but oppressive regimes. A legitimate government, it said, had to reflect the will of the people broadly considered while at the same time protecting the rights of everyone, including unpopular minorities—a decidedly tricky business.

Jefferson's arguments were given added weight by their grounding in everyday reality. In the summer of 1776 there was no power on earth above the Continental Congress except Nature and God. Americans had come together as free beings and worked this out among themselves. The Declaration was living proof of its own principles.

The War of the Revolution

It might be argued that the war itself, and the way it was fought, contributed as much to the American Founding as the Declaration of Independence did. A different war fought in a different way might have led to a different political society.

American Revolution: http://www.dell.homestead.com/revwar/files/INDEX2.HTM#time

The Politics of War Making

Historian John Shy applied insights gained from a study of the Vietnam War to understand the military history of the American Revolution. For the two experiences were by no means unrelated. In both, a formidable power attempted to suppress outmanned and outgunned insurrectionary forces—and lost. In both situations, moreover, it was political rather than military factors that made the difference.

But Is It True?

In his freshman seminar on American politics, Professor Charles R. Kesler, at Claremont McKenna College, asks his students a question they have trouble answering. Is the Declaration of Independence true?

"Invariably," he reports, "there are loads of bright freshman who know all about it and who can discourse plausibly on the Declaration's roots in the Enlightenment, John Locke's influence on its authors, the economic interests that actuated its signers, and so forth." But then comes the question about truth. "*Are* men created equal and endowed with certain unalienable rights?"

The question never fails to floor them. What does he mean? they ask one another. What is he talking about? They look at each other and grin sheepishly. For in the world of the college freshman of the 1990s, things are never true or false; they are useful, or handy, or relevant, or the opposite of those three. "In *their* [the Founders'] opinion, it was true," a freshman will finally volunteer. That is about as far as a discussion of truth can go these days.

So, we might stop and consider the same question ourselves. *Is* the Declaration of Independence true? Or is it only one of those expressions of belief which have their little niche in history? And if the latter, what is it, exactly, that we stand for as a people—and why do we think it is so wonderful that Communism is falling apart?

From America: A Study in Heritage, *Kendall/Hunt 7th edition, p. 80*

In the American Revolution, Great Britain possessed not only a world-class fighting machine but also the economic strength and the political will to make it run. The British had seasoned officers, professional soldiers, the finest hardware available, undisputed control of the seas, and a handful of truly gifted generals. On the battlefield, they quickly proved what every colonial already knew—that in a conventional encounter they could defeat any force the Americans could bring against them.

But the British were far less effective politically. They did not understand Americans very well, and as a result they routinely misassessed the political situations they encountered. In this kind of war, that was a serious liability.

To begin with, they failed to gauge the depth of American feeling. The colonists, they supposed, already had more liberty than was good for them, and it was absurd to bewail its threatened loss. Conversely, all the talk about conspiracy and slavery sounded like so much paranoid raving. To the British, in short, the American case seemed weak and muddled—the work of a few rabble-rousers.

As a result of these misperceptions, British commanders invariably overestimated the support they could count on from "good Americans" and underestimated support for the rebels. For some officers, in fact, this was not even a real war, only an insurrectionary disturbance, and the proper way to deal with it was through police action.

So conceived, British strategy passed through several phases. The first reflected a police-action mentality and sought only to isolate and punish. This approach fizzled, as we have seen, for the Americans perceived it as an attempt to divide and conquer. The British next pressed their advantage in conventional warfare and endeavored to fight textbook-style engagements. But the Americans would rarely go along. Occasionally they might take on the redcoats openly—as on that memorable Christmas night in Trenton, New Jersey when they dealt Colonel Rall's Hessians a stunning blow—but mostly they settled for rearguard harassment actions and the like. The British couldn't lose this way, but they couldn't really win either.

American Revolution: http://www.historyplace.com/unitedstates/revolution/index.html

And so a third change of strategy, to that of comprehensive pacification. The concept was to move into strategically important areas with a convincing presence and try to get the machinery of government up and running again. Here was where the British were the most effective. And here was where the crucial issue of the Revolution came to the fore. For the Revolution, as Adams said, became a contest for the minds and hearts of the ordinary people, people who were essentially apolitical and uninvolved, who were mainly interested in getting on with their lives. Without the active commitment of this majority, it was not possible for the patriot side to win—much less to found a successful republic.

In some areas, especially in the South, the pacification strategy worked well enough. Still there were difficulties. In the first place, the British had to operate everywhere, while the rebels could pick and chose, and this stretched their forces thin. And in the second, they were obliged to operate with great sensitivity, luring the undecided to their side. Of this task they seemed almost utterly incapable.

For virtually everything the British did was wrong. If they were lenient with rebel sympathizers, they embittered the loyalists. If they were harsh and vindictive, they angered the very people they were trying to win over. If they dug in, the enemy would operate with impunity all around them. If they headed off in pursuit, their friends were abandoned to reprisals. And try as they might, they couldn't seem to teach their own soldiers the difference between "rebels" and "noncombatants"—who were sadly treated alike.

Militias and the Continental Army

American Revolution: http://www.fordham.edu/halsall/mod/modsbook12.html#American Independence

Where the British got the politics of the war all wrong, the patriots got important things right. And what they got most right was the militia.

Colonial militias, as every schoolchild knows, were regarded as a joke on the battlefield. Yet they often proved to be an effective military force, especially at providing timely reinforcements. The militia presented something that most British officers had never dealt with: a great spongy mass that could be driven from the field but could never be isolated or decapitated. It was, in Shy's words, "a people numerous and armed."

The militia was also a political educator. For where the Continental Army was professionally staffed, military in tone, always a bit foreign to the local scene, and not infrequently on the run, militias were unprofessional, down home, filled with neighborhood good old boys, and never far away—a sort of military version of the little red schoolhouse.

And their schooling was in appropriate community behavior. The problem for neutrals—and this included a large segment of the population—was how to behave prudently with a revolution going on. One could cozy up to the British army and gain protection that way, but what if its commander suddenly marched his troops out of town? It was more prudent to be on terms with the local militia. But the militia was likely to require something. Aid and comfort. Supplies and equipment. The affiliation of male family members. With the militia around, in other words, it was almost impossible to continue sitting on the fence.

The British did their own part to keep the fence clear by persecuting any and all who joined a militia, even if only for prudence's sake. So, for better or worse, neutrals did have to choose. And once they chose the militia, they were enfolded in its camaraderie and anointed with its esprit de corps. They were also treated to serial sermons about virtue, honor, and patriotism. If their group saw action—and most of them did—these words took on a meaning they might otherwise lack, for nothing wins converts like bloodshed.

It would be hard to imagine a better way to politicize the populace. Hundreds of thousands of Americans who hadn't cared one way or the other about taxation or jury trial not only found themselves enlisted in battle groups but drinking in the revolutionary spirit. Their turbocharged tenure in the militia laid the foundations of the American Republic.

Something similar could be said for the Continental Army. In its ranks, young men from all over the colonies rubbed elbows with their contemporaries and got to know one another. Regional stereotypes often melted away. The loyalty of the men to one another and to their common cause was repeatedly tested under fire. Privation, sacrifice, defeat, betrayal all had psychological effects. Many leaders of the new republic gained their American sense of patria right here.

The dark side of the Revolution was epitomized by the deprivations suffered by George Washington and the Continental Army at Valley Forge, Pennsylvania during the winter 1777–78.

The Yankee Doodle memories enshrined in popular culture have obscured the dark side of the Revolution. Most of it was exceedingly grim. The revolutionary armies were on short rations and dwindling supplies much of the time. Physical privation was made more poignant by the willingness of some Americans to engage in profiteering or take advantage of flexible loyalty. Desertion and turncoatism were common. So were corollary misbehaviors ranging from alcoholism to psychosis. Battlefield medical services were poor to nonexistent, and care of the sick and wounded was often simply appalling.

Prisoners of war commonly suffered an unspeakable fate. Since rebellion enjoyed no legal protection in the British Empire, there were no rules governing the treatment of combatants. Many of those who were captured were summarily tossed into prison compounds, the holds of ships, or damp stone cellars and left to rot. They were fed irregularly and given no other attention whatever. Conditions in several of the "prisons" were little short of horrifying.

Out of the suffering and sacrifice came an unusual kind of heroism. Some of the indifferently committed easily fell by the wayside. Others found inner sources of strength and a new access of meaning. The language of the Revolution and the significance of its cause were seared into their souls.

Nailing Jelly

In the end, the patriots won the day. With no victory in sight after seven years of conflict, the British finally despaired of their pacification strategy and switched back to conventional warfare. They continued to win major battles, but they could not seem to win the war. One disgruntled officer likened the American quagmire to "nailing jelly to the wall."

Just when the jelly seemed to stick, the bedraggled insurrectionists would rise above themselves and deliver a crushing counterblow. One such coup fell in October of 1777. A British plan to slice the colonies in two by driving one force up the Hudson Valley and another down from Canada went completely awry. Gen. John Burgoyne and his entire army were forced to surrender after the resulting Saratoga campaign. Even worse for the British, France was so impressed by Saratoga that it decided to come into the war.

And it was French participation that made Yorktown possible and brought the war to an end. Lord Cornwallis, one of the most successful of the British commanders, allowed his army to become trapped on a narrow spit of land between the York and James rivers in Virginia, on the assumption that the British fleet would arrive in Chesapeake Bay to relieve him. But it was Admiral de Grasse's battle squadron that got there first—George Washington literally jumped for joy when he saw the French flag fluttering from the frigates. Cornwallis held out under siege for a few days and decided to surrender.

The effort might have continued but to no good purpose. Great Britain, as it turned out, had been fighting on the strength of a false belief about the empire—a belief holding that the American colonies were essentially British and therefore that the points at issue were resolvable. They were not resolvable. They had grown and festered into a full-scale revolution. And the colonies of almost two-centuries standing had struggled free.

The Revolution as a Revolution

Historians have long debated the revolutionary nature of the American Revolution. What precisely was "revolutionized?" some have asked. J. Franklin Jameson and others began compiling a list. Several items on that list are of relevance to the Founding of the American Republic.

Social and Economic Changes

Quitrents: Money substitute for produce or labor owed to a landowner; held over from feudalism and still practiced in colonial America.

In the first place, a way of life that had focused on patronage relationships and psychological dependence was undermined by the revolutionary experience. In time the old mode of thinking would give way to a new one, and Americans would speak ever more insistently of independence and self-reliance.

Land ownership, already broad, was broadened still further by the Revolution. Loyalist lands were confiscated, some of them in expansive tracts, and distributed to veterans, small farmers, and postwar immigrants. Aristocratic patterns of landholding were further weakened by the abolition of quitrents, primogeniture and entail.

Primogeniture: The eldest son's exclusive right of inheritance.

Entail: To restrict inheritance to landowner's lineal descendants; kept colonial estates from being divided and sold in smaller parcels.

Official churches were disestablished in several of the new states, notably Virginia, where the Anglican establishment had provided the Revolution with a few extra bones of contention. Americans became accustomed to the extraordinary idea of separating church and state altogether.

Slaves were freed throughout most of the North—the practice seemed wholly unrepublican—and there was a marked decline in indentured servitude. The democratic feeling spread elsewhere. New colleges were founded. New voices were heard in literature and journalism. New ideas began to circulate about the education of women and their role in society.

Even in the economic realm there were changes. Credit expanded. Capital started to accumulate. New manufactures were set up and new patterns of trade established. Farmers began to feel the effect of systematic agricultural improvements. And a host of new businesses sprouted along Main Street.

Political and Constitutional Changes

The elimination of the king and exile of the loyalists—many of an aristocratic bent—altered the complexion of American politics. The "popular element" was now without rivals. While that element still did not grasp for the reins of power, voting privileges were further broadened and legislatures came to represent a wider array of interests.

Not only the structure of politics but its conduct became more democratic. For example, the argument of no taxation without representation cut against American, as well as British, practice: if people were to be taxed by their legislatures, they wanted to be represented. Soon there were voter instructions and other forms of accountability, even public galleries for the onlookers. Those who ran for office were asked deeper and more difficult questions than ever before, and once elected they began hearing a lot more from their constituents.

As with democracy, so too with liberty. The word, which was heard so often, could be applied to everything from the freeing of slaves to the discarding of outworn contentions. Liberty, for many, became liberation: cutting away from a European past and forging into an American future. There was something breathlessly exhilarating about it.

All said and done, the most truly revolutionary thing about the American Revolution was its creation of successful republican government—which is getting a little ahead of the story.

Revolutionary Symbols of Nationhood

In the mythology of the Revolution, incidents were transformed into legends, objects into symbols and men into monuments. Any schoolchild could supply a list. "Indians" throwing tea into Boston Harbor. Paul Revere's midnight ride. Lexington Green. Concord Bridge. A bell cracked from joyous ringing. The winter encampment at Valley Forge—those bloody footprints in the snow. Nathan Hale's last words: "I only regret that I have but one life to lose for my country!"

None of these exactly encapsulates the Revolution as a founding event, but one comes close. The image of the "minutemen" responding to the call of duty, turning out of warm beds and into the dark night, expresses much about the way Americans have come to see themselves and their national birth. The inspiration, of course, is classical antiquity once again—Cincinnatus, the farmer who left his plow in the field and took charge of the army that would save the Roman Republic, only to resign his command afterward and return to farming. It is a nice synthesis of all three republican ideals—classical, liberal, and democratic.

After the war, Washington's officers formed themselves into the Society of the Cincinnati for the purpose of getting together and remembering the old times. While any of them might have resembled the Roman warrior in some way, it was their commander, George Washington, who fit Cincinnatus to a tee. And it was George Washington who came to symbolize the entire Revolution. As *the* Founding Father, Washington's importance could hardly be overstated.

Lacking Jefferson's intellect or Lincoln's common touch, Washington was once thought to require image enhancement. Accordingly, in successive editions of his biography, Mason Locke Weems added such humanizing touches as the story of the cherry tree. Weems needn't have bothered: the real George Washington was well up to the mark. After all, he held the Continental Army together by the sheer force of his personality, kept it in the field as a power to be reckoned with, bucked staggering odds, endured a thousand setbacks, brooked all kinds of intermeddling, won enough battles to keep the American people hanging onto hope, and in the end cornered the redoubtable Cornwallis at Yorktown.

The Protestant clergy, who had done so much to suffuse the conflict with meaning, had no trouble picturing Washington as a demigod. Nor was it difficult to find supporting imagery. Washington crossing the Delaware on a foggy Christmas night. Washington kneeling in the snow at Valley Forge. Washington groping for his spectacles at the great mutiny—"Gentlemen, I fear I have gone blind as well as gray in the service of my country." Washington refusing the laurels of a Caesar and the offer of a crown.

George Washington: http://gi.grolier.com/presidents/ea/bios/01pwash.html

Americans cherish these images to good purpose. For they depict George Washington as the very soul of classical virtue. If the liberated colonists were truly to believe in such virtue, and base their own republic on it, they needed to have at least one shining exemplar before their eyes. Washington was that. He was one of the few to take virtue so seriously as to deliberately cultivate it. As a young man, he made lists of desirable attributes and systematically worked to perfect them one at a time. He wanted to be a successful planter, a daring innovator, an imaginative entrepreneur, a graceful horseman, a winning poker player, a loving husband, an inspired leader, a victorious general, and a wise statesman. Small wonder that our idea of him has become marbleized.

George Washington, the indispensable Founder, in life and legend.

The Revolution and the Founding

Seldom in history had an entire populace transferred its allegiance from one sovereign to another. An enormous infusion of energy—indeed a full-scale shooting war—was required to accomplish it. The American Revolution made the inhabitants of thirteen separate colonies ask themselves who they were and what they were all about. It forced the American people, as a people, into being.

As denizens of an empire, Americans had taken many things for granted. Now they must think for themselves. They had to assess the relative merits of peace, security, freedom, prosperity, dignity, power and nationhood. And important questions underlay the choice making. Were the Americans *a* people? Did they have a common purpose? A common identity? A common destiny? Given the difficulties faced by national states in the modern world, was an American nation as such really feasible? If so, what sort of entity ought it be? A loose confederation of sovereign states? A single unity? Some sort of compromise between the two? Since any American nation must in some sense be "invented"—something rare in human history—what were the possible materials of its composition, and how could they best be brought together?

While the Revolution provided no answers as yet, it had provided a new set of realities. The colonies were independent of Great Britain. They could more or less govern themselves. They were recognized by the European powers as a bona fide nation state. They had a place on the map and by European standards one of breathtaking size. They possessed a loose-jointed confederal government that had proved it could at least make war. And they had a founding document committing them to life, liberty and the pursuit of happiness.

America was off to a reasonably good start.

CHAPTER 4 STUDY AID

1. Mercantilism:

- Closed economic network regulating trade between a dominant 'center' and subordinate markets and producers on the periphery.
- Trade between the 'peripheries' and foreign markets restricted.
- Production of finished products in peripheries proscribed.
- Peripheries guaranteed market for raw materials at the center. Some trade among peripheries may also be protected.
- The center bears the cost and responsibility of administration, regulation, enforcement, and protection.
- If successful, mercantilism concentrates wealth at the center while peripheries remain subordinate. This provides incentive for the center to refine and maximize efficiency of the system.

2. Myopic Attempts to Reform the British Mercantile System:

- Focused on rationalizing colonial administration to assure the accumulation of wealth at the center.
- British protection invaluable to North American colonies until French expelled from Canada in 1763.
- British administrators and Parliament sought to increase revenues and control over the colonies through a series of 'reforms.'
- By the second half of the eighteenth century, the North American colonies were politically, socially, and economically mature enough to respond vociferously to perceived encroachments upon their assumed rights and privileges.
- British attempts at reform, though generally well intentioned, were consistently ill devised and ill timed. Colonists consistently misinterpreted their intent and British administrators consistently misinterpreted colonial reaction resulting in a classic and profoundly consequential failure of communication.

3. Hearts and Minds:

- The mind-sets, experiences, and interests of the North American colonists diverged from their British roots and fostered the creation of a distinctive consciousness and identity.
- The Country party's identification of 'virtue' with the 'country' and 'corruption' with the 'court' closely paralleled the Calvinist-Puritan distinction between regenerate saints and an unregenerate world.
- Colonial reaction to imperial reforms coalesced into a pattern of evasions, arguments, and confrontations that defied imperial authority in the name of British constitutional liberties.

- British policy remained focused on the concentration of power and the enhancement of revenue.
- Though the process of confrontation, escalation, and crisis the American colonies were gradually transformed—rallying around the principles of virtue, liberty, and democracy.

4. Forging a Nation:

- Identity—Thomas Paine's *Common Sense.*
- Boundaries—Geographical, social, economic, political and cultural.
- Authority—Colonial assemblies and continental congresses.
- Legitimacy—Functioning institutions and organizations.
- Sovereignty—Foreign alliances and creation of an army, etc.

5. Declaration of Independence

- Vision—Republicanism, in all its aspects and writ large.
- Laws of nature and nature's God.
- Life, liberty, and happiness (property).
- Consent of the governed.
- Slavery unresolved.
- Blended three forms of republicanism and virtue, identified liberty with personal freedom, linked legitimacy and authority to the consent of the governed, and thus vested sovereignty directly in the people (however defined).

6. The Conduct of the War:

- A founding influence.
- British effort shifted from punishment to confrontation to pacification.
- Colonial militias—source of political education.
- Continental Army created shared experiences, suffering, and vision.
- British misunderstanding of the political nature of the war, occasional defeats, colonial perseverance, and foreign alliances contributed to an improbable colonial victory that turned the world upside down.

7. Consequences of Revolution:

- Tangible social and economic changes occurred.
- Political and constitutional changes occurred.
- Archetypal national myths and symbols created.

CHAPTER 4 STUDY GUIDE

NAME: ______________________ **SECTION:** ____________

Key Terms

Mercantilism

The Stamp Act

Patronage

Boston Tea Party

"Common Sense"

Bunker Hill

Declaration of Independence

Colonial Militias

Multiple Choice Questions

1. Which of the following phrases best represents the idea of patronage?
   a. "Pull yourself up by your bootstraps."
   b. "It's not what you know, it's who you know."
   c. "Work hard and you can get ahead."
   d. "The best things come to those who wait."
2. The text lists three "doors" the crisis with Britain opened before the war. Which of the following was not one of those doors?
   a. Democracy
   b. Liberty
   c. Virtue
   d. Equality
3. Which of the following was not a bond of nationhood mentioned in the text?
   a. Unity of purpose
   b. Sovereign government
   c. Demystified monarchy
   d. Common suffering
4. The Revolutionary War has been compared to which other American war?
   a. Vietnam
   b. Korea
   c. World War I
   d. The Gulf War

Review Questions

1. Why have some argued that American independence came by evolution instead of revolution?

2. Why did British interests and actions change in America in 1763? What were some of those changes?

3. How did the Revolutionary War affect the colonies economically?

4. How did the creation of a national military force forge a bond of nationhood?

5. Why did the British exclaim that fighting the Americans was "like nailing jelly to the wall"?

Review Exercises

1. Explain how the Battle of Bunker Hill affected the American psyche at the start of the war.

2. Briefly explain the role the Declaration of Independence played within the war itself.

____

____

____

____

3. The text mentions the Minuteman as a lasting symbol of the revolution. What does the Minuteman symbolize?

____

____

____

____

Big Essay Question

Think about the key reasons why the colonists won the Revolutionary War. Come up with a list of your top three reasons and explain them, in descending order, in a one-page essay. Feel free to use material in the text or in other sources.

CHAPTER 5

The Confederation and the Rule of Law

With the defeat of Cornwallis' army at Yorktown, Americans could turn attention to the fact of their independence, not just the idea, and begin to think about an American republic.

Or was it republics, plural? There were reasons for supposing the latter. Indeed, by the war's end, the colonies-turned-states had had republican governments up and running for more than five years. They had their historic names, of course, and other marks of individuality. Each had its own government, its own constitution. And while their citizens loosely spoke of being "Americans" or "continentals," they very much regarded themselves as separate and self-contained polities.

All of these entities—including the blurry "America"—saw themselves as Good Societies. From the discussion given in the introduction, we recall that a Good Society:

- Extends the benefits of social interaction beyond a single favored group or class to society as a whole.
- Strives for a reasonable balance between liberty—defined as the right of individuals to seek their own happiness in their own way—and order.
- Offers its citizens "goods of the soul"—such things as political participation, a sense of individual empowerment, and a sense of human dignity.

Borrowing from antiquity, the American states used the term *republic* to designate the type of Good Society they had in mind. It came from the Latin *res publica*, "the public thing." And although the concept was a bit amorphous, most versions of it included a unifying moral code, a strong sense of public good, a reliance upon reason, a concern for justice, and a notion that such things were possible because of human virtue.

Within these parameters there was plenty of room for variety, and the American republics had no certain course marked out for them. Would they evolve into modernized classical republics, Lockean liberal republics, or some other unglimpsed form? Would their governments lean toward Aristotle's rule of the one (monarchy), rule of the few (aristocracy), or rule of the many (democracy)? What role would religion play in their political life? What role would commerce and industry play? How would they relate to one another and to the rest of the world? Would the feeling of nationhood come to reside in the whole or the parts—or both? And most importantly, would they rely upon virtue, as had the ancient republics, or upon more modern ideas as the key to their happiness? It all had to be worked out.

Virtue Versus Structure

While the new states already possessed working governments, some of those were only hastily revised colonial governments, and most Americans realized they must begin thinking about governance in a new way. The design of a truly republican regime ought to reflect human intentionality, they supposed, not just historical circumstances. Indeed, in some of the states constitution making inspired a heady enthusiasm.

The Failure of Civic Virtue

A beginning assumption held that virtue was the crucial element in constitutional design, far more important than anything else. There were solid reasons why. Americans often attributed their miraculous victory over Great Britain to God on the one hand and virtue on the other. Furthermore, history had convinced them that virtue must be the foundation of *any* republic. And "Jean Jacques," as they called him, did his part. For Rousseau's widely read *Social Contract* argued that modern equivalents of classical civic virtue were both possible and vitally necessary.

Some believed there were stronger claims for virtue in the New World than the old. After all, several of the colonies had been founded for specifically religious purposes, suggesting a high moral tone in the people. Life in the wilderness had required dogged determination and enormous strength of character. And during the Revolution, as we have seen, there were many acts of courage on the battlefield and much old-fashioned patriotism at home. Besides, virtue *had* to work. Americans had burned their ships, politically speaking, and there was no turning back. They had better be as virtuous as they claimed to be—or the future looked grim.

What did a republic of virtue look like in the modern world? How did it run? What set it apart? Americans weren't sure. They knew only that its life would be marked by simplicity and austerity—as different from London's Mayfair as they could make it. It would be governed by "aristocrats," to be sure, but they would be *natural* aristocrats, men of talent and virtue, not those whose only asset was a title. Small, simple, ascetic governments had to reflect virtue, Americans believed, and they had to respect freedom.

During the Confederation period this faith eroded considerably. Even before the war ended, there had been numerous stories of profiteering and disloyalty, but these were usually chalked up to *la guerre*. Harder to dismiss were instances of corruption, jobbery and other forms of misbehavior in high places—and they didn't end with Yorktown. Postwar America was a place of disillusionment, even cynicism, and it showed.

Of particular repugnance, at least to James Madison, was the interest-driven, me-first politics of state legislatures. The discourse there was all about power and advantage, he fumed, and no one so much as mentioned the public good. Worse, legislators were logrolling and horse trading with scant regard for justice, and when it came time to vote they could never see beyond their own little factions. Human nature, it seemed, had put on an altogether different face.

Shays Rebellion in Massachusetts was a kind of last straw. Disquiet among farmers in the western part of the state became so intense that angry bands began mustering in the townships and shutting down the local courts. In the winter of 1787, an "army" of these malcontents, many of them veterans, marched on the arsenal at Springfield and dispersed only after cannon fire had left four of them dead in the snow. Some Americans began wondering if the Revolution had been fought in vain.

Shays Rebellion: http://www.sjchs-history.org/shays.html

The Failure of Structure

At the same time, there was a growing interest in the use of political **structure** as a corrective to such difficulties. Aristotle, we recall, had taught that civic virtue was not constant in human affairs and could be influenced by a variety of institutional and situational factors. For example, citizens were more orderly and more mindful of the public good in a mixed regime, where the aims of one group were balanced against those of another, than in, say, a simple democracy. Hence, the intriguing idea of "structure."

If virtue was the mainstay of the ancient republics, structure came to be that of the modern ones, as we saw in chapter 1. If the right structure could be found, that is. For all structures did not work equally well. The much-admired English monarchy, though it mixed and balanced Aristotle's one, few and many in a dozen ways, had proven itself only too capable of tyranny. Americans had to try to do better.

In field testing the effects of structure, the new state governments presented thirteen experimental laboratories, for each of them had different structural features in its constitution. One could observe, among other things, how virtue and structure worked together—or perhaps failed to work together—and hazard some guesses as to the cause.

While some state constitutions were merely revised colonial charters, others had been written from scratch and with great care. The latter were the more interesting. For in several of them, owing to the strong faith in civic virtue, the framers had thrown caution to the wind and shaped the constitution solely to accommodate the will of the people. If the voice of the people was the crucial element in a republic, they had concluded, by all means let us hear that voice loud and clear.

Pennsylvania was the best example. Here the framers designed the constitution around a strong, democratic, unicameral (one house) legislature. They made the executive office weak, for they feared the inclination of executives to work mischief, and even placed the judicial system under the legislature's control. There ought to be no authority, they believed, to challenge that of the people's elected assembly.

Structure Undermining Virtue

Shrewd observers began to see that certain kinds of structural weakness played directly into the lapses of civic virtue. Take Pennsylvania again. The omnipotent assembly may have represented the people, all right, but it also represented the clash of competing interests among them. Large and vocal interest groups found that it was not too difficult to muster a legislative majority, and when they did so they could pass self-serving laws with impunity. The outcome was democratic, perhaps, but it was not conducive to political morality. For, as James Madison had noted, there were only winners and losers in such a game and it was hard for either of them to think about the public good.

In the postwar world, moreover, depression and dislocation added sharp urgency to the assembly's actions. Legislators saw themselves as fighting for their constituents' very survival, and a no-holds-barred mentality gripped the state house. There was another problem. Civic virtue turned out to be far less restraining on organized groups than on isolated individuals. For the dynamics of group interaction tended to nullify the role of conscience and convince everyone they were doing the right thing.

When other branches of the Pennsylvania government dissented, the assembly demonstrated another weakness of the structural situation. It could often find means, both legal and extralegal, to undercut competing authority. There was no executive to speak of, and the judiciary could be brought to heel by simply firing the judges and appointing new ones. John Adams pointed out that Pennsylvania was not really being "governed" at all. The strongest faction in the state was simply ruling according to its will—much in the way that tyrants in the ancient world had done.

Legal Tender Laws: Legislation which declared that US currency must be accepted at face value for the payment of all debts and obligations.

Stay Laws: Prevented foreclosure of mortgages, etc.

Ex Post Facto Laws: Laws retroactively applied.

Pennsylvania was the worst case, but many other states experienced similar difficulties. The tendency was for power to drift toward the newly democratized assemblies and there to be misused. The pattern went something like this. Indebted farmers, facing closed markets and other postwar adjustments, pushed hard for paper currency, legal tender laws, stay laws and the like, the practical effect of which was to cancel the debts they had incurred before the war. The farmers were strong enough as a political interest to get their way. Indeed, they were so strong they could roll right over the constitution, as in Pennsylvania, and those who failed to cooperate found themselves at risk. Their authority could be redefined. Their power could be emasculated. Their very offices could be wiped out. And constitutional prohibitions proved to be scant protection for them—mere "parchment barriers."

There were troubling implications in all this, and some Americans began groping to define them. "Majority tyranny," in Madison's phrase, was certainly one of them but there were others. The legislation the farmers typically demanded was both special interest and ***ex post facto*** in character. The laws singled out specific groups—such as moneylenders—and basically defrauded them of their property. The targets could not know in advance that this might happen to them, of course, so they were not only being punished, they were being punished after the fact.

Similar issues had had much to do with bringing on the Revolution. The security of property, for instance, had been a primary bone of contention between the colonies and Great Britain. Oddly enough, Americans were begin-

Events during the French Revolution demonstrated just how tyrannical a 'majority' could become when not checked either by virtue or structure.

ning to see that their own legislatures were now doing more or less what the British Parliament had done earlier, and for much the same reason.

Equally troubling was the question of expectations. All business—indeed, all affairs of society—must be conducted in a climate of certainty. A banker who places depositors' savings out on loan has to have some expectation that the money will be coming back with interest—or there will be no banks, no loans, and no business. Without the stability of settled expectations, the American experiment would be in dire straits.

Many of those who found this situation alarming were among the revolutionary leaders themselves. They had fought a terrible war to secure the Good Society. The idea of property being wrested from one faction by another, more numerous and rowdy, was not the earmark of any Good Society—it was more like ancient Athens run amok.

The problem was not that virtue had failed, precisely, though in some ways it certainly had. It was that virtue and structure were not working together. A legislature structured in such a way as to allow one interest to run roughshod over another was simply not an inducement to civic morality. The structure, let us say, was pushing the wrong buttons in human nature.

Americans were not ready to despair, however, at least not yet. Analysis of the problems convinced many of them that the difficulties were not beyond resolving. But much thinking and rethinking were clearly in order.

The Confederation

According to the eighteenth-century doctrine of ***sovereignty***, only a single sovereign authority could exist in any given political society. We can think of sovereignty as the place where the buck stops—the ultimate, complete and final authority to make political decisions. In a monarchy it is the king who is sovereign. In a democracy it is the people as a whole. And so on.

By such thinking, the political society now calling itself the United States of America had to ground its sovereignty *either* in the states *or* in the Confederation as a whole—there was no middle ground. Given that choice, it was a foregone conclusion that sovereignty lay in the states. After all, each state had its own separate identity and historical experience. Relations among them (as colonies) had always been difficult, sometimes even troubled, and a sort of rugged individualism had come to be the rule. And until the passage of the Stamp Act, all attempts to unite the colonies in some sort of larger dominion had dismally failed—even in the face of dire necessity.

Republican theory supported this state of affairs. For it strongly submitted that modern republics must be as small, local, and separate as their ancient ancestors had been. Only in such a setting, as the argument went, could citizens exercise proper vigilance and could civic virtue operate as it should. By its very nature, a republic required political intimacy.

We have forgotten how appealing this idea was to many Americans—probably most. Small republics, each in command of its own affairs; popular participation in the political process; a town-meeting style of interaction with one's own friends and neighbors; sensitivity to local needs and problems; respect for the distinctive culture of a given locale; the ennobling aspects of a hands-on democracy—these and other benefits were assumed to derive from this modern rendition of the classical republic.

In consequence, the Confederation could hardly be imagined as anything but what it was—a loose-jointed alliance of sovereign republics for the purpose of common defense. Its organization and powers sufficed only to that end. It had a unicameral legislature in which all states were represented equally. (The equal voice bespoke state sovereignty more than anything else, for it represented states as individual entities, not the people as a whole.) There was no executive as such, nor was there a national judiciary. And the government lacked the all-important authority to raise revenue, the very hallmark of sovereignty, relying instead on humble requisitions to pay its bills. This meant that any other powers the Confederation had were largely illusory.

The Confederation was particularly hobbled by its inability to regulate interstate commerce. This was another sovereign power, for nation-states of the time commonly used trade regulation as a weapon against their neighbors. By taxing strategic imports, forbidding strategic exports, and either subsidizing or monopolizing key industries, governments attempted to promote their own national well-being at the expense of others, and not rarely, such behavior led to actual war. This situation had characterized Europe for hundreds of years—and now it characterized America.

For these and other reasons, thoughtful Americans were deeply concerned about the Confederation's prospects. Yet few supposed that this league or any other could advance much further toward true sovereignty. Its constitution, called the **Articles of Confederation**, had been written by a committee of the Continental Congress during the Revolutionary War. The work of drafting and revising the document had begun in 1777 and had drug on for nearly four years before final approval. And even then John Dickinson, who had played a central role in the process, regarded the adoption of the Articles as miraculous. This tells us much about the reluctance of American states to part with the meagerest scraps of their sovereignty.

Articles of Confederation: http://www.earlyamerica.com/earlyamerica/milestones/articles//

The problem wasn't merely one of sibling rivalry. It was centered in the republican idea itself. For any powers given to a supergovernment like the Confederation were by definition "imperial" rather than "republican" in character. *Empire* was the only word in Thomas Jefferson's dictionary to describe an entity larger and more extensive than an individual state. Did they want to create an "empire?" To verify the answer, the framers of the Articles drove a final nail in its coffin. They stipulated that there could be no amendment of the document's provisions except by the approval of *all* members.

The behavior of the states was entirely consistent with their muscle-flexing sovereignty. They slammed tariffs on each others' goods, levied huge transit fees, and wielded monopolistic assets—such as port facilities—like a bludgeon. They sent their own ministers to foreign capitals and conducted their own diplomacy with the Indians. They went around the Confederation at every turn and subverted it whenever expedient.

Several of the states maintained their own military establishments—quite apart from the Confederation's anemic forces—and made use of them in settling interstate disputes. Indeed, in the six years between the cease-fire at Yorktown and the Convention in Philadelphia, there were no fewer than four separate brushfire conflicts involving state militias. The worst of these was between Pennsylvania and Connecticut over rival claims on the western frontier. The Pennamite War, as it was called, racked up scores of casualties and dramatized the high cost of state sovereignty. For in the absence of a national court system, there was no other way for states to settle their differences.

Foreign governments also took advantage of the Confederation and played off one state against another. Great Britain was the worst offender. Its diplomats maneuvered fast and loose among the Americans precisely because there was no central government to rein them in. They pressed the advantage of their own merchants and drove merciless trade bargains. They continued to occupy their forts in the Northwest, too, and conspired with the Indians to harass the American frontier. When the US Minister in London complained that this was in explicit violation of the peace treaty just signed, the British replied that American states violated the treaty every day, especially in the matter of repaying English debts. But, then, most borrowers couldn't repay the debts for a good reason: the British wouldn't grant them trade privileges. It was all a hopeless tangle.

Nationalists came to depict this state of affairs as confused, dangerous, self-defeating and bordering on outright crisis. Their concern may have been overstated, perhaps, but the Confederation clearly did lack the power to deal with significant difficulties, and its impotence cast a long shadow.

To begin with, given the essential anarchy among them, states could never be sure of the good faith or ultimate intentions of their neighbors. Thus, they had no choice but to adopt strategic behavior, pressing the advantage when they could and seeking defensive measures when they must. A vague sense of hostility came to suffuse their world.

The Jerseys: Colonial New Jersey was once divided by its two proprietors into East and West Jersey.

Clusters of states began to consider regional confederacies as an alternative to going it alone. New England, for example, comprised a rough-and-ready unit with its Calvinist background and common interest in trade. In the South there was a common interest in staple agriculture and the pervasive practice of slavery. The Middle States—New York, Pennsylvania, the Jerseys—had less in common than New England or the South, but they were beginning to share a growing cosmopolitanism. And beyond the Appalachians there was the West, already vibrant with distinctiveness. Might the future lie in such groupings?

Thoughtful Americans discussed the possibility. Since regional confederacies would be stronger than solitary states, and since they would also be faced with strong rivals, each would have to think about defending itself against the others. This might lead to standing armies, military budgets, high taxes, and an imperial style of leadership. Parallels for this situation could be found in places like the Balkans: rival principalities, well armed, often at war with one another, living in constant uncertainty. None could exist as a republic.

The problem, once again, was not simply a failure of virtue. Indeed, virtue of a certain kind might make *this* situation worse. For remember that virtue's classical ancestor, *areté*, had a lot to do with deeds of honor on the battlefield and aggressive leadership at home—precisely what could (and in ancient Greece did) lead to a free-for-all. The problem, rather, was that virtue and structure were not working together toward a Good Society.

John Adams and the Rule of Law

Like many of the Founders, John Adams was a perceptive observer of these events. He was also—and not coincidentally—widely read in history and political thought. Adams had trained as a lawyer and was an accomplished practicing attorney. It was the law, he often observed, that might well resolve America's present difficulties. In order to take in his full meaning, we need to examine a connection of crucial importance to the doughty New Englander—the connection between law and liberty.

The connecting argument was set forth in Adams' well-thumbed copy of Locke's *Second Treatise*. "Freedom is this," Locke had written,

> freedom of men under government, to have a standing rule to live by, common to every one of that society and made by the legislative power erected in it, a liberty to follow my own will in all things where the rule proscribes not, and not to be subjected to the inconstant, uncertain, unknown, arbitrary will of another man.

There was, Adams knew, a long and tortuous history behind those words.

In the early days of the ***common law***, as the English legal system would be called, the king heard and decided controversies himself, judging them according to custom, reason or whatever rules might seem to apply. Eventually, when the "caseload" became too heavy, the king delegated his judicial function to others, and these became the first common law judges. If a dispute required some new principle, or "precept," for its resolution, the judges carefully spelled it out so that all future cases involving similar questions could be resolved the same way. Precedent embodied precept and became binding.

John Adams understood the essential role the rule of law would need to play in the emerging American Republic.

A few common law precepts had political implications. One of these, evolving through the mists of time, was called *habeas corpus*. It held that a person ought not be arrested and detained unless there were grounds for believing that he or she had violated some specific law. If the individual in question was merely, say, a political undesirable who had been arrested in order to be silenced, he could request a writ of *habeas corpus* from a judge, and armed with that he could require the arresting authority to show that he had actually broken the law. If no such evidence could be shown, the accused had to be set free.

The important feature of *habeas corpus* and similar devices was that in theory at least they applied to *everyone* in the realm—the king and his ministers not excepted. That was often a tough point to make, especially in the face of royal prerogative, but the judges kept hammering away. They argued over and over that for law to have meaning, the king himself must be bound by it.

When the point was finally made, centuries hence, a subtle transformation came about in English life. The government became obliged to *govern* the people not just *rule* them. That is, the king could no longer order his subjects around and make them submit to his will. The existence of known, standing laws and their application to *all* members of society changed everything. England had achieved the ***rule of law***

Of the rule of law's many benefits, one of the most important was stable expectations. For wherever the law truly ruled, the laws of society became in a sense like the laws of nature—steady and reliable. Just as one could count on the law of gravity to propel water through an aqueduct, one could count on the law of contracts to insure the repayment of loans. And in both cases, predictability made society work better.

But the rule of law's greatest benefit was liberty. That was Locke's point. For knowing in advance what the law would or would not condone, people could calculate their situation to best advantage—precisely as athletes do when playing by the rules—and make the most of it. And when they could do so, what a change in the atmosphere. Citizens could get on with their lives unafraid.

Libel: Publication of knowingly false or malicious statements.

Sedition: Acts that incite rebellion against established government.

They could discuss public issues with candor. They could even stand up and criticize the government. A member of Parliament, for example, could make a hard-hitting speech against the ministry, and as long as he avoided **libel** or **sedition**—there were laws governing both—he could say whatever he pleased. It was the rule of law, more than anything, that made the English feel free.

Understandably, then, John Adams, who prized liberty, equally prized the rule of law. To him it represented the cardinal principle of a Good Society. If the rule of law existed, other things tended to fall into place: freedom, order, benefits for everyone and not just the few. And if it did not exist, nothing else seemed to work well either.

In 1770, some five years before the Revolution, Adams encountered a dramatic opportunity to demonstrate his commitment. After the so-called Boston Massacre, the soldiers accused of perpetrating it were arrested by the local authorities and charged with murder. After all, they had loaded their muskets and fired into a crowd—never mind what the crowd was about to do to them. What patriot leaders had in mind was a "show trial" aimed at discrediting the royal government. They wanted to use the law, in other words, to "get" a political adversary, precisely as the law was often used to get *them*.

Adams wanted none of that. He was as patriotic as the next person and certainly no friend of the royal government, but he knew the law of homicide, and knowing that knew that "murder" had not been committed on King's Street by the accused. He agreed to defend the soldiers in court—outraging fellow patriots in the process—and see to it they were given a fair trial. For he knew the rule of law was on the line. If lives and liberty could be made pawns in a political chess match, the law would be finished.

Now, a decade later, Adams became convinced that the rule of law was the key to unlock present difficulties. The governments of the American states were republican in a technical sense, perhaps, but many of them had failed to establish the rule of law. The government of the Confederation had a similar failing, though for a different reason. In both cases, to be sure, laws were being passed by legitimate legislative authorities. But in the one case, the laws were perceived as unjust, and their effect inclined toward tyranny. And in the other, the laws were perceived as unenforceable, and their effect inclined toward anarchy. Uncertainty was the bottom line in both situations: people never knew what to expect because the law was not steady and reliable.

Later on, when Adams spoke of the American Republic, he often used the phrase "empire of laws." In time, many of his colleagues came to see the Founding that same way. The notion that began to take shape in the early 1780s was to make the empire of laws not just a hopeful ideal but a working reality.

Five Principles of the Rule of Law

In John Adams' day the rule of law was more often felt than understood. For when applied to specific cases, the concept could abruptly dissolve into puzzling

The Trial of the Accused Soldiers

All Boston was scandalized. The soldiers accused of murder in the Boston Massacre were to be defended in court, not by the king's attorneys—there was not a one of them who would touch the case—but by native Yankees: Robert Auchmuty, Josiah Quincy, and, most astonishingly, John Adams himself. None of the three had volunteered. Adams, for one, took the job only because of a dramatic personal appeal by one of the soldiers' friends.

Adams' soft-heartedness cost him dearly. His cousin, Sam, leader of the Liberty party, suddenly turned cool toward him. Rocks were thrown through his windows and a mudball splattered his cheek. There were whistles and catcalls wherever he went, and a voice behind a fence asked him why he left his red coat at home. He found solace from a passage in Beccaria: "If, by supporting the rights of mankind and of invincible truth, I shall contribute to save from the agonies of death one unfortunate victim of tyranny . . . his blessings and tears of transport will be sufficient consolation to me for the contempt of mankind."

The defense attorneys prepared their case well. Of the three, Adams was best versed in the English Common Law, and he grilled his colleagues for days on the distinctions between homicide justifiable, homicide excusable, and homicide felonious. To many Bostonians such distinctions seemed irrelevant. They sided with the *Boston Gazette* when it piously thundered: "Whoso sheddeth Man's blood, by Man shall his blood be shed!"

That was tommyrot, and John Adams knew it. The soldiers had been cornered by a wild mob spoiling for their blood. One of the casualties himself, lingering before death, forgave the soldiers and confessed that the fault had been on his side. But to go too far into that, as John Adams also knew, would be to put the town of Boston on trial and harm the cause of liberty. So the defense strategy would be to stick to the law, and to the law's venerable distinctions. They would argue that this was a case of homicide justifiable.

The trial itself was the event of the decade. People drove in from the countryside and picnicked on the common while awaiting it. The courtroom was packed to overflowing, and the spillover thronged the steps outside. The defense began with lengthy interviews of prospective jurors, throwing out one after another, and including in the final roster not a single Bostonian. Then the charge was read, the pleas entered, and the trial formally commenced.

Auchmuty and Quincy acquitted themselves well. The latter's impassioned rhetoric on the soldiers' behalf made the jurors stop and think. But John Adams was the trial's real hero. Standing before the court in his barrister's wig and gown, he seemed to personify the law and its ancient traditions.

He stepped near to the jury and addressed it earnestly as he spoke. "Place yourselves in the situation of Killroy or the sentry, with the bells ringing—and you know well there is no fire—the people shouting, huzzahing and making the mob whistle as they call it. The people are crying, Kill them! kill them! knock them down!—and heaving snow balls, oyster shells, clubs, white birch sticks three and a half inches in diameter. . . . Consider yourselves in this situation and them judge if a reasonable man would not consider they were going to kill him." The law, said Adams, did not require a man to stand still for that. The law allowed him the right of self-defense—even if the result was homicide.

In the closing arguments, the defense appealed to good, hard, common sense. That and the law. Adams quoted Algernon Sidney that "The law no passion can disturb. 'Tis void of desire and fear, lust and anger. 'Tis written reason, retaining some measure of the divine perfection. 'Tis deaf, inexorable, inflexible." Them he added his own wisdom. "The law," he said, "on the one hand is inexorable to cries and lamentations of the prisoners. On the other it is deaf, deaf as an adder to the clamors of the populace."

The jury considered the matter for two and a half hours and returned a verdict of not guilty for six of the soldiers, guilty of manslaughter (a lesser charge) for the remaining two. The punishment was a mere branding on the thumb. Then and there, with jurors and spectators alike craning there necks to see, an iron was heated, the two soldiers stretched out there hands, and the requirements of justice were met.

As Adams walked over to the prisoners, they seized his hand in gratitude. "God bless you. Mr. Adams!" they said. "We owe our lives to you and Mr. Quincy." Matthew Killroy, reputedly the toughest of the eight, sat weeping in his chair. Montgomery, the other brandee, held up his thumb as one might exhibit a trophy. "A small price to pay for our lives, Sir," he said with a smile.

From America: A Study in
Heritage, Kendall/Hunt
7th edition, p. 102.

ambiguity—and even today we often grasp it imperfectly. But people *felt* that something was wrong when the rule of law was violated. The words "unjust," "unfair," "unreasonable" or "arbitrary" were used to convey their displeasure. We have all been in situations described by these terms. They usually occur when we find ourselves confronting authority. If we perceive that a parent is being heartless, a bureaucrat mindless, or a boss soulless, it is often because the rule of law is being shorted.

The rule of law is not law itself. It is an abstract ideal, a set of philosophical principles, guiding the way we think about law and make use of it. Accordingly, the rule of law can never be passed by a legislature or encoded into specific language—it truly must be "felt" in the hearts of the people. And, like any ideal, it can never be perfectly realized.

We have seen that an instinct for law is embedded in the human personality. Still, in some societies, there is an exceptional sensitivity to the principles we are discussing here, and when those principles are violated, the people cry out. Colonial America happened to be such a society. When Parliament passed the Stamp Act or the Townshend Duties, many Americans could not explain why they found those measures so odious. They heard all the reasonable arguments about paying their own way, enjoying relative prosperity, being "virtually" seated in Parliament—and came back all the louder with "No taxation without representation!" They were voicing a rule of law objection without fully realizing it.

Conversely, in other societies, there may be very little appreciation for the rule of law, and when its principles are violated, people are easily mollified with explanations. Hitler's Germany is a case in point. Although technologically proficient and highly civilized, Germans nevertheless failed to apprehend transgressions of the rule of law that became more or less systematic with the Nazis. If a given edict was "official," they supposed, it must be okay.

Hitler and the Nazis' disregard for the rule of law enabled them to achieve extraordinary power for a relatively brief period of time before Germany came to be regarded as something of a renegade state. The consequences for the world were catastrophic.

This fact tells us something else about the rule of law. It exists quite independently of the visible machinery of government. Nazi Germany had a legislature, a legal code, a court system, and many other appliances of the law. All it lacked were those philosophical principles.

If John Adams had made a list of such principles, it would probably look something like this:

1. *Generality*

John Locke railed against the fact that in Stuart England there was one legal code for somebodies and another for nobodies. The laws, he said, are "not to be varied in particular cases, but to have one rule for rich and poor, for the favorite at court and the countryman at the plow." In modern-day terms, this means that the laws must apply to broad categories of people. They must not single out specific individuals or groups for special treatment. If there is any singling out, it should be done by people themselves *after* the laws are passed. That is, I place *myself* in the category of "drivers" when I climb into my car and start the engine.

We would sense there was something wrong with a law addressed to the drivers of red Toyotas, for such an ordinance could also single out Mormons or Southerners or liberals. Those who make the laws should never know in advance to whom they apply. Lamentably, the British Parliament lacked that sort of blindness when it passed laws expressly for Americans and then began laying special taxes on them.

But in practice we too create exceptions. Some of these truly violate the principle of generality and ought to be viewed with extreme mistrust. Others are a practical necessity. And there is no hard and fast way to tell which is which.

A few categories of singling out, however, have become absolutely taboo in democratic polities. These involve ethnicity, gender, religion, political affiliation, and increasingly, sexual preference. Above all, we abominate separate laws for the rich and the poor—somebodies and nobodies—knowing these could throttle a Good Society. "Equality under the law" has become a watchword.

Birmingham, Alabama police respond to a 1963 civil rights demonstration. Laws which discriminate against specific segments of society violate which aspects of the rule of law?

2. *Prospectivity*

This means that the laws must apply to future action, not past. We would feel as uneasy with a law punishing those who presently drink soda pop as with the singling out of red Toyotas. For the potential violator must always be able to decide *in advance* whether or not to obey.

Tyranny, of course, disagrees. For the tyrant wants to say in effect "I don't like what you did," and punish accordingly. The last thing a Hitler or Stalin would desire is for intended victims to evade sanctions by altering their behavior.

In practice, most violations of generality also violate prospectivity and vice versa. Together, in fact, these two abridgements of the rule of law account for most of the tyranny the world has known.

3. *Publicity*

This means that the laws must be both known and certain. They must be well publicized, and their enforcement must be reasonably reliable. Laws that are capriciously enforced—or else not enforced at all—do no favors for the rule of law. On the contrary, they sow public contempt.

But police apparatus is expensive. So governments occasionally trim back by funding a mere token enforcement of the laws, hoping that the possibility of punishment, no matter how remote, will suffice to deter lawbreakers. Many people will accept such gambles and bet they won't be caught. If government isn't serious about the laws, citizens won't be serious either.

Capricious enforcement was an aspect of British policy before the Revolution. It led to widespread smuggling on the part of colonials and to customs racketeering on the part of the authorities. Later on, when the British sought to reform the system and enforce the Navigation Acts rigorously, Americans were outraged.

4. *Consent*

In a single catchphrase, consent comes down to "No taxation without representation!" It means that the laws must be generally acceptable to those who must live by them. Electing the lawmakers and upholding the constitutional system are examples of the way we give consent.

Yet not all law making is subject to this sort of approval. Customary law, judge-made law and natural law are not usually submitted to the voters for endorsement. What counts, however, is that the people, if they truly *dis*like a certain law, have the means at their disposal to wipe it off the books, as they did with the Jim Crow laws of the South or the Volstead Act. For the theory of consent holds that the people as a whole would never consent to injustice—because they themselves would be its victims. The theory only works, of course, when the laws are *also* general and prospective.

Jim Crow Laws: Segregation laws which were widespread in the South.

Volstead Act: A 1920 law enforcing Prohibition.

Where consent becomes absolutely mandatory is in the matter of taxation. Once the people themselves consent to the amount they will be taxed, government undergoes a seismic transformation, and the distinction between rulers, on the one hand, and the ruled, on the other, virtually disappears. Put another way, if lawmakers themselves have to come up with the taxes *they* determine, all incentives to be unfair go out the window.

It was the absence of consent that sent the British Empire to its doom in 1776. For where one group, Parliament, is allowed to make laws for another group, Americans, and is not accountable to those laws itself, there is nothing whatever to restrain it. Why *not* pass a Stamp Act? asked the MPs. *They* didn't have to buy the stamps.

5. Due Process

The phrase *due process of law* simply means that the laws must be administered impartially. Justice, as the saying goes, has to be blind. It must consider nothing but guilt or innocence. If the accused person is black, or poor, or a Communist, or a Christian Scientist, the law must say: so what? And there must be an established set of procedures to insure that everyone before the bar is given a fair trial, an adequate defense, and if guilty an appropriate punishment.

Due process accounted for yet another sore spot in imperial relations. For the British authorities had few qualms about revoking jury trial and other traditional rights in the interest of tighter law enforcement. If Americans were a bad risk, they said, then by all means close up those procedural loopholes. We often say the same thing about "bad risks" today.

We have come to take these five principles so much for granted that they seem no more than simple common sense. Yet they have been willfully ignored by those we would least expect. For example, back in the days of the Stuart monarchy, not only did the king breach the rule of law, so did Parliament. In the notorious bills of attainder, a political offender—usually one of the king's henchmen—was singled out by name in a parliamentary act, accused of a crime, and pronounced guilty. Think how many of the five principles such a law violated.

Subtler abridgements of the rule of law are still committed almost everywhere. For governments with the worthiest of motives see a problem and take action to solve it, unmindful that they may be aiming legislation toward a specific group, penalizing past action, or ignoring the rights of possible violators. In so doing, they come close to "administering" the lives of their citizens the way they administer park maintenance or road repair.

The same could be said about the controversy leading up to the American Revolution. Americans viewed British violations of the rule of law as darkly conspiratorial, as we have seen. But records show that the policy makers simply thought they were solving problems.

During the Confederation period, similar violations were common. Either the law was being used for political purposes—to advance one group's agenda at another group's expense—or else the law was simply incapable of enforcement. There was often no conscious thought of tyranny or anarchy, but the damage was the same as if there had been.

Virtue and the Rule of Law

While John Adams often alluded to civic virtue, he was one of the Founders who privately expressed doubts. "You will be very sensible," he wrote to James

Warren on the eve of the Constitutional Convention, "that our Countrymen have never merited the Character of very exalted Virtue. It is not to be expected that they should have grown much better."

Many of the Founders had reason for such skepticism. Their own hard experience, their reading of history, and their acquaintance with the latest theories of psychology had convinced them that virtue was highly problematic. Unquestionably, some individuals *were* capable of the *areté* described by Plato. They could place public good ahead of private advantage. They could face stark sacrifice. They could find deep meaning and high purpose in politics. And they could love their country with a profound and selfless patriotism. The Founders themselves answered this description, and it broadly bespoke their aristocratic world.

But the day-to-day reality of a commercial society was simply otherwise. Most Americans were less motivated by high idealism than by old-fashioned self-interest. Comfort and convenience counted a great deal with them, and they measured life in dollars and cents. Accordingly, they weren't overmuch concerned with most public issues, especially the abstract or remote ones, and whenever they became so, it was usually with an eye to private advantage. The psychology of the day—notably that of David Hume—questioned whether the human mind was even capable of the moral thinking virtue required. What people took to be morality, Hume contended, was often just custom in disguise.

There were other problems in basing the Good Society on virtue. As the ancients had defined it, virtue appeared to be both unstable and situational. Soldiers were capable of great sacrifice in time of war—yet as veterans they thought nothing of engrossing themselves at public expense. A Benedict Arnold could be a national hero one minute and an archtraitor the next.

Gulag: The labor camp penal system of the USSR.

Virtue had a frightening side as well. Deeds of honor were one thing when the colonies were fighting for their lives—they were something else when the new states began jostling one another over rival claims. And empire wasn't the only danger Americans faced. *Areté* also stoked the fires of fanaticism, as history plainly showed, and while Stalin's gulags and Hitler's storm troopers were still far in the future, the French Revolution was just around the corner.

In pondering such matters, the Founders were grappling with one of philosophy's most elusive problems—the fundamental nature of man. They didn't claim to have final answers. At the same time, it was critically important that they get certain things "right," and human nature was one of them.

Wisdom and Virtue

As the Founders revised estimates of American virtue downward, the word *virtue* continued to appear in their correspondence—but it was increasingly connected to the word *wisdom*.

Wisdom was also a kind of code word. It represented an array of values, attitudes and specific insights which taken together meant that life was not simple. When *wisdom* and *virtue* were linked together, the republican picture that Plato had painted in broad strokes and blazing colors took on depth and shadow, and for the Founders it began to look something like this. A modern republic, even in America, could not be based on Spartan-style civic virtue, for liberal individualism simply couldn't deliver such an intense and single-minded public-spiritedness. Yet Americans did have their own forms of virtue—and clearly *some* forms were essential to *any* republic.

The questions posed by wisdom, then, were these. What kind of virtue *could* human nature deliver? What kind could liberal individualism deliver? What kind was necessary for a modern republic? And finally, how much virtue was "enough?"

We might think of the rule of law as representing a new approach to such questions. Probably none of the Founders, including John Adams, worked out the relationship between law and virtue in so many words, but Adams seemed to grasp the main idea. The rule of law offered a different way of focusing moral energy. Instead of asking citizens to put the good of society ahead of their own personal interest or to perform great and lofty deeds, they are asked instead to do two things:

> *Cherish the law.* People must commit themselves to observe the law with an almost religious devotion. Indeed, the law takes the place of the king as the symbol of nationhood and gains its own sense of mysterium. Obeying the law is not about fearing to get caught—but desiring to live virtuously.
>
> *Cherish the rule of law as an ideal.* This is a bit trickier, for it requires a sophisticated understanding of what law is and how it works. To revere the *rule of law*, rather than just law itself, one must know something of politics, something of life, something of philosophy perhaps, and assuredly something of wisdom. One must know, for example, how the rule of law is often undermined by seductive political temptations.

Alexis de Tocqueville: French observer of American life circa 1835.

Machiavelli: Italian philosopher best known for his treatise on the attainment of political power through any means.

The rule of law required its own kind of virtue, to be sure, but it was a virtue within human reach. Indeed, it was the *kind* of virtue—orderliness, accountability, self-restraint—that a liberal society might well be able to deliver. (Insightfully, **Alexis de Tocqueville** dubbed it "enlightened self-interest.") The rule of law also required its own kind of wisdom, that of both the wise leader and the wise electorate. Shrewdness, cunning, the Machiavellian traits so often associated with politics, did not work well in a rule of law society, nor did go-for-broke activism or flashy charisma. What the rule of law society called for, rather, was statesmanship.

The Rule of Law and Constitutional Design

Virtue could not establish the rule of law by itself; it was also a matter of constitutional design. But what were the specific design principles involved? Why did some constitutions do a much better job of it than others?

Separation of Powers

One clue—and it would turn out to be an immensely important one—came out of the English constitutional experience. Once the king, the Parliament and the

The Noble Experiment

What happens to respect for the law when masses of citizens cannot bring themselves to obey it? Before the days of Prohibition, no one thought to ask. After all, among a law-abiding people, laws were supposed to be respected automatically. As Prohibition Commissioner John F. Kramer said of the Volstead Act: "This law will be obeyed, and where it is not obeyed it will be enforced." But it turned out that Prohibition was neither obeyed nor enforced, for American drinkers simply continued their drinking.

At home, in their basement, they brewed hard cider, cheap wine, and makeshift brandy, and in their bathtubs they literally concocted bathtub gin. For the making of high octane beverages, they rigged up their own distilleries, some of them fairly sophisticated. One still, steaming away in a New York cemetery, had produced fifty-one barrels of moonshine by the time Prohibition agents got onto it.

Most people, however, simply bought their bootleg ready-made. This could be done at the local speakeasy, a club or café with a secret back room, or at a variety of other establishments. Near Cornell University it was soda fountain. In lower Manhattan it was store that sold "sacramental wine." In Baltimore it was a sidewalk fish market, with tiny bottles of booze in the cash register drawer. In Atlanta it was a confectionary, and in New Orleans it was a taxi—*any* taxi. One woman bought a milk shake in New York's Van Cortlandt Park and wrote indignantly that by mistake it was filled with gin. (Bad gin, too, she reported.) Another complained that in a grocery store in Harlem, a clerk "charged me two dollars for a can of tomatoes, and when I got home I found there was nothing in it but a lot of nasty-smelling water." And nosing around in Brooklyn pawn shops, agents uncovered then one thousand dollars' worth of liquor wrapped in the clothing left as pledges.

Nor was the illicit trade in alcohol confined to lowbrow establishments. In New York's tony Half Past Nine Club, a large stock of contraband was found in a stuffed grizzly bear. In a fashionable Madison Avenue delicatessen, the beribboned baskets of fruit concealed small bottles of rye. Before the "noble experiment" was over, Manhattan lost some of its most exclusive nightspots: Jack's, Shanley's, the Beaus Arts, Reisenweber's.

People regarded the mass evasion as good, clean fun. At the football game, they pulled flasks out of hip pockets, garter belts, and raccoon coats, and had a wonderful time. Only dimly were they aware of names like Johnny Torrio, Frankie Yale, and Al Capone; and the connection between big-time gangsters and the good, clean fun seemed tenuous. Then, on St. Valentine's Day of 1929, two carloads of Al Capone's hit men battered their way into a Chicago warehouse and gunned down seven members of "Bugs" Moran's gang. Americans were puzzled to learn that the massacre took place in a bootleg distillery. Somehow, the country was becoming downright lawless.

From America: A Study in Heritage, *Kendall/Hunt* 7th *edition, p. 112.*

court system had accepted one another as fully legitimate entities, there was a division between the rule-making and the rule-enforcing authorities of the realm. The rule maker was Parliament. The rule enforcer was the king and the court system working together, royal officers bringing suspects to trial and the courts deciding their guilt or innocence.

It was only a rough-and-ready division, to be sure, and sometimes it broke down, but in the main it worked. As long as the rule makers couldn't directly enforce the rules, they had to make them blind, as it were, and general. And as long as the rule enforcers couldn't make the rules, they too had to operate on a general basis. The law couldn't be used for extracurricular purposes—to "get" someone—because it was under no *single* authority's control.

Enlightenment thinkers had expanded the theory of these mechanisms somewhat, positing that there were three functions with respect to the law, not just two. The legislative branch made the law. The executive branch implemented the law. And the judicial branch applied the law to specific cases. But it still came down to a structural separation between rule making and rule enforcement, and that separation worked to keep the law "blind."

It is also instructive to note what had happened to this system in the early 1700s. With Parliament on a roll after the Glorious Revolution, it took over some of the executive operations of government. Parliamentary officers did not arrest lawbreakers or decide their cases in court, to be sure, but with the emergence of the ministerial system (which is still in use today), these officers did acquire certain enforcement responsibilities. The separation between rule maker and rule enforcer grew blurry—and opened golden opportunities for political jobbery. Indeed, much of the "corruption" alluded to in the commonwealth ideology had its origins here. Powerful ministers and their friends could identify specific goals, marshal support for these in the House of Commons, secure the measures needed to further them along, and implement them in the precise way they had in mind. No one had to operate "blind" any longer—for the powers got along with each other only too well.

Several of the American Founders began to grasp the difficulty. From their reading of Montesquieu, they had picked up the idea that there was something to be gained by keeping the powers apart. For when rule making, rule implementing, and rule applying were all in different hands, officials were more likely to be aboveboard, or at least so held the growing belief. The idea—it was not yet a full-fledged theory—had gained enough currency to find its way into a few state constitutions. "The legislative, executive, and judicial powers of government ought to be forever separate and distinct from each other," was the way it was generally put—a sort of little sermon in the preamble.

The trouble was, this strategy had almost never worked. The powers of government did *not* remain separate and distinct, as we have seen: the moment a muscle-flexing legislature set out to solve a problem, it regarded the rest of the government as an impediment to overcome—and the little sermon as a mere nuisance. James Madison shook his head at such developments. Constitutional prohibitions, he said, posed but a "parchment barrier."

But there might be ways of designing the structure so that the powers really would force each other apart in the course of their routine operation. The question was, how?

Sir Thomas More and the Law

Nowhere was the power of the law to thwart tyranny depicted more clearly than in the struggle between Henry VIII and Sir Thomas More. King Henry, having married his late brother's wife, Catherine of Aragon, found himself with no son and, more important, no heir. He thus determined to divorce Catherine and try again with someone else—someone more comely. But there was a hitch: the king was Catholic, and the pope would not sanction his divorce. Henry plunged ahead anyway. He went through with the divorce, married Anne Boleyn, and repudiated the authority of Rome.

Sir Thomas More was the king's chancellor, a high officer in this government. He was a wise, a saintly man, devoutly religious, and respected all over the realm. Had he ratified the king's bold actions, it would have gone far toward making them acceptable. But this his conscience would not permit. Of course, he would not renounce the divorce either, nor condemn the king's break with Rome, for that would amount to treason. He simply said nothing. A master of English jurisprudence, More knew exactly what the law would and would not condone. He also knew that English respect for the law was such that the king himself feared to break it.

Just where that left him was aptly illustrated by his trial in 1535, imaginatively recreated in *A Man For All Seasons*:

> *Thomas Cromwell* (the prosecutor): "Let us consider now the circumstances of the prisoner's silence. The oath was put to loyal subjects up and down the country and they all declared his grace's title to be just and good. But when it came to the prisoner, he refused. He calls this 'silence.' Yet is there a man in this court, is there a man in this country who does not know Sir Thomas More's opinion of this title? Yet how can this be? Because this silence betokened, nay this silence was not silence at all but most eloquent denial."
>
> *More*: "Not so. Not so, Master Secretary. The maxim is 'qui tacet cosentire videtur.' The maxim of the law is 'Silence gives consent.' If therefore you wish to construe what my silence betokens, you must construe that I consented, not that I denied."
>
> *Cromwell*: "Is that in fact what the world construes from it? Do you pretend that is what you wish the world to construe from it?"
>
> *More*: "The world must construe according to its wits. This court must construe according to the law."

In the end, tragically, More was condemned and executed anyway, but only because his enemies found a witness willing to testify against him falsely. Short of such outright skullduggery, the law of England steadfastly shielded him, even from the wrath of a tyrannical king. It was an example that was not lost on the authors of the American Constitution.

From America: A Study in Heritage, *Kendall/Hunt 7th edition, p. 109.*

Constitution of Massachusetts (1780):
http://www.vi.uh.edu/pages/alhmat/masscons.html

The Model Constitution

John Adams was one of those who thought he might have an answer. On the basis of his own wide reading in political theory and the recent experience of the several states, Adams wrote a landmark booklet titled *Thoughts on Government*. Although he wasn't thinking precisely about separation of powers, he hit upon an important idea. The antidote to too much power in the legislature, he submitted, was to strengthen the power of the executive.

To some, this seemed wrongheaded. Americans had fought a revolution because of, among other things, too much power in the hands of executives—the king's appointed governors. Were they now to return to that selfsame abuse? Yet there were arguments for doing just that. Strong executives were dangerous only when there was nothing to counterbalance them, Adams argued. Strong legislatures were dangerous for the same reason. But when *both* were strong, it might be a whole new situation.

The new situation became manifest in 1780 with the Massachusetts "Model Constitution." The original Massachusetts Constitution, drafted in 1775 at the beginning of the Revolution, had been as problematic as any. In the Model Constitution, the Massachusetts fathers hoped to learn from their mistakes—and from the best thought and reflection available. Many of Adams' ideas found their way into it.

To begin with, the office of the executive was strengthened in several ways. His term of office was lengthened. His appointment power was increased. And he was given a conditional veto over acts of the legislature. He could strike down any law he didn't like, but by a two-thirds vote his negative could be overridden.

The veto was the hardest innovation to abide. After all, it was by means of the "negative," as it was called, that the British governors had done their worst damage. Detractors envisioned a dark future with the veto back in place, everything from executive blackmail to governmental gridlock. Besides, political theorists—and especially those who liked separated powers—pointed out that the conditional veto in effect made the governor a legislator, at least in the negative sense, and thus *compromised* the separation of powers.

In operation, however, the Model Constitution had none of the predicted ill effects. On the contrary, it altered the psychology of power among the branches and promoted a new interactive spirit. We recall that the main defect in state constitutions was their vulnerability to a legislative power-grab. Once the governor was rearmed with the veto, he could ward off attempts of the assembly to usurp power from the other branches (including his own). Conversely, when the legislature was counterarmed with the override—which colonial legislatures had never had—it could similarly protect itself from the governor. Each branch continued to eye the other warily, of course, but each knew it was safe from a truly tyrannical purpose. Both could relax and cooperate.

Above all, the Model Constitution performed wonders for the rule of law. By forcing the powers of government to work independently (without working at cross-purposes), it posed a real obstacle to those special-interest measures and cozy inside deals for which state regimes were becoming sadly famous. Bills lucky enough to be approved by both the legislature *and* the governor passed the rule of law test, as it were, with significantly higher marks. For no governor could make the legislature do his bidding—or vice versa.

The general lesson might be stated this way. In designing the structure of government, anything that could be done to insure that the laws were general, prospective, public and so on was good for the republican cause. The separation of powers did just that.

The Rule of Law and the Good Society

At the beginning of the chapter, we suggested that there were many possible visions of the Good Society. Here is the vision suggested by the rule of law.

In a rule of law society, the government must restrict itself to governance—the creation and enforcement of general rules—not solving problems and certainly not administering citizens' lives. It could promote the *general* welfare by providing defense or building highways, but it couldn't fret about the welfare of farmers who had borrowed money—or bankers who had loaned it to them. This

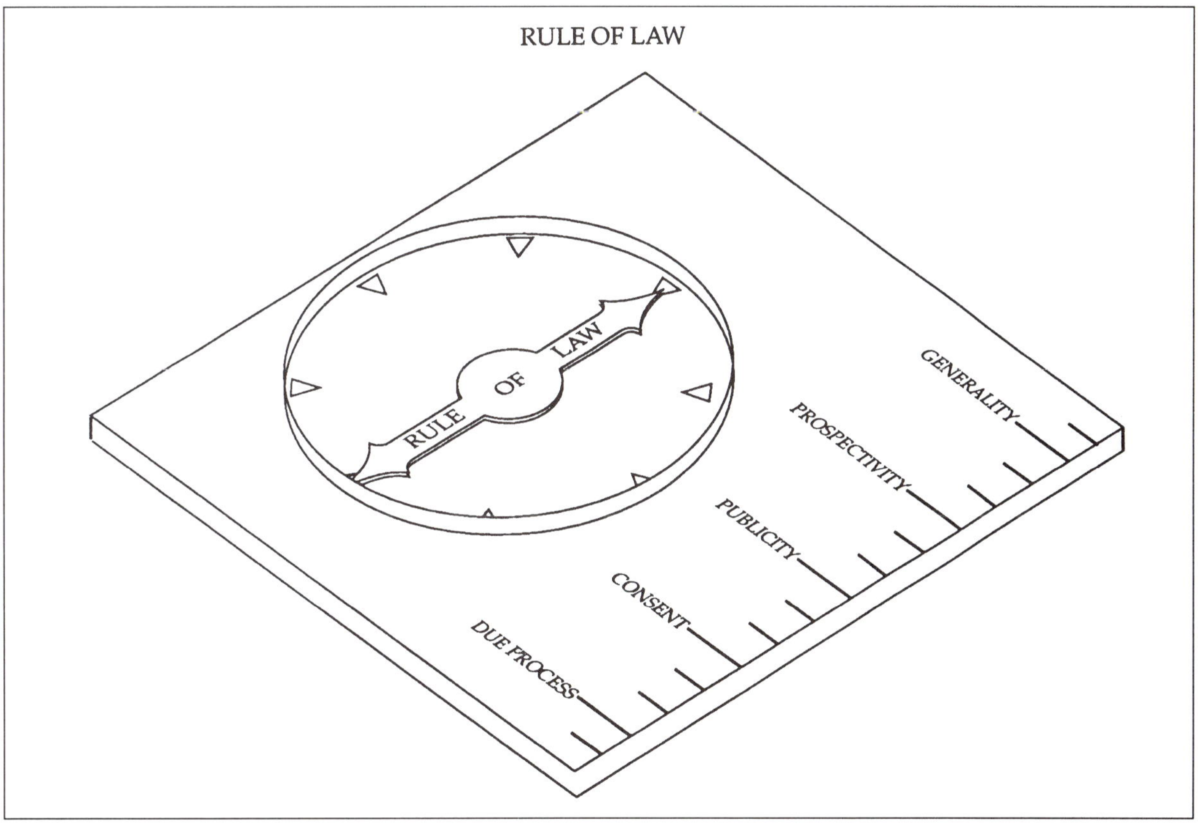

The concept of the "rule of law" represents the philosophical ideal that government may not act except in the impartial enforcement of a known general rule. The actions of such a government are predictable, much like the laws of nature. Those living under the rule of law can chart the course of their destiny in predictable self-determination.

is a far more restrained role than government played in the ancient republics. It is also far less interesting. When government is not conceived as the answer to every problem, it is no longer uppermost on everyone's mind.

Similarly, the rule of law society would be little inclined to save the world. The grand revolutionary schemes of modern times wouldn't appeal very much to the citizens. For they wouldn't be able to work up the turbocharged "virtue" such schemes require—theirs would be more along the lines of ordinary human decency. The citizens would not be the *zoons politikons* described by Aristotle either. The meaning of life for most of them would be found elsewhere than politics—in the home, in business, in the trades and professions, in leisure-time activities. Politics might be engaging at certain points, even exciting at election time, and, yes, there would always be "professional politicians," but the consuming intensity of ancient public life simply wouldn't be there.

In a rule of law society much of the problem solving would take place outside of government. There would be more in the way of private initiatives, community efforts, and voluntary organizations (churches, service organizations, lodges, clubs) to address social needs. Where government did become involved, it would tend to be on the local level.

The moral sense of the people might need to adapt somewhat to a rule of law society. What the ancients called virtue, for example, might have to trade public venues for private ones—fewer John F. Kennedys, let us say, and more Martin Luther Kings. And the individual pursuit of happiness might need to become a little less concerned with the condo in Malibu and a little more concerned with the plight of the homeless.

There would be a role for schools and churches in the society of law, but it would not be the role they played in the ancient republics. There the school and the church, like almost everything else, had been part of the government and in charge of promoting civic virtue. In a rule of law society, religion, like all personal matters, would be outside of public scrutiny, and learning would be only slightly less so. Ironically, however, if schools and churches failed to promote moral sensitivity and ethical standards, the Good Society might well turn bad.

In theory at least, the rule of law provided a way of bringing classical, liberal, and democratic republics together. By transferring initiative from the public sphere to the private individual and the local community, it might redefine and redirect classical civic virtue. By requiring citizens to play by the rules in their pursuit of happiness, it might use that pursuit to benefit the whole of society. And by presupposing an entire populace committed to high (though not impossibly high) ideals, it might base its foundation on the *demos.*

But *only* in theory. In reality, there was no rule of law society anywhere in the world, and especially among the fledgling American republics. There were only a few ideas, a few experiments, a few significant intellectual correspondences—and a whole raft of practical difficulties.

CHAPTER 5 STUDY AID

1. Proto Nation

- Alliance for protection; allegiance to one's state
- Republicanism based on virtue and structure
- States retain sovereignty
- States retain effective power and ultimate authority

2. Problems of Virtue

- Civil virtue declines
- Corruption
- Interests = factionalism & partisanship
- Shays Rebellion

3. Problems of Structure

- Structure undermining virtue
- Pennsylvania (weak executive, easily swayed judiciary, strong unicameral legislature)
- Majority tyranny, elective despotism
- Special interests, ex post facto
- Uncertainty

4. Un-sovereign Confederation

- Sovereignty in the States
- Small is better—republican theory
- Powerlessness
  a. Revenue—no power to tax
  b. Commerce—no power to regulate
  c. Adjudication—no judicial power
  d. Diplomacy—no authority to conclude

5. Rule of Law

- John Adams
- English common law; habeas corpus
- No rule of will
- Predictable and reliable laws = liberty

6. Five principles of the rule of law

- Generality
- Prospectivity
- Publicity
- Consent
- Due Process

7. Virtue and the rule of law

- Wisdom and enlightened self-interest
- Cherish the law
- Observe the law
- Exercise self-restraint

8. Rule of law and constitutional design

- Separation of powers
- Rule making separate from rule enforcing
- Model Constitution—Massachusetts

9. Rule of law and the Good Society

- Government governs—enforces general laws
- Private sector—problem solving

CHAPTER 6

The Constitutional Convention I
The Framers

The Confederation experience underscored the difficulty of creating successful republican governments—a difficulty stretching back to the ancient world. But the Founders took heart from the Enlightenment notion of the social contract, free men coming together and creating their own government through conscious deliberation. That element—conscious deliberation—had often been missing heretofore. The Founders believed that it was possible to do much better. Through "wisdom and reflection," as Hamilton put it in *Federalist* 1, they trusted that the difficulties before them could somehow be resolved.

In a way, it was not a new challenge. Americans had been tinkering with constitutions of one kind or another for almost two centuries. They had learned that certain mechanisms in a colonial charter inclined in a rough way toward certain outcomes in the colony's life. There was a kind of "science" there—dim and shadowy though it was.

The trick was to bring the right people together, first of all, and then get them to deliberate as carefully and creatively as possible. It was not, alas, an easy trick to perform.

The Idea of an American Nation

The story begins with a small group of Americans generally referred to as "**nationalists.**" Most of them were personally acquainted with one another, although they came from distant parts of the continent, and they tended to have a common background in the Revolutionary War.

Nationalist: Supporter of strong central government.

The important quality the nationalists shared was a vision of America as a unified nation. In their eyes, the America of possibility somewhat resembled the ancient Roman Republic—large, powerful, stable and "honorable to mankind." It would be respected in the world and capable of performing great works of public benefit. Not least, it would provide a veritable cornucopia of prosperity. For the nationalists, the very meaning of the Revolution required such a polity as its outcome.

Alexander Hamilton of New York was probably the best known of the nationalists. Certainly he was the most energetic. He had been urging a national constitutional convention ever since the guns had fallen silent. Eventually, others had taken up the cry, and by 1785 they were corresponding enthusiastically with one another. For with each passing month there was fresh evidence that the Confederation simply wasn't working.

Even so, it required the interest of George Washington to move matters off dead center. Like Hamilton, Washington had been a nationalist from the beginning. As commander in chief of the Continental Army, he had known firsthand the impotence and frustrations of the Confederation, but he also shared Hamilton's vision of American nationhood. Washington had once had a strong personal friendship with Hamilton, who had been his aide-de-camp during the war, and they had mutually reinforced one another's nationalist sentiments. Indeed, it was Washington rather than Hamilton who took the first step.

The place, appropriately, was Washington's backyard. Mount Vernon was on the Virginia side of the Potomac River, looking across the water toward Maryland. Friction between the two states had made navigation of the river difficult, and Washington was eager for a settlement that would please both parties. Accordingly, at the suggestion of fellow Virginian James Madison, he decided to host a conference at Mount Vernon in March of 1785 for the purpose of talking things out.

The Mount Vernon Conference succeeded well enough to prompt a wider effort. The Maryland legislature suggested inviting Pennsylvania and Delaware to a trade convention at Annapolis the following year. Once again, Madison became involved, and the invitations were expanded to include all the states. The Annapolis Convention did not lay commercial issues to rest—delegates from only five states appeared—but it accomplished a broader purpose. The nationalists spent several days reinforcing one another's convictions and lamenting the state of affairs. In the end, the delegates adopted a report by Alexander Hamilton calling for yet another convention, to meet in Philadelphia in May of 1787. The subject was no longer trade. It was the impotence of the Confederation.

If these developments confirmed the leadership of Hamilton, they also pointed to the emergence of his friend and fellow nationalist James Madison. Occasionally called the Father of the Constitution, Madison was known to his contemporaries as a Virginian's Virginian. He was the scion of a tidewater family whose fortune stretched back for generations, and his plantation, Montpelier, radiated patrician charm. Yet Madison felt uneasy about his wealth and even uneasier about the slaves who wrung it from the soil. He compensated by approaching life with a volcanic energy that belied his slight stature and tinsely voice. After deciding, reluctantly, against the priesthood, Madison read law at Princeton and buttressed his learning with a wide study of history, the classics, and contemporary thought.

Madison was a nationalist for essentially two reasons. First, his service in the Continental Congress had given him that vision of American nationhood that so many of the nationalists shared; and second, his experience in the Virginia House of Burgesses had convinced him of the perfidy, duplicity and "mutability" of state governments. Up close, republican politics hadn't been about the public good at all, Madison concluded, but about rabid partisanship and factional interest. The cure for the disease, as he gradually came to see it, was to find constitutional means of bringing virtue to bear more effectively. A strong, energetic, national sovereignty, led by the wisest and most talented in the nation, was for Madison the only solution.

Madison the politician had an intuitive grasp of the situation. He realized that if any convention were to succeed, it would do so only with the active support of George Washington. Madison the scholar understood that a knowledge of political theory and historical precedent might come in equally handy. Both personae swung into action. The politician wrote to Washington and began clarifying the issues for him. The scholar wrote to his friend Thomas Jefferson, then serving as minister to France, and asked him to round up whatever books he could find on relevant topics and send them with dispatch. A hefty crate of volumes arrived shortly before Madison left for Philadelphia. He studied them literally night and day.

However serious the nationalists' preparations, much depended on factors beyond their control. How would the Confederation government respond to this call for a convention? And how would the *states* respond? (After all, no matter how bad things might be, some vested interests would prefer to leave well enough alone.) The most worrisome question was that of Washington's attendance. He was presently enjoying his retirement to the full and seemed little inclined to rejoin the political fray. Besides, it might be risky to commit his prestige to a project that might prove abortive. As March turned to April, letters from Mount Vernon appeared to banish all hope that the general would be on hand. Madison set out for Philadelphia on May 1 in a spirit of deep misgiving.

The Convention and Its Delegates

In the spring of 1787, Philadelphia was America's largest and most prosperous municipality. With a population of 45,000, it was larger than any city in England except London. At the intersection of Broad and High Streets, on the town square, stood the State House, "Independence Hall," an imposing brick edifice with Palladian windows and a towering cupola—although this had recently been taken down for repair. Philadelphia was the home of eight newspapers, several magazines, a dozen libraries, and at least one authentic museum. Among its narrow cobbled streets were to be found more than a hundred inns at which travelers could rest and patrons could drink and dine. Several of these would become central to our story.

James Madison was the first delegate to arrive, nearly two weeks in advance of the scheduled opening on May 14. He took a room in a small pension on Market Street, tucked among the red brick townhouses, and returned to his heroic reading schedule. By the time the Convention gathered, he had become a walking disquisition on government.

On Sunday, May 13, Madison caught the first inkling that his hopes had not been in vain. To the boom of cannon and the rattle of musketry, General Washington had arrived in Philadelphia—the Convention had a chance! In the days following, as coach after coach rattled in with distant passengers, Madison's excitement grew. Almost all the states had sent delegates, and from appearances they had sent the very best.

Father of the Constitution

Delegates to the Constitutional Convention who did not know James Madison could not understand how such a small, quite unimposing man could wield so much influence. The secret, if there was one, was preparation; he always did his homework. Preparation, in fact, remained the key to Madison throughout his long and highly accomplished life. He was the epitome of the scholar in politics.

Both his mother and father came from established Virginia families, and the Madison estate, Montpelier, might be mistaken today for a well-endowed college. James himself was uneasy about his wealth—always playing it down—and doubly uneasy about the nearly two hundred slaves who produced it. (When his trusted valet ran away and was recaptured, Madison, instead of selling him off into the West Indies, set him free and then hired him.) It was, in fact, "to depend as little as possible on the labor of slaves" that the young man took up the study of law.

And when James Madison took up anything, he did so with single-minded devotion. Under Donald Robertson, his first tutor; under Reverend Martin, with him he later studied the classics; and finally under John Witherspoon at Princeton, Madison worked so diligently that he seriously endangered his health. More than once he confided that he did not expect to live long.

Once settled in such habits, Madison was loath to alter them. Whether it was working on a problem in the Virginia House of Burgesses, guiding the Revolution in the Continental Congress, or hammering out the new Federal Constitution, he addressed his tasks with a fearsome intensity. Much of his subsequent career was foreshadowed by his first term in Congress, where he did not utter a single speech for six months. When he finally did take the floor, however, he was so well prepared that he immediately rose to leadership.

Curiously, though, everything about him belied his volcanic drive. His delicate stature, his reserved manner, his thin, tinsely voice combined to put adversaries off their guard. It was only upon looking closely that one noticed the exquisite tailoring, the finely powdered wig, the bright, scintillating eyes, the confident, bouncy step. "In the affairs of the United States," wrote one of the delegates, "he perhaps has the most correct knowledge of any Man in the Union." It was true. He was prepared on everything and for everything. Every argument, every objection, every wrinkle in the tapestry of ideas he knew with loving intimacy and could address the same with cogency and conviction.

Madison reached the great watershed in his life with his work in Revolutionary Virginia. A passionate believer in virtue, he lost his bid for reelection because he refused to canvass actively or ply the voters with rum. But virtue, he found, didn't count for much in state politics. Indeed, after battling the factions and their special interests through the long years of the war, he gradually concluded that republicanism could never survive without a strong—and virtuous—central government. "The advice nearest to my heart and deepest in my convictions," he wrote, "is that the Union of the states be cherished and perpetuated."

There were ideas besides Madison's in Philadelphia, of course, and many of them found their way into the Constitution. But the heart of the document was his: it represented the strong and virtuous national republic that had shimmered before his eyes. When all else was said and done in the constitutional debates, there was just no gainsaying the boyish little Virginian with the amazing store of knowledge.

From America: A Study in Heritage, *Kendall/Hunt 7th Edition, p. 154–155.*

Madison the politician was as interested in the delegates as Madison the scholar was in theories of republican governance. The small Virginian undoubtedly assessed his colleagues from every possible angle as they bundled their luggage into various lodgings.

James Madison has been referred to as the Architect of the Constitution.

Collective Portrait

Altogether, seventy-four delegates had been appointed by twelve of the thirteen states. (Rhode Island, always the maverick, remained aloof.) Fifty-five of them actually showed up. Most of those who failed to appear "smelt a rat," as Patrick Henry put it, and chose not to affiliate with "consolidationists."

In background if not in temperament, the fifty-five were remarkably similar. Thirty-four of them were lawyers. Eight were judges. And literally all of them had held high public office. Some were governors or former governors of their respective states, and most had been members of Congress. Even so, their public careers were coterminous with private careers in the law, mercantile enterprise or agriculture. A quarter of them owned commercial plantations. They were men of practical affairs and counted themselves as such. If they had private libraries and well-thumbed books, they also made decisions and expected to abide by them.

Virtually to a man, they were representatives of the American elite. Some came replete with Honorables and Esquires by their names, and several were much at home in London drawing rooms. Still, a fair number were lowborn and self-made, often by means of commercial enterprise and occasionally by means of the right marriage. One delegate, Roger Sherman of Connecticut, had grown up in a hardscrabble farmhouse in which there were three beds for a family of nine. Self-creation was an American theme, and its representation here was significant.

The Founding Fathers:
http://www.nara.gov/exhall/charters/constitution/confath.html

Something else of significance was the instability of the delegates' fortunes. Robert Morris could boast that he had superintended the finances of the Revolution, but his reach often exceeded his grasp and his financial empire would eventually come crashing down. James Wilson would suffer a similar fate. George Washington, who was one of the wealthiest men in Virginia, was perennially hard pressed for cash, as was his Virginia colleague George Mason. Many of the delegates were involved in some way with land speculation or with government securities, both of which were gamblers' games, and they were used to living life on the edge.

If they were not stuffy, neither were they doddering. A list of fifteen key participants at the Convention ranges in age from twenty-nine for Charles

Pinckney to sixty-six for Roger Sherman. Benjamin Franklin at eighty-one was by far the oldest in attendance. So debilitating was his gout, he was carried to the State House every morning on a sedan chair borne by four prisoners from the Walnut Street jail, and he favored the group with an endless stream of jokes about it. Their average age was forty-seven—a little over half that of the sage Dr. Franklin.

Finally, the delegates were men of the Revolution. Eight of them had signed the Declaration of Independence. Thirty had served in the army, and roughly half of those had seen action. Several were authentic battlefield heroes. They had not convened here to undo 1776, as historians have occasionally imagined. On the contrary, what brought them together was an almost mystical sense that something more was needed, something elusive perhaps yet crucial, to bring the Revolution to full fruition.

Intellectual Influences

Several of the delegates were true scholars. James Madison certainly was that, and so was his friend and ally James Wilson. Wilson was a Scot, recently immigrated, and he still spoke with a lowlands burr. He had a razor-sharp legal mind and an encyclopedic store of knowledge. Only a cut below the two Jameses were the two Georges, Wythe and Mason, both from Virginia and both of a studious, scholarly bent; and along with them William Samuel Johnson of Connecticut, soon to take his father's place as president of Columbia College. Elbridge Gerry of Massachusetts had advanced degrees from Harvard and an interest in political philosophy. Alexander Hamilton had been a child prodigy in New York, writing sophisticated political tracts while still a teenager, and he enjoyed awing people with his facile brilliance. Yet he had nothing over Gouverneur Morris, a stout Pennsylvanian with a peg leg and a curmudgeon's temperament, who was one of the most learned and accomplished men in America. The scholars had an advantage at the Convention, especially when the voting was close, and they made the most of it. It was one of the few occasions in American politics when the central players were willing to heed the voice of learning.

Most of the delegates had some semblance of a classical education. In the process of reading their Greek and Latin, they had been exposed to the experience of the ancients, and to the doleful, cautionary lessons that experience seemed to convey. They were also schooled in the European Enlightenment. They had read their Hobbes and Locke, their Voltaire and Rousseau, and most of them had at least a passing acquaintance with Montesquieu's *Spirit of the Laws*, which impinged directly on their work in Philadelphia. Grotius, Vattel, Burlamaqui, Puffendorf all popped up from time to time in their discourse, and there were repeated references to Blackstone and English constitutionalism.

A number of the delegates had some exposure to the Scottish Enlightenment as well. They knew the works of Adam Smith, Francis Hutchinson, and Lord Kames. Of particular significance were the writings of Thomas Reid and the so-called "common sense" school. Reid had argued that abstruse philosophy was not the means by which mankind apprehended moral truth—but rather by simple, everyday common sense. That point of view squared well with Americans.

One more Scot is worth mentioning. A few of the delegates—an important few as it turned out—were familiar with the writings of David Hume. His work had ruffled feathers throughout Europe, for it came down on the skeptical side of every question from human nature to divine miracles. But it was hard to question Hume's logic, especially when it came to understanding political society. Hume brimmed with ideas for controlling human behavior, reining in power, and turning the dark side of man to moral profit. To the seekers of Utopia he offered a trenchant piece of wisdom. What *is* established, he said, is just as important as what *ought* to be established. A dynamic interrelation of the "is" and the "ought" was to become a subtext of the US Constitution.

Still, philosophy as such would play but a minor role at the Convention. There was only a single discourse upon Enlightenment political theory—and it fell flat. For the delegates generally believed that truth was learned through observation, not speculation. "Experience must be our only guide," said John Dickinson, adding "Reason may mislead us."

It has occasionally been said that the American Founders were irreligious. A few may well have been **Deists**. Gouverneur Morris was a secular humanist and unabashed skeptic. Most, however, were more-or-less committed churchmen. The vast majority were mainstream Protestants. Two were Catholic. A few were Quakers. Some form of Calvinism lay in the background—or described the behavior—of several central players: William Paterson, George Mason, Elbridge Gerry, Roger Sherman. And the Convention's truly central players, James Madison and James Wilson, had both pondered careers in the ministry.

Deism: An18th-century movement denying interference of the Creator with the laws of the universe.

Ideology and Interest

According to Karl Marx, there is a close relationship between "ideology" and "interest" in human affairs. In any given society, so the argument goes, there will always be a certain group in control of things, and the specific interests of that group will always be expressed in a system of governing ideas. Thus, for example, if religion is revered in a society, look for priests to be the "master class" in question. And so on.

The American Founding is sometimes presented as a showcase for the Marxian point of view. After all, the Framers of the US Constitution came together in a dramatic way, and they were white, Protestant males representing a well-defined social elite. Half of them held government securities bought up at bargain prices and roughly a third of them owned slaves. What else could they write but a constitution that protected property, acknowledged slavery, and required the government to redeem its securities at par.

At Par: Redeemed for value stamped on face of note.

So, why isn't this a good explanation? Because it is too simple. It is true that the Framers made no bones about interest—they constantly discussed it in their deliberations. However, interest came in an assortment of forms and cut in many different directions. The interests of the various states prompted one course of action—highly conflicted of course—while the interests of the regions prompted quite another. If there was a class interest among the Framers, it was at odds with both of the above, for the rank and file of Americans had no titles by their names and knew only a farming life. There was a commercial interest that favored low tariffs, an agrarian interest that favored high tariffs, a city and town interest, a back-country interest, a debtor interest, a creditor interest, a large-scale plantation interest, a small-scale family farm interest, pro- and anti-slavery interests, pro- and anti-British interests, even a western interest whose main theme was suspicion of the East. These were all represented at the Convention. It was virtually impossible to bring them together in a single "controlling interest"—as we will soon see.

Ideology needs a closer look as well. As men of affairs, the Framers distrusted systems of ideas as such. Still, they had at least three broad commitments that could in some sense be called "ideological."

First of all, they distrusted power. It was dangerous by its very nature, they believed. All power tended to engross itself and invade other domains. All power inclined to curtail freedom. And all power disposed toward corruption. The Framers talked endlessly of ways to restrict it, tame it, civilize it and render it responsible.

Second, they upheld property. When they used that term, however, it was not quite in the way we would today. *Property*, for the Framers, was a kind of freedom—as it had been for the commonwealthmen of England. Ownership gave meaning and dignity to life. It made it possible to stand on one's own feet and be responsible for one's own actions. Fear of power and respect for property went hand in hand.

Third, they were committed republicans. A few of them leaned toward classical republicanism. William Paterson, for instance, had a great deal to say about the waning virtues of old, and George Mason wanted the Constitution to restrain private luxury. The vast majority were liberal republicans, however, and it was liberty rather than virtue that focused their concern. And an important few were democratic republicans. Benjamin Franklin, Charles Pinckney, and one or two others belonged in this camp, along with James Wilson, its leader. Perhaps the most interesting of the "democrats" was Madison himself. He viewed the political world in terms of popular sovereignty, and sought practical means for breathing that sovereignty to life.

In a word, then, ideology and interest didn't jibe. For time and again the Framers demonstrated they would vote against their "interests"—state, regional, class, personal—in the name of some transcendent higher good.

Getting Started

Interesting facts about Constitution: http://www.nara.gov/exhall/charters/constitution/conqa.html

As the delegates greeted one another, there was a whiff of apprehension in the air. This was a risky business, they knew, and it might well result in failure. And high reputations were on the line.

At least the symbolism felt right. They were convened in "Independence Hall" and seated in the selfsame room in which the vote for separation had been taken. It was a spacious chamber, done up with Georgian flourishes and yet quietly dignified. At one end, before a classical pediment, stood an elevated dais on which the presiding officer sat at an imposing walnut table. The rest of the seating was at smaller circular tables covered with green baize cloth. Delegates sat grouped together by states, so they could confer with one another before casting their state's vote.

Independence Hall in Philadelphia was the site of the Constitutional Convention.

The hall was cool in the mornings, but as sunlight poured through the slatted blinds, the air grew fetid and stale. With the windows closed and locked to protect privacy, it was not long before beads of sweat began to stand out on foreheads or trickle beneath woolen waistcoats. It was, said the old-timers, an exceptionally hot summer, the worst in half a century.

As credentials were read out and special instructions voiced, the precarious nature of the Convention's mandate became clear. The delegates had been sent not to create a new government but to strengthen the existing one. Thus, their project must always be referred to as "the Convention," or perhaps "the Grand Convention," never "the *Constitutional* Convention." For similar reasons, the nationalists had to bury their intentions in cryptic phrases and speak only of creating "a more perfect union." *Innovation* was not a happy word in the 1700s, especially when applied to politics.

On May 25, George Washington was officially elected president of the Convention and solemnly escorted to the dais. He made a short speech of acceptance, notable for its humility, and took his place in the president's chair, which was carved with a gilded sunburst. Washington had an air of magic about him, variously described as personal charisma, deep strength of character, or a sense of spiritual wholeness, and placing him at the head of a proceeding that promised to be fractious made practical good sense.

The first of the Convention's committees set to work on a list of operating procedures. Among the rules to be followed were:

> Rise to speak, addressing the president.
>
> Refrain from conversing, reading, et cetera while a member is speaking.
>
> Refrain from walking in front of a speaker.

Some of the rules seemed legalistic and nitpicky. Yet on at least two occasions critical votes would have gone the other way but for the invocation of one small rule or another.

Benjamin Franklin and the Ambiguous Sun

Ben Franklin seemed to view the world in similes. The comment in his *Autobiography* that "Keimer stared like a pig poisoned" was just the sort of thing he would come out with, and similes filled *Poor Richard's Almanac* from beginning to end. So it was only natural that the old diplomat was thinking in similes during the long hot summer of 1787.

Franklin knew that settling the sovereignty issue would be monumentally difficult. As it happened, he had tried to settle it himself some thirty years earlier, at the time of the French and Indian War. Then, the much younger and less-experienced diplomat had conceived a plan for uniting the colonies in a common defense. The need for such unity had been real enough—almost critical—but the plan had been scuttled by the colonies' legendary jealousies and rivalries. Americans, it seemed, would literally rather face their enemies alone than join themselves together.

Why was that? Franklin asked himself. He supposed it was because as colonies they had all been so very different. They had been founded at different times by different groups and for different purposes. They had come of age individually, proud of their self-reliance, and more than a little snooty toward one another. A New Englander traveling in Virginia remarked that the reputation of that place for hospitality was grossly exaggerated. A Virginian had the same idea about Pennsylvania, adding that its women were "remarkably homely, hard-favored, and sour." And, to complete the circle, Pennsylvanians lost no love on New England, a place, said one, of "low craft and cunning." It was that way all over.

It was not a happy condition, Franklin admitted, but it did explain some things. It explained why the states so cherished their independence. It explained why they sent their own diplomats abroad and why nine of them now boasted of having their own navies. New York was presently engaged in a shooting war with Vermont, as was Pennsylvania with Connecticut. No wonder the British had failed to take the "United States" seriously.

It was with such gloomy thoughts in mind that Franklin, sitting in the infernal heat, swatting flies, fanning himself with a sheaf of papers, began to ponder the sun disk painted on the back of General Washington's chair. It was a stylized sun, with rays fanning outward, a decorative motif often used. The painter, thought Franklin, wasn't very good. He had failed to make it clear whether the sun was rising or setting.

Franklin's own sun was setting. At 81 he was far and away the oldest man here, and though he was revered as the elder statesman, the tough issues were being tackled by younger and more vigorous men. He had seen a great deal in his lifetime. He had seen America grow from colonies, to provinces, to independent nations. Would he live to see it become a single nation—one capable of holding its head up in the world? He wondered.

And suddenly there was the perfect simile. Was that sun on the chair back rising or setting? Was it casting its rays over the beginning of something important or the end? What did the American Revolution mean, after all, a mere colonial rebellion that happened to succeed—or a new epoch in human history?

Ah, yes, that was it, the ambiguous sun. Rising or setting?

From American: A Study in Heritage, *Kendall/Hunt 7th Edition, p. 144.*

Beyond formal procedure, there was a great deal of old-fashioned courtliness. It was always *Colonel* Mason, *General* Pinckney, *Judge* Wythe. And in addressing George Washington, it was always *Your Excellency*. A wonderful anecdote recounted that Hamilton and some friends made a wager with Gouverneur Morris that he, Morris, would not dare to lay a hand casually on Washington's shoulder and address him as a personal friend. According to the tale, Morris took the dare one evening at the City Tavern, and for his pains received a glare so icily forbidding that he slunk from the great man's presence. Decorum had its value, though, especially when stakes were high and divisions deep.

The Convention's most remarkable rule was to conduct all proceedings in secret. To that end, the windows were kept tightly shut, sentries were posted at the doors, and members were not allowed to copy from the daily journal. Benjamin Franklin, ever garrulous, was judged to be the number-one security risk, so delegates were careful to be at his side at all entertainments, steering the conversation away from sensitive topics. At one point, a set of notes was found on the floor, dropped there by one of the delegates in his haste. Washington treated the matter with such gravity that the notes were never claimed.

Secrecy played a crucial role. If all utterance was off the record, members would be free to speak their minds and change their minds. Nor would it be possible for news media to seize upon a scrap of information and whirl it into a major controversy. All the same, the hush-hush was disconcerting.

As veteran parliamentarians, the delegates knew all the special jargon and procedural devices of political assemblies. Among the latter was the ad hoc committee, which came in two varieties. Grand committees, usually composed of one delegate from each state, were appointed to work out a compromise when everything else had failed. Special committees were given particular tasks. The Committee of Detail, for example, took the broad understandings hammered out in general assembly and worked them into meticulous prose.

One committee was a legal fiction. The Committee of the Whole House, as it was called, was a parliamentary device enabling the General Assembly to reconstitute itself as a "committee." When it did so, Washington would step down from the dais and take his place at the Virginia table, and Nathaniel Gorham would replace him at the head of the room as committee chairman. Why the charade? In the Committee of the Whole House, votes could be taken again and again as members heard new arguments, rethought their positions and possibly changed their minds. In General Assembly votes were binding.

Voting was by states, as in the national Congress, not by individual delegates. As long as a state had at least two delegates residing in Philadelphia, its vote could be cast by any of them in attendance. This created an odd situation. If a state had, say, five delegates in residence and four happened to be indisposed, the sole member in the hall could cast the state's vote any way he chose. Suppose, on the other hand, two delegates were in attendance and they happened to disagree; they cancelled one another and the state's vote was nullified. Months of persuasion might come to naught if the right parties weren't present when a critical vote arrived.

The big question of the opening sessions was where everyone stood on the question of a stronger union. A majority of the delegates were nationalists, plainly enough, but there were all sorts of gradations among them—and some

Serious Business

In its first item of business, the Constitutional Convention wisely elected George Washington to be its presiding officer. Solemnly escorted to the dais, the General made a little speech of acceptance, confessing his want of qualification and calling upon God for help. After that, though a lively interest never left his face, he took no further part in the proceedings. But his presence on the dais, day after day, through the heat, the flies, the sometimes bitter debates, meant everything.

Washington was deeply concerned—sometimes painfully distressed—about the condition of the American Republic. It faced a host of difficulties, and there was no promise that it would ever surmount them. Had the threats been military, the old general would have known precisely how to deal with them, but they were political, intellectual, even spiritual in nature; and Washington felt ill equipped in those areas. Accordingly, he did what he believed he could do best: make the delegates understand how serious a business this was.

Take the matter of secrecy. The delegates resolved early that their proceedings must be kept closely under wraps. For, if the public were looking over their shoulders, they could neither speak their minds freely nor work out delicate compromises.

How did Washington regard the secrecy rule? Early in the Convention, he was handed a paper dropped accidentally and left on the floor. That afternoon, just before adjournment, he rose from his seat. "Gentlemen!" he said, in a voice he might have used at the Battle of Germantown, "I am sorry to find that some one member of this body has been so neglectful of the secrets of the Convention as to drop in the State House a copy of their proceedings. . . . I know not whose paper it is, but there it is [throwing the paper down on the table], let him who owns it take it." William Pierce, alarmed, shot his hand into his own breast pocket to make sure he was not the culprit. "It is something remarkable," he penned in his memoirs, "that no person ever owned the paper."

One for whom seriousness came naturally was James Madison. He was the first to arrive in Philadelphia, almost two weeks before the Convention got under way, and he used the time to full advantage. He spent days and nights poring through hundreds of volumes, writing out lengthy notes and even penning an essay on comparative government. Small—"no bigger than half a piece of soap," someone said—slight of figure, and so quietly spoken that he was repeatedly asked to speak up, Madison nevertheless dominated the proceedings. He sat on the front row and took notes far more complete than those of the Convention's paid scribe. For him, too, constitution writing was a serious business.

There were, of course, moments of pleasure. Everywhere, it seemed, there were teas, formal receptions, and dazzling balls. If it was the hottest summer in Philadelphia's memory, it was also the most brilliant.

By day, however, the Convention was invariably back on task. And in its own perverse way, the heat may actually helped out. Slatted blinds kept the sum from the Assembly chamber, but with the windows closed, the air inside grew lifeless and oppressive before midday. The New Englanders, who had come ill provided with woolen suits, stole envious glances at Washington, who seated on the dais in his light camlet coat and breeches, seemed imperturbable as a Gilbert Stuart portrait. But sitting through the suffocation, day after day, served as a reminder, if one were needed, that the American republic was in peril—and that the business of rescuing it was serious business indeed.

From America: A Study in Heritage, *Kendall/Hunt 7th Edition, p. 122.*

held their cards very close. Then, too, a small but determined minority were as far from nationalism as one could be. So who, precisely, was who?

• The Nationalists

No nationalist was more deeply committed than Alexander Hamilton. But the New York he represented was badly divided politically, and the other two delegates, John Lansing and Robert Yates, made sure to come down against Hamilton on every question—including nationalism.

It was the same way with Virginia. Madison was an ardent nationalist, but there were six other members of the state's delegation and none quite matched his zeal. Washington came close, perhaps, but the role of Convention President made it difficult for him to get down in the trenches. There were two other prominent Virginians, Edmund Randolph (the Governor) and George Mason, but their nationalism wasn't always beyond question.

Washington believed that the Convention's president ought to be above politics. Subsequently, scholars have shown that various delegates believed he did not embrace a clear position.

In the Pennsylvania delegation, James Wilson was as ardent a nationalist as Hamilton, and such was the force of his personality that he often carried the others along. But most weren't hard to carry. Gouverneur Morris believed in strong, can-do government and wasn't all that hostile to monarchy. And Benjamin Franklin could be described as nationalism's patriarch.

In the small states, nationalism depended on how individual delegates felt about their state going it alone. In the Deep South, for instance, with the Spanish down the coast and Indians on the frontier, going it alone was not an appealing prospect, so nationalist feeling was strong. Elsewhere some small state delegates matched the strong union sentiments of a Madison or Hamilton—George Read of Delaware was willing to see the states "swallowed up entirely"—while others were more iffy. A few of the delegates converted to the nationalist cause in the course of the Convention. Rufus King of Massachusetts, for one, was rather skeptical at the outset, but as events moved along, he became an enthusiastic supporter.

• The Confederalists

Those who liked things the way they were—with the states fully sovereign and the Confederation weak—were generally termed "confederalists." Luther Martin of Maryland was the Alexander Hamilton of that persuasion. As far as he was concerned, American nationhood was a bad idea and the program advocated by the "monocrats" an ill-disguised attempt at empire. Martin frankly attended the Convention for the purpose of obstructing it, and believed that the best way to do so was by delivering long and ponderous harangues. It was not an effective approach.

Next to Maryland, the state most likely to opt out was New York. Both Lansing and Yates were cool to the idea of a strong union, and when the Convention began leaning that way, they pulled up stakes and left for good, promising to fight against any plan that ensued.

Beyond these firebrands, a number of other delegates were in sympathy with state sovereignty, even while they lamented the present situation. Randolph and Mason of Virginia, Paterson of New Jersey and Gerry of Massachusetts were among the deeply conflicted. They leaned first one way and then the other.

The confederalists reflected American thinking at the time far more accurately than the nationalists did. Most Americans were simply not ready for an American nation as such—indeed, they could scarcely conceive such a thing.

The Virginia Plan

When the day scheduled for the opening arrived, only two delegations had presented themselves. Those two, Pennsylvania and Virginia, began holding informal afternoon talks. What came out of these meetings was a set of fifteen resolves, mainly the work of Madison, constituting a general design of government. The Virginia Plan, as it would be called, moved boldly for an American union.

As soon as the Convention's preliminaries were out of the way, the Virginia Plan was officially presented to the General Assembly by handsome and debonair Edmund Randolph, who spoke for a full three hours. This wasn't at all what the delegates had expected—a full-dress proposal right off the bat. Nor could there be any doubt about the polity it proposed. There would be a national legislature elected on the basis of proportional representation, with larger states choosing more representatives and smaller states choosing fewer. The government, Randolph urged, must represent *the people*, not *the states*, and it must act upon them directly. Thus, everywhere there were, say, 40,000 voters within a contiguous space there would be an elected representative to the national assembly. Each district would have its own separate interests, and these would contend with one another for preference in the political arena. States as such would become irrelevant.

What must have turned in the minds of Randolph's listeners. A single government for the entire continent. Only people to be represented. States irrelevant. This was no minor revision of the Articles of Confederation—this was a revolutionary overthrow.

The Grand Convention was under way.

The Deliberative Process

By June, things had settled into a semblance of routine. The delegates met six days a week and generally took Sundays off. They convened around ten o'clock in the morning and continued until five or six in the afternoon. Not everyone attended all the sessions. On most days, there were no more than thirty members on the floor at a single time.

There were late arrivals and early departures too, some for political reasons and some for personal. By the time the New Hampshire delegation finally pulled in on July 22, the New Yorkers had all gone home, so there were never more than eleven states present. And by the end of August eleven of the original fifty-five delegates had left Philadelphia for good.

The proceedings were formal by our standards. Many of the speeches were written out beforehand and delivered rather stiffly. Where the discourse became interesting was in the more off-the-cuff exchanges, where the delegates batted theories and authorities back and forth like badminton birdies. On the whole, however, they worked out their own approach to the intellectual problem before them—an approach favoring their own experience. Never ask whether something *might* work but whether it has worked, does work, or is likely to work, was the way they might have put it.

A 1799 engraving of George Washington presiding over the Constitutional Convention at Independence Hall in Philadelphia.

In this spirit, the Framers continually scrutinized their ideas by the light of human nature. And here they were often dismally pessimistic. Given the opportunity, almost everyone would lie, cheat, steal or try to beat the system—or so one might gather from listening to the debates. In this, of course, the Framers were laying out worst-case scenarios. If a mechanism seemed to function properly in the direst situation they could imagine, chances were it might function in the everyday world.

Decision making was despairingly tedious. The delegates would often decide an issue, place it on the shelf, and return to it in a day or two, discovering that it had to be decided all over again. Resolving one knotty problem required no fewer than sixty separate ballots, while the wording of a single clause in Article III had to be reworked seven times. Typically, the moment a question was finally "settled," a new objection would pop up.

Participation was lopsided. Some twenty percent of the delegates accounted for eighty percent of the dialogue. Gouverneur Morris was heard most often with 173 speeches. James Madison was a close second with 161, and James Wilson and Roger Sherman were in third and fourth place respectively. On the other extreme, six delegates—Blount, Blair, Gilman, Bassett, Few, and Ingersoll—never uttered a single word. Some of the disparity was due to a difference in political style. Morris' style was open, aggressive and forensic, whereas John Rutledge preferred to work behind the scenes. "Care not who reigns," he often cited, "care only who rules."

Core Framers

A careful study of the Convention's day-to-day process suggests that roughly fifteen of the fifty-five delegates—a quarter—were of truly commanding importance. These "Core Framers," as we will call them, made a variety of contributions. Some were good at crafting arguments or persuading the undecided. Others came up with creative innovations. Still others knew how to build a coalition or negotiate a compromise. In chapter 7 we will focus on the core delegates and see how they worked together to tackle a specific problem.

For now, our list of Core Framers, in three descending tiers of importance, looks something like this:

James Madison
James Wilson
Roger Sherman

Gouverneur Morris
Charles Pinckney
William Paterson
George Mason

John Dickinson
Edmund Randolph
Benjamin Franklin
Oliver Ellsworth
John Rutledge
William Samuel Johnson
Elbridge Gerry
Rufus King

The list is entirely subjective, of course, and it offers only a rough approximation, but few historians would probably quarrel with the general sense of it. Personality had a heavy impact on the Convention's outcome.

Conditions and Circumstances

Finally, we might keep in mind the physical conditions and social circumstances under which the delegates labored, for these too worked a subtle influence. The summer's Bengal heat, already mentioned, was accompanied by swarms of pestilent flies. These were a "veritable torture," as one visitor recalled, constantly "stinging everywhere and turning everything black [with their] filth." Mosquitoes were bad that year too and several members complained of bedbugs. And Philadelphia was anything but a model of cleanliness: hogs and chickens wandered the streets freely, and when the wind shifted, the stench of the slaughterhouse wafted in.

Orrery: A mechanical model of the solar system.

Yet Philadelphia was also America's showcase. But a few blocks from the State House the delegates could visit Charles Willson Peale's museum, with its fossil bones, stuffed animals, and gleaming portraits of American heroes. Or they could pay a call on David Rittenhouse and see his famous orrery, or venture out to Bartram's Garden with its amazing botanical collection. Down along the riverfront one could wander along the miles of wharves and warehouses, and find shops selling everything from Chinese silk to French soap. There were bookstores and stationery shops and the Library Company at Carpenters' Hall, with its models of new-fangled machinery. There was even a College of Physicians, reputed to be on the cutting edge of medical science. The message? This was no moribund society falling into tatters. America was going places.

Self-portrait of Charles Willson Peale welcoming the public to his museum.

Most of the delegates roomed within a few blocks of the State House. Madison took lodgings with Mrs. Mary House, celebrated for her hospitality, at the corner of Fifth and Market. Elbridge Gerry and his new bride rented a house on Spruce Street. Several Convention members checked into the Indian Queen, a hostelry on Fourth Street, where boarders sometimes crowded two to a room. Washington was a summerlong guest of Robert Morris, who despite his tottering financial empire lived as sumptuously as a European prince.

Morris was only one of those lavishing entertainment. He was joined by fellow Philadelphians (and fellow delegates) Jared Ingersoll, Thomas Mifflin and George Clymer. Dr. Franklin laid in a cask of porter for his own socials, which were noted for their conviviality, even though a glass or two made him dangerously voluble on Convention matters. Nor were the matrons of society to be outdone. They hosted an endless parade of afternoon teas and evening suppers, often provided with musical accompaniment, to the extent that some delegates began to feel a surfeit. Crusty George Mason, for one, complained of all the "etiquette and nonsense so fashionable in this city."

The entertainments, however, were often only extensions of Convention intercourse. Groups of delegates would gather nightly at the Indian Queen, the George, the Black Horse, the City Tavern, or the London Coffee House to sip brandy, chat about current difficulties and, not infrequently, conduct back-channel negotiations. Indeed, the cozy, candlelit parlor of a comfortable inn provided an agreeable atmosphere for the Convention's spadework. Here it was not a matter of declamatory rhetoric but of engineering compromises, building coalitions and devising creative solutions.

For, first and foremost, the Founding needed to be a *political* undertaking—the forging of a common identity and sense of nationhood—and no document laying out a plan of government could perform that feat. Somehow the working politics of patria had to be assembled from the raw materials of discord.

An Assembly of Demigods?

Jefferson, writing from Paris, pronounced the Grand Convention "an assembly of demigods." Hyperbole aside, the Framers of the Constitution were exceptional by any standard. Several of them could be counted among the most brilliant men of their time. A dozen others might be described as first rank. And all were of conspicuous talent and achievement. Even more striking for such a gathering, they possessed an almost uncanny ability to reason together in ways that surpassed their reasoning as individuals. They were men of politics rather than politicians, the difference being that unlike our own world, their names would have been high on any list of the most admired. Some of them aspired to the highest renown possible for statesmanship—to be founders of a great nation. That aspiration would be sorely needed in the trials that lay ahead.

CHAPTER 6 STUDY AID

1. Overcoming weaknesses:

- Enlightenment (social contract)
- Experience
- “Science”
- Creativity

2. Nationalism:

- Unified nation
- Hamilton, Washington, Madison
- Mount Vernon Conference & Annapolis Convention
- Philadelphia

3. Delegates:

- Elites and statesmen
- Revolutionary experience
- Classical education
- Enlightenment influences
- Religion
- Interest (How influential?)
- Ideology (Distrust of power; property; republicanism)

4. Convention:

- Washington
- Rules of order
- Behind closed doors
- Nationalists vs. Confederalists
- Virginia Plan
- Processes and conditions
- Core framers

CHAPTER 7

The Constitutional Convention II
Compromise and Creativity

The Constitutional Convention was four long, sweltering months of politics. Members recalled the experience as exhausting, frustrating and often disheartening in the extreme. Ben Franklin, his gout-swollen legs propped on pillows, fanned himself continually and came out with sprightly one-liners about the ordeal of it all—it was virtually the only comic relief.

Politics meant compromise. Almost everything had to be compromised—not once but over and over again. There were so many interests, agendas, motives, and points of view to consider that not infrequently resolution appeared to be hopeless. And as in all compromises, no one liked the result. Indeed, at summer's end, some delegates concluded that their work had been compromised to death.

Politics also meant creativity. This was a bit more mysterious. It is one thing to bargain or horse trade, and quite another to come up with constitutional mechanisms that really work. "A free government," John Adams observed, "is a complicated piece of machinery, the nice and exact adjustment of whose springs, wheels, and weights, is not yet well comprehended by the artists of the age, and still less by the people."

Opposing Forces

The Debate:
http://www.yale.edu/lawweb/avalon/debates/debcont.htm

The story of the Constitutional Convention can be told in terms of two sets of opposing forces. What we might call the "centripetal" forces were the cohesive ones that tended to draw the states together to form a strong union. And the "centrifugal" forces were those that tended to drive the states apart.

Centripetal Forces

Had there been no cohesive forces drawing the states together, the project of union would have soon faltered. What were those "centripetal" forces?

Centripetal forces:

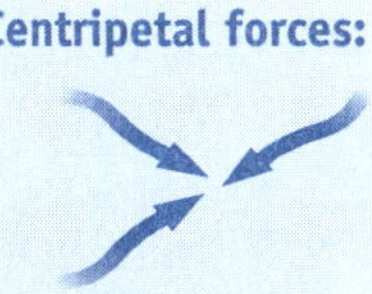

- Impotence of the Confederation and the Threat of Anarchy

We have already seen something of the Confederation's impotence. It was much on the minds of the Philadelphia delegates. The problems went well beyond difficulties in obtaining public credit or the free-for-all in interstate commerce.

Martial law had been declared in Georgia. A group in the New York legislature was allegedly negotiating with the Viceroy of Canada. And in Massachusetts, the worst case, Shays Rebellion seemed to confound the American Revolution itself.

Problems in the West may have been even worse. Spain had closed the lower Mississippi and was inciting Indian tribes to harass the frontier. The British were fostering similar intrigues in the Great Lakes. Some disgruntled westerners had begun to question affiliation with the East on any terms. "It will not be an easy matter to bring the Americans to act as a nation," gloated Lord Sheffield, adding, "They are not to be feared as such by us."

Anarchy was not an attractive prospect. Thirteen independent states, fighting and bickering among themselves, forming alliances and counteralliances, conspiring with foreign powers, using force to settle their disputes—was something like the nationalists' worst nightmare. "Look at the public countenance, from New Hampshire to Georgia!," cried Edmund Randolph. "Are we not on the eve of war, which is only prevented by the hopes from this Convention?"

- A Vision of American Nationhood

American nationalism was a by-product of the Revolution. It had sprouted in various places and for various reasons, but none was more important than in the ranks of the Continental Army, where young men had come together from all around the colonies to fight for a common purpose.

What the vision of an American nation looked like might be pasted together from scattered references. For some it was a peaceful, happy Arcadia stretching across the continent. For others it was a realm of promise and prosperity. Classical republicans described it as a people devoted to republican principles—justice, dignity, the rule of law—and to great human achievements. For the democratic republicans it was a place of equality, without rich or poor, where ordinary citizens counted. Virtually all the nationalists portrayed it as a power respected in the eyes of the world. Above all, it was a land of liberty.

That this was to become a compelling vision is an understatement. We have seen so many representations of it in patriotic songs and poetry, in election broadsides and celebratory posters, in symbols and effigies and Miss Liberty herself that we easily forget it was *created* not *revealed.*

- Strong Leadership and Esprit de Corps

While most of the delegates at Philadelphia leaned toward a stronger union, most Americans did not—at least not yet. Because they were swimming against the tide, the nationalists enjoyed unusually strong leadership. They were well organized, determined and politically savvy, and they had a knack for bringing things off. The battle they were now fighting was reminiscent of the battle they had once fought against the British. The mood in the Pennsylvania State House was often one of patriots called together at a time of crisis.

- George Washington

Washington may well have been the Convention's most important asset. Certainly he was a centripetal influence of the first order. He had become a symbol of the Revolution and a living embodiment of American nationhood. He was among the earliest and most energetic of the nationalists, writing poignant laments from Mount Vernon over the Confederation's sorry state. "Weak at home and disregarded abroad is our present condition, and contemptible enough it is," he fumed.

Our image of Washington as being above the fray of politics obscures how politically adept he was. The merest handful of US Presidents can compare. He realized that his principal benefit to the Convention would be in his physical presence rather than positions he might take or theories he might expound. All the same, no one doubted where his heart lay. And when they saw him, day in and day out, anxiously engaged in the cause, all those images of the Revolution came back to mind. Washington held the Convention together and kept it on task as surely as he had done with the Continental Army.

Centrifugal Forces

Centrifugal forces:

The obstacles to be overcome were daunting, to say the least. They were not quirks in the minds of individual delegates but deeply rooted ideas, interests, and opinions in American political society. The delegates only mirrored what was out there.

We must also remember that the American people had to ratify any proposal the delegates worked out. Surmounting difficulties among fifty-five men seated together in a room was one thing—winning the support of a diverse and often conflicted public opinion was quite another. The Framers worked, as it were, with skeptical constituents looking over their shoulders. If they gave away too much at the bargaining table, their work would surely come to naught.

In a manner of speaking, then, the story of the "centrifugal forces" and how they were overcome is the real story of the Constitution.

- The Confederation

The Confederation resolved the problem of territorial expansion by ceding all territorial claims to the central government.

Impotent though it was, the Confederation was recognized as America's legitimate government. It had won the war against Great Britain and proven itself capable of at least some governance. There were those who spoke of the Articles of Confederation with deepest respect, and several of the Philadelphia delegates had been among its authors.

Public opinion favored the idea of repairing the Confederation, shoring it up in crucial areas and plugging the main leaks, and most Americans seemed to think this was truly possible. Some historians have argued that the Confederation was not beyond hope and might indeed have been put into working order.

The moment the Philadelphia delegates went beyond their instructions to rehabilitate the Confederation, they entered a legal and ethical limbo. They had been sent by the states but under the authority of the Confederation to accomplish a prescribed task. Would the American public sustain their decision to toss a working government onto the trash heap? Doubts about this question dogged the Convention's heels from the beginning.

- The States

Unlike the Confederation, the American states were robust republican sovereignties. Each had its own identity, its own history, its own social structure and

Virginia Plan/Resolves: Called for states to be represented by population in national legislature.

economic life, its own feeling of wholeness, its own culture and sense of style. And each was in the grip of powerful local interests who were not about to let go.

What the **Virginia Plan (Resolves)** proposed was essentially to put the states out of business. Under the strong national government envisioned by Madison, state governments would be reduced to administrative units, much like counties today, their sovereignty but a memory. Needless to say, not a single one of them contemplated such a step.

There was more than simple jealousy at work here. Political theory held that only a small, cohesive and culturally homogeneous society could exist as a republic. People had to know one another in order to exercise the responsibilities of citizenship, and they had to participate in a government close at hand. The convenient term for what the nationalists had in mind was not *republic* but *empire*.

Accordingly, many of the delegates had been sent to Philadelphia with explicit warnings not to tamper with the states. At the mere thought of such tampering, several delegates packed up and went home.

THE VIRGINIA PLAN OF REPRESENTATION

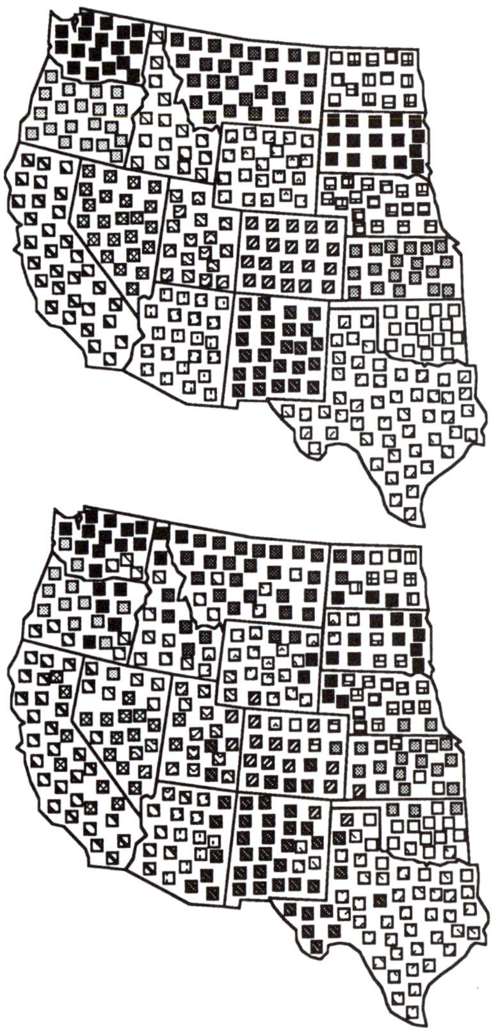

The Virginia Plan was to have representation in a national legislature based on the size of respective state populations. Small states feared that this would make for domination by large states (top). In practice, however, representatives were more likely to represent local or regional interests, which did not group within state lines (bottom). Actually, it would be the people of the United States, *not individual states, that such a body would really represent.*

- The "Convention Breakers"

The last two centrifugal forces deserve a place of their own in this chapter. For these "convention breakers" very nearly did that—broke the Convention. One of them seemed odd and out of place. The other was only too predictable. How the delegates dealt with them takes us into the inner dynamics of how the Convention worked. These issues were, in turn, the question of large states versus small states, and the question of slavery.

Large States Versus Small

Going into the Convention, few delegates worried about a political difficulty in which a couple of large, populous states would be on one side and more numerous small states would be on the other. But just such a conflict awaited them. For there was a perception on the part of ten smaller states that Massachusetts, Pennsylvania and Virginia were bullies and troublemakers.

The charge was not wholly imaginary. The Big Three, as we might call them, were much larger and more populous than their neighbors.

Tyranny of the Majority

Were there actual situations in which people were simply voted to death? Indeed there were. In the French Revolution it occurred several times. The incident the world remembered best was the execution of the king.

In truth, Louis XVI posed a difficult problem for the Revolution. Now he would accept the new order and now he wouldn't. He was docile one moment and hostile the next. It occurred to the Jacobins that as long as Louis Capet was alive and dithering, he would be a rallying point for royalism within the country and intrigue without.

But how, exactly, to get rid of the king? Most of the deputies favored some sort of trial. After all, if law was really to govern the new republic, then bring Louis before a high court, charge him with treason, present evidence, and carry out the sentence.

At that point, however, the constitutional scholars began scratching their heads. The new constitution specified "inviolability" for most of the king's actions. Besides, what were the specific crimes he could be charged with, and where was the evidence for his guilt?

But Robespierre, the leader of the Jacobins, cut through the tangles cleanly. Addressing the Assembly on December 3, 1792, he recast the issues in the most practical terms. "This is no question of a trial," he said, "Louis is not a defendant; you are not judges, but statesmen, and the representatives of the people. You have not to give a verdict for or against an individual, but to adopt a measure for public safety. . . . We talk of a republic, and Louis still lives!" The logic seemed convincing. Call it a trial if you will, and call the deputies judges, but don't worry about specific laws or details of evidence. Just vote.

On January 14, the voting began. The first question was whether Louis was guilty. (No crime was alleged, just guilt.) By a vote of 683 to 65, it was decided that he was. The second question was the choice of punishment. There, the vote was closer: 361 for immediate death, 72 for temporary reprieve and imprisonment, and 288 for imprisonment or exile. If the votes for reprieve and imprisonment were added together, they came to 360. Louis lost by a single vote.

The sentence was carried out in the Place de la Revolution on the morning of January 21, 1793. Like all affairs of the guillotine, it was short and simple. And at the end of it the people shouted: "*Le Roi est mort: Vive la nation!*"

One vote!

From America: A Study in Heritage, *Kendall/Hunt 7th Edition, p. 104–105.*

Their histories went back further, and they sometimes regarded sister colonies as offshoots or afterthoughts. They did indeed throw their weight around occasionally and believed they had a right to do so. And they wondered why it was that in the affairs of the Confederation, the vote of a single Rhode Islander was equal to that of eight or ten Virginians.

For Madison and the nationalists, this injustice simply had to be corrected. Any republic worthy of the name must represent the people *as* people, they said, and it must act upon them the same way. If a government gave equal representation to states, then it represented states and not people. By the same token, if it could not command (and of course punish) individuals, it could never be sovereign. If, for example, smugglers violated commercial laws in the sovereign-state sort of polity, the national government could never arrest and try them directly; it could only "request" that they be apprehended by some state—and perhaps be told to go to hell. For the nationalists, this was unacceptable.

The code word for their philosophy was *proportional representation*. That is, government must represent the people proportionally and act upon them as full citizens. The thinking may have been sound enough theoretically but it ignored that whole history of large-versus-small-state friction. So when the small state delegates looked over the Virginia Resolves, they saw something entirely different from the people being given a voice: they saw large states dominating the union—and small states getting shorted. That was not Madison's intention. What he envisioned was a withering of *all* states, small *and* large. Put another way, he imagined the American Republic as a single consolidated entity in which state boundaries were irrelevant and human beings were all that mattered.

But how did it look to the New Jersey delegation? It looked like going home and announcing that their state had become a colonial possession of dreaded Pennsylvania. Delegates from the small states saw themselves as fighting for their very existence.

James Madison was not often politically blind, but in this instance he came close to it. While he and the other nationalists kept insisting that the problem was artificial and would soon go away, quite the reverse was true. It got worse and worse. Given the fact that the small states outnumbered the large ones eight to three, and that the voting was equal among all eleven, it was remarkable that the nationalists were able to push the Virginia Plan toward tentative adoption. They were able to do so by building a coalition between the three largest states and the three southernmost ones—which believed they too would someday be populous. (At the time, Georgia had a smaller population than tiny Delaware.) It was a tribute to the political skills of Madison and his colleagues that they were able to keep this alliance together and operating.

New Jersey/Paterson Plan: Representation equal for each state regardless of population.

In the face of the large-state juggernaut, the small states, led by William Paterson of New Jersey, asked for time to regroup and come up with their own plan. Given the realities of the situation, the **Paterson**, or **New Jersey**, **Plan** was necessarily a moderate revision of the old Confederation. It would delegate to the national government some increased authority, but it stuck to the principle of equal representation of the states.

The New Jersey Plan was no more appealing to the nationalists than was the Virginia Plan to the confederalists. As long as representation was based on the states, sovereignty would remain fragmented and the American nation a chimera. In a series of speeches beginning on June 16, the New Jersey Plan was debated by the Committee of the Whole and more or less demolished. "We must remember the language with which we began the Revolution," pleaded James Wilson: "'Virginia is no more, Massachusetts is no more, Pennsylvania is no more. We are now one nation of brethren, we must bury all local interests and distinctions.'"

On June 19 the small state plan was voted down seven states to three, with Madison's coalition holding firm. But that was not the end. The small states would never sign onto the Virginia Resolves, they said, and if push came to shove they would leave the Convention. The question for the large states was, should they let the small states leave?

THE NEW JERSEY PLAN OF REPRESENTATION

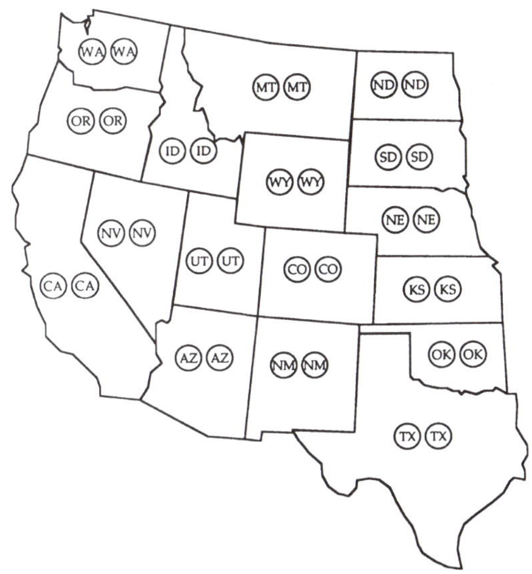

The New Jersey Plan was to have equal representation of states within a national legislature. What was to be represented was the states, as sovereign bodies, not the people.

Not without a try at compromise, said Connecticut's Roger Sherman. He had been thinking about the problem since the beginning of the Convention, and he had worked out the only solution possible. If there was to be a two-house legislature, he submitted, let each side claim one house for its own. Proportional representation in the lower house would preserve popular sovereignty, and equal representation in the upper house would preserve state sovereignty.

Because we take this solution for granted today, we overlook how bizarre it seemed at the time. Far from being a unified legislature representing the American people, Congress would be the strangest hybrid imaginable. And the rest of the country—would it be a single sovereignty (as indicated by the House) or multiple sovereignties (as indicated by the Senate)? The question seemed unfathomable.

For if political theory had one clear and settled idea, it was the idea of *sovereignty*. By its very definition, sovereignty could be vested only in a single authority. Accordingly, either the states must be sovereign, as they were in the present Confederation, or else the national government must. It was simply unimaginable that *both* could be sovereign at once.

However, this analysis missed an important point. Since any piece of legislation would have to pass *each* house independently, both principles could in fact be applied. A measure unacceptable to the American people would fail in the House, and a measure unacceptable to the American states would fail in the Senate. Perhaps it really *was* possible to have "parallel sovereignties" at work.

Lacking our hindsight, both sides saw Sherman's compromise as a defeat and dug in their heels. But the small states dug in less vigorously. Indeed, given the fact they were outvoted, small state delegates came to see the compromise as a rescue—the way a losing team might view the sudden hope of a tie—and reluctantly gave it support. They would vote for proportional representation in the lower house, they said, if Madison and his cohorts would vote for equal representation in the upper house.

But Madison balked. If states were the problem and union the solution, how else to view this ungainly proposal than as a disaster, he asked? And if the small states wanted to walk, let them do so. In the end they would have no choice but to come around.

So began the long hot summer. It commenced with the two halves of Sherman's compromise framed as separate votes, and both taken on June 19. Proportional representation passed for the lower house. And equal representation *failed* for the upper. The nationalists had held their alliance together and their forces in line. The small state delegates cried betrayal. Then the Convention began slugging it out.

Day in and day out, the slugging continued. The large states threatened to unite on their own. The small states threatened to seek out foreign powers. The delegates grew weary. Tempers flared. "I . . . declare," shouted William Paterson, "that I will never consent to the present system. . . . Myself or my state will never submit to tyranny or despotism." James Wilson answered him icily. "The gentleman from New Jersey is candid in declaring his opinion," he said. "I commend him for it. I am equally so. I say again I never will confederate on his principles. If no state will part with any of its sovereignty it is in vain to talk of a national government." There was a great deal more.

As June wore into July, the deadlock continued. Neither side was willing to budge, for both saw too much at stake. The speeches grew tiring and repetitious. The heat made everything worse. Some delegates went back to their homes for a respite—a few never to return. And hope for a settlement began to ebb away. It was, wrote William Few on June 28, "an awful and critical moment. If the Convention had then adjourned, the dissolution of the union . . . seemed inevitable."

While the dream of an American nation teetered on the abyss, the facilitators of compromise mustered their forces. Benjamin Franklin, for example. While he stood squarely with the nationalists, he also saw the need for a resolution. So, after listening to the umpteenth large-state-versus-small-state speech, he facetiously offered to chop Pennsylvania into small states if that would solve the problem. At other times he lectured like a schoolmaster. "When a broad table is to be made and the edges of planks do not fit," he intoned, "the artist takes a little from both, and makes a good joint. In like manner here both sides must part with some of their demands, in order that they may join in some accommodating proposition."

This was when "the doctor," as he was often called, made his historic plea for divine aid. He reminded the delegates that during the Revolutionary War the Continental Congress had often prayed for guidance in the very room where they now sat. "I firmly believe," he went on, "that without [God's] concurring aid we shall succeed in this political building no better than the builders of Babel. We shall be divided by our little partial local interests; our projects will be confounded and we ourselves shall become a reproach and bye word down to future ages."

Franklin's motion was not adopted, most likely for want of a chaplain, yet his plea may not have been without effect. They did stand at the judgment bar of history, some realized, and destiny did sit among them in this hall. And Franklin's appeal to a sense of the universal couldn't help but strike an American cord. God, he suggested, had favored their cause in order to favor that of mankind. "And what is worse," he added, "mankind may hereafter from this unfortunate instance despair of establishing government by human wisdom and leave it to chance, war, and conquest."

Whether by divine intervention or otherwise, events did take an unexpected turn. On July 2 the question of representation was voted on again—this time with a different result. Both Maryland and Georgia had been voting with the nationalist majority to produce a 7-to-3 result. This time, however, Maryland delegate Daniel of St. Thomas Jenifer was strangely absent when the vote was taken, and that swung the state's precarious delegation in favor of Sherman's compromise. Madison was still reeling from that blow when a second one fell. Abraham Baldwin of Georgia switched sides and nullified the Georgia vote entirely. The 7-to-3 majority was suddenly a 5-to-5 tie.

Now at a complete loss, the delegates appointed one of their Grand Committees (one member from each state) to try to work through the impasse. This group labored through the Fourth of July holiday to have a compromise proposal ready on the fifth—but the nationalists simply rejected it. They might not be able to go forward but they weren't going back either. That evening Luther Martin wrote that the Convention was on the verge of collapse.

And nine more days of stalemate still lay ahead. In the records, this period resolved itself into a wearisome blur. The summer heat was now record breaking. The flies buzzed and stung. The debates spun around and around. On July 10, Robert Yates and John Lansing Jr. departed for New York, promising to fight against any document the Convention might produce. Hamilton had already left. That a state as important as New York was now completely unrepresented further deepened the gloom. Washington wrote that he despaired of finding a way out—and regretted his own involvement.

On July 16 the heat wave finally broke. There were storms the night before, and with the dawn came cool breezes. The final vote was scheduled for the Connecticut Compromise, and Madison, although bone weary, was confident that it would be rejected once and for all. But another surprise was in store for him. When the vote was taken, two of the Massachusetts delegates had unaccountably changed their minds—the vote split and Massachusetts was out. Mary-

The city of Philadelphia (pop. 45,000, then America's largest city) became part of the story of the Constitutional Convention: record-breaking heat, social gatherings, the bustling urban environment and a stimulating intellectual community all contributed to the ambience surrounding the convention.

land was back in the nationalist fold, a repentant Daniel Jenifer gazing birdlike over the rims of his bifocals—but now North Carolina defected. Madison's gasp was almost audible.

There was a climactic, mud-spattered entry by Alexander Hamilton, who had hastened back from New York for this very moment of truth. He dramatically cast the winning vote with the nationalists. But, alas, the rules of procedure forbade his participation. At least two delegates had to be in residence for a state to be officially represented, and the other New Yorkers had gone home.

Accordingly, the Connecticut Compromise passed, five states to four.

For those who consider the turn of cards in history, there was much to ponder. Daniel Jenifer's mysterious absence. Franklin's appeal to Providence. Abraham Baldwin's sudden change of mind. Votes switched at the very center of the nationalist coalition. The appearance of a procedural rule so obscure that the delegates had to stop and look it up. These small things had literally made an American nation possible.

Not least to ponder was the attitude of James Madison. He had watched his majority shrink, then falter, then die. He was numb with despair. When the small states had lost early on, they threatened to abandon the Convention. Madison made no such threat now. He returned to his lodgings and perhaps to his books. Later that evening, at the Lion Cafe, he still wasn't smiling, but he was grudgingly ready to propose a toast. This truly was a new kind of government, he allowed, one for which there was absolutely no precedent in history. It was on the one hand a government of the sovereign people of America, and on the other hand a government of the sovereign states. "Gentlemen," he said, "to the United *States* of America."

North, South and Slavery

The compromise over state sovereignty was an example of what we might call "constitutional serendipity," where an agreement forged in direst necessity was later revealed to be a source of great benefit. The Constitution eventually abounded in such windfalls. The compromise over slavery was not one of them, however. Quite the opposite.

Some scholars believe that agreements worked out over slavery at the Philadelphia Convention made the Civil War inevitable. At the very least it legitimized a practice far out of keeping with republican principles. The question becomes, then, why should such a "monstrous contradiction" be woven into the very fabric of the Constitution?

An answer of recent popularity holds that the Framers could not help but act out of ideological interest. After all, they were white Protestant males—and five of the Core Framers (not to mention George Washington) held slaves themselves. Hopefully we have learned to be wary of such tidiness. The real situation at the Convention was far more complex, far more ambiguous, and in hindsight far more tragic. It wasn't a simple case of master-class ideology.

Slavery, like state sovereignty, was a historical accident. It had made sense in colonial times, perhaps, but appreciably less sense now. By the end of the Revolution most Americans were at least mildly apologetic about it and some were downright hostile. Still, as David Hume had pointed out, what *is* established has its own validity, and slavery was established only too well.

Colonial slavery had been no one's particular fault—and at the same time everyone's. Labor was scarce. Conditions were harsh. And the wholesale kidnapping and selling of Africans had been a thriving enterprise for centuries. Oddly, as it seems to us now, most colonists never really thought about its moral implications one way or the other.

Eventually though, opposition to the practice did develop. It appeared first in Quaker Pennsylvania, and from there spread to Puritan New England and elsewhere. By the time of the Revolution, slavery had more or less died out in most northern colonies. Its lack of economic viability was as important as moral repugnance in its demise—it simply didn't work very well. The Revolution also played a role. How, Americans asked, could someone be willing to fight and die for liberty if that person owned slaves?

At the same time, however, slavery got a second wind in the South. While the cotton gin had not yet been invented, tobacco, rice and indigo cultivation were all growth industries, and other forms of slave-driven enterprise were creating new promise. By the summer of 1787 slavery was not the "dying institution" it has sometimes been characterized.

We should also be clear about what was and wasn't conceded at the Philadelphia Convention. Literally none of the delegates imagined their mandate to include the possible elimination of slavery no matter how out of keeping it might be. Their task was to work out a better government—not work out a better society. Besides, they were already operating on the margins of what was politically possible, and taking on slavery would have put them far over the line. Whatever else it was, slavery was a system designed to insure white domination over an alien, stigmatized and resentful populace whose concentration in some areas inspired terror. For some whites it was a matter of self-preservation.

In parts of the Carolinas, the ratio of slaves to whites was as high as 50 to 1.

The real question dealt with the future. Many northerners (and some southerners) believed that slavery was on its way to eventual extinction, so why make a fuss over it? The answer in three words was: *the slave trade.* With a steady supply of new slaves, all sorts of eventual possibilities opened up. New kinds of crops might be cultivated. New uses for slave labor might be developed. And the new lands of the West were out there beckoning.

So, what about the slave trade? We recall that one of the Convention's central concerns was the regulation of commerce. It promised enormous benefits but was fraught with thorny side issues. The power to regulate commerce might become a weapon of destruction in the rivalry of the regions, for example. If one region gained control over the central government, it might seek to stifle another region's interests simply by burdening certain kinds of commerce—such as the slave trade.

Here was a possible scenario. The North gains control over Congress and immediately throttles slave imports. The effect, of course, would be to weaken the South, both economically and politically, which some northerners might

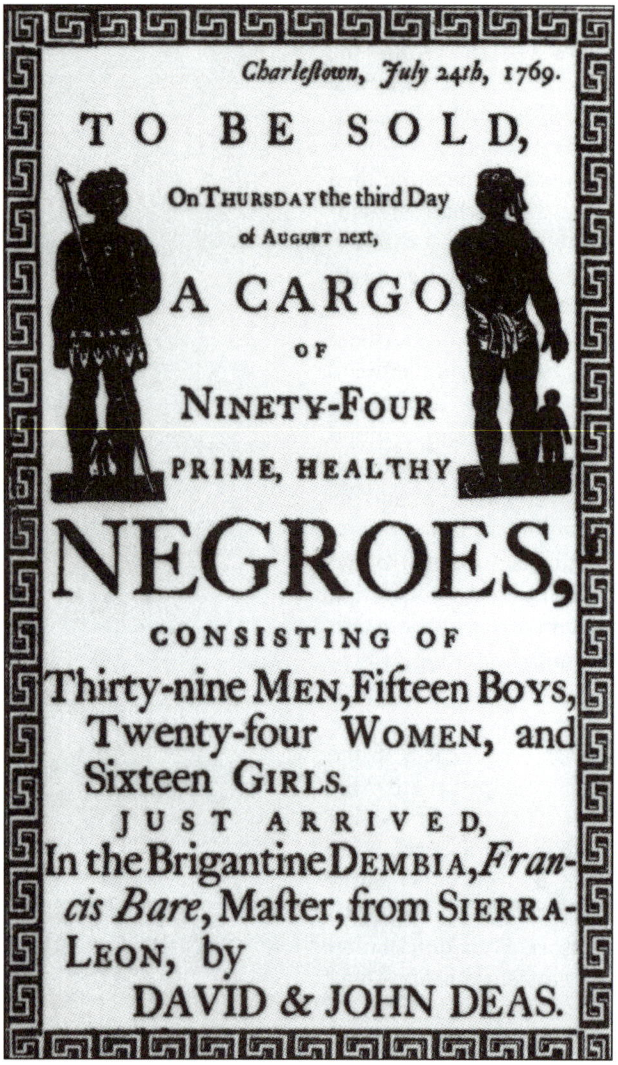

The question of slavery remained one of the great unresolved issues which confronted the delegates to the Constitutional Convention. Compromise temporarily deflected consideration of implications inherent in the 'peculiar institution,' for several decades before the conflagration of the Civil War irrevocably settled the matter.

not particularly regret. At least there would be no further expansion of slavery nor contradiction in American ideals—for slavery really *would* be bound for extinction.

The question for the southern delegates, then, was whether they wanted to take such a chance? Some of course didn't; it was just too risky. Others did, though, and they had their reasons. Some of them harbored their own qualms about slavery. (George Mason of Virginia, for one, stood up in General Assembly and gave a rousing abolitionist speech.) Others saw the institution as a liability. Sooner or later, they knew, a slave-owning society was bound to experience servile insurrections, and these in turn could lead to foreign intermeddling and even worse complications. With slavery, they had a wolf by the ears, as Jefferson put it, and wondered if they ought not turn around and run.

There was potential here for a compromise, and there were delegates capable of effecting it. Two of them were William Samuel Johnson of Connecticut and John Rutledge of South Carolina. They shared an important commonality. Both states were entirely dependent on exports for their income, and thus neither wanted to see taxes placed on exported goods. The Connecticut delegates were masterful politicians, too, and they spent much of their leisure time with the charming Carolinians.

Over port one evening, Johnson and Rutledge worked out the following agreement. Southerners would support the New England position on commercial regulation—that is, no power to tax exports, only imports—and New England would support the southern position on the slave trade. The Constitution would bar any interference with such trade for the next twenty years. That would give the South a hedge on the future.

This agreement touched only one aspect of the slavery problem. There were two others. The first was the question of how slaves should be counted, both for representation and taxation purposes. The southern position was clear if self-serving. Slaves ought to be counted as "persons" in calculating representation and as "property" in calculating taxes. In other words, let the slave population amplify the South's voice in Congress without increasing its tax liability. This the northern delegates couldn't abide. It would give the South a fixed and permanent political advantage, while at the same time boosting the traffic in slaves. One Yankee, Elbridge Gerry of Massachusetts, drily observed that slaves in the South were used to perform the labor of horses and mules in the North. Why not allow horses and mules representation? he asked.

This time compromise was proposed by James Wilson. Peering through the thick lenses that were always perched precariously on the end of his nose, Wilson dug up an old fraction used in the Articles of Confederation. Suppose the slave population were to be counted for three-fifths of the free population both for taxation and representation, he said. It was a shade better than half a loaf both ways. After a great deal more discussion, the southerners reluctantly agreed. Without the "federal ratio," Hamilton said later on, "no union could possibly have been formed." But there was another hidden cost. Now slavery was both recognized and protected in the Constitution—not for twenty years but forever.

The final element of the slavery compromise involved the recovery of fugitives. There were not a few northerners who would prove sympathetic to the cause of runaway slaves, as the Underground Railway would someday show, and it would be easy for local governments to throw stumbling blocks in the way of recapture. The Constitution, said the southern delegates, must forbid such legislation absolutely.

Underground Railway: Support system that assisted fugitive slaves.

Here was yet another painful dilemma. Northerners knew there would be opportunities for gross abuse, and indeed, literally tens of thousands of free blacks would be kidnapped into slavery under the terms of Article IV, Section 1. Yet one purpose of the national union was to protect property, all kinds of property, and the slaves sadly were that.

We can see these compromises as the requisite price of union. Yet we can also see them as a missed opportunity of tragic proportions. For while the abolition of slavery was beyond the Convention's capacity, closer limits on slave expansion might not have been. The South would not have been pleased by such limits, but it might well have accepted them.

As it was, in the twenty years of grace given to the slave traders, more than a million *more* Africans would be imported into the South. The abundant supply would keep prices low, and low prices would fuel rapid expansion. Slavery would go on to accomplish everything the prophets of doom had foretold.

Creative Chemistry

So much for the politics of half a loaf. There was another kind of politics at work—a politics of innovation. On the surface, the two looked much the same. Both required haggling. Both entailed maneuvering and arm twisting. Yet with the politics of innovation there was a different end in view: to come up with ideas that worked.

We have already seen that a group of some fifteen Core Delegates led out in the Convention's work. A useful way to understand how these central players interacted is to identify three separate roles they tended to play.

- Visionaries

We can think of the visionaries as political philosophers. They were the last-ditch nationalists like Madison, Hamilton, Wilson, the two Morrises and Charles Pinckney. They were "visionary" because they believed, without much proof, that a republic could be fashioned of imperial size and placed under a strong central government. Their role was to convince the others that it could really be done—and that *they* could figure out how.

- Innovators

If the visionaries were political philosophers, the innovators were political scientists. They were the ones who had to come up with the working dynamics of the new system. Sometimes they could draw upon the experience of the states for ideas, and sometimes they could borrow from Hume or Montesquieu. In a significant number of cases they had to figure things out for themselves.

The Convention's primary innovators were James Madison, James Wilson, and Gouverneur Morris, with an important secondary in George Mason, Edmund Randolph, Oliver Ellsworth, Elbridge Gerry and Charles Pinckney. They worked together well, for the most part, and often in a characteristic pattern of interaction, as we shall soon observe.

- Facilitators

If the visionaries were political philosophers and the innovators were political scientists, the facilitators were old-fashioned politicians. It was their job to make sure that any idea the innovators came up with would be workable, practical, and above all acceptable to the people.

Some of the facilitators were masters of compromise. They could take antithetical positions and figure out a way to reconcile them. However, some of the facilitators performed the opposite function—steadfastly *refusing* to compromise. This too had a purpose, for if the Framers had compromised on some kinds of issues their work never would have been ratified. And there were other implements in the facilitators' toolbox. One was a knack for building alliances. Another was the art of persuasion. And a third might simply be called brokerage—the ability to work out political bargains.

Among the facilitators were Edmund Randolph of Virginia, Rufus King of Massachusetts, John Dickinson of Delaware, and John Rutledge of South Carolina. Both Paterson of New Jersey and Madison of Virginia refused to compromise, famously so, when much was at stake, but on the other side, Benjamin Franklin spoke often and movingly of the need to accept half a loaf.

The best facilitator of all was Roger Sherman. The Connecticut delegation, of which Sherman was the leading member, was an engine of facilitation, with both William Samuel Johnson and Oliver Ellsworth playing roles similar to Sherman's. But Sherman handled the tough cases. He was one of the Convention's lowborn and self-made delegates, of a square-set, no-nonsense countenance. He was the only member who might be called a politician by profession, having served in one post or another for the whole of his adult life—and he truly was a pro.

Designing the Presidency

The chemistry of the Convention made some rather startling innovations possible. We might trace in detail how one of these was developed in order to gain a clearer picture of the Convention at work. The creation of the Presidency is a good example. It shows how visionaries, innovators and facilitators interacted with one another in complex and sometimes surprising ways, and how together they created something entirely new.

On the subject of the executive, the Framers began with nothing but questions. What is an executive? What do executives do? What constitutional mechanisms would promote appropriate executive behavior? Or *mis*behavior? There was a whole science to be learned.

As a beginning assumption, most of the Framers supposed that a weak executive—or executives, plural—would suit the American purpose best. This was the model used in several states and it had been proposed in the Virginia Resolves as well. The thinking was that since legislatures reflect the will of the people, executives ought simply to do the lawmakers' bidding.

When the issue was first taken up on June 1, however, one delegate had a markedly different idea. James Wilson proposed that the executive be a strong office, not a weak one, and be occupied by a single individual of "energy" and "dispatch." The motion received no second. "Energy" in an executive was no blessing, scolded Roger Sherman; it was a dangerous prelude to tyranny. Edmund Randolph agreed. A single executive was nothing but a "fetus of monarchy," he said.

But Wilson came back with a surprising argument. He had a skeptical, Socratic mind that cut like a razor through specious reasoning. He also had a strong bias toward democracy, which had something to do with the rugged, egalitarian world of his youth in the Scottish lowlands. Think about it, he began, a single executive, *if elected by the people*, might be just the opposite of the bugaboo tyrant—a safeguard *against* tyranny. Think about the dynamics involved. The single executive, with his own accountability, might provide a sort of counterweight to the legislature. If the latter got out of line—as it had in Wilson's own state—the executive might be inclined to resist it, for he would be mindful of the people who elected him. Wilson may have been pondering Adams' *Thoughts on Government* or the Massachusetts Model Constitution, both of which supported strong executives. Yet he was reasoning his way toward an entirely new constitutional idea.

The following day, two of the South Carolinians, John Rutledge and Charles Pinckney, declared themselves persuaded by Wilson's arguments and echoed his call for a single executive. But Randolph dissented yet again. There ought to be an executive committee of three, he insisted, whose members would be drawn from different parts of the country, and the three would have to work out agreement among themselves.

Wilson had an answer for that proposal too and he submitted it the next time the question came up. A three-man committee would simply bicker among itself and never decide anything, he said. Besides, single executives were found in several state governments, and *they* hadn't morphed into monarchs. Did Wil-

Executive Tyranny

With the benefit of hindsight, we see the Constitution today much as its supporters (the Federalists) did: as a preserver of republican freedom. But how did it look to the doubters? In a single word, threatening. Not benefiting from our hindsight, the Anti-Federalists could see only tyranny in the Constitution's provision for a strong and vigorous central government.

James Lincoln of South Carolina was typical: "From a well-digested, well-formed democratic you are at once rushing into an aristocratic government," he cried. Pennsylvania's James Warren agreed. "The Constitution will result in an immediate aristocratic tyranny," he said, "that from the difficulty, if not the impracticability of its operation, must soon terminate in the most uncontrolled despotism!" George Mason of Virginia had still another set of doubts. "The executive and the Senate form a combination that cannot be prevented by the representatives," he fretted. "The executive and legislation powers thus connected will destroy all balances."

The Anti-Federalists were missing a fine but nonetheless important point. A government with the right structure could be powerful without being tyrannical. Indeed, it must be powerful in order to maintain the rule of law. "A weak constitution," said Madison, "must necessarily terminate in dissolution. . . . Tyranny has perhaps oftener grown out of the assumptions of power [under] a defective constitution, than out of the full exercise of the largest constitutional authorities." That, as it turned out, was the view sanctioned by hindsight.

From America: A Study in Heritage, *Kendall/Hunt 7th Edition, p. 140*

son's eye twinkle when he said this?—for the doubting Randolph happened to be one of them.

Roger Sherman, who the week before had scorned the idea of a single executive, now swung around and came on board himself. However, the executive ought to consult a council before acting, he added. Wilson demurred yet again. A council would *cover* malpractice, not *prevent* it, he argued. If something went wrong, the executive would blame the council and the council members would blame each other. All sense of accountability would be lost.

The Convention voted seven states to three in favor of a single executive and left the council question up in the air. Wilson was making headway. As discussion now switched to other features of the office, the delegates reflected again and again on the curious idea of a strong executive and strong legislature playing against one another to create balance. At least one of them, Gouverneur Morris, found the idea preposterous and spoke forcefully against it. But others began to come along. In the end, they voted to kill the council idea and go with a single executive.

Most of the delegates still favored his election by the legislature, however. There was a way to accomplish it, they thought, that would still let the executive function as a counterweight. If he were limited to a single term, for example, there would be no incentive for him to be the legislature's yes-man in order to insure his reelection. But, then, that single term would have to be a long one—maybe seven years or so—in order to keep from constantly changing executives, and with a seven-year term, what would keep him from straying into corruption? Well, there would have to be some mechanism of impeachment. The problem was getting a bit sticky.

Besides, Wilson still had his own idea. He had won the day on the question of single versus plural—perhaps he was feeling lucky. Why not, he urged, seriously consider election by the people instead of the legislature? The question must have been greeted by blank faces. Leaving aside the advisability of such a practice, how on earth could it be brought off? How could three million people scattered over a continent have the slightest idea how to nominate candidates, how to winnow them down to a manageable number, and how to make the final choice among them?

Wilson was undaunted. He volunteered to sit down and see if he could work something out. And the following day, July 2, he returned as promised with a set of concrete proposals. Let the states be divided into large districts, he said, and let the voters in each district cast ballots for electors. Let these electors then meet together in some designated place and elect the executive. Wilson's colleagues were not impressed with this contraption of springs and wires, and they voted it down forthwith. They opted instead to have the executive chosen by the Congress and be ineligible for reelection.

The matter was shelved for the time being and the Convention turned its attention elsewhere. When it took up the executive issue again, the Great Compromise between large and small states had been achieved, and it was clear that subtle balances had shifted. For one thing, several delegates had been pondering

Wilson's counterweight mechanism and its possible value. And now that the small states had a national government they could live with, their representatives viewed the strong executive concept more kindly. Gouverneur Morris was among those who had undergone a change of heart. Wilson was right, he said flatly: the executive ought to be chosen by the people.

Hard-headed Roger Sherman was still unpersuaded, however. The people would never be informed well enough to make a wise choice, he lectured, and even if they were, they could never muster a majority for any given candidate. So Wilson thought it all through once again. Maybe the popular vote wouldn't reach a majority, he allowed, but there was still a way to make the system work. In the absence of a majority, let the election be decided by the legislature, choosing among the strongest candidates. That way, the people would in effect do the nominating and the legislature would make the final pick.

Parties now emerged. Pinckney of South Carolina and Mason of Virginia led the fight against election by the people, while Wilson and Morris of Pennsylvania pressed it forward. The Convention listened to both sides and voted once again for election by the legislature. Only this time they struck out the ineligibility clause. If the executive was doing a good job, why shouldn't he be able to run again?

But that brought back the old problem of kowtowing to the legislature. James McClurg of Virginia, a delegate who had been sent to the Convention on short notice and had taken little part up to now, suddenly announced that he had the answer. Let the legislature elect the executive, he submitted, but let the executive *serve for life*. That way, there would be no need to curry favor with the lawmakers. Some delegates took the idea seriously. Gouverneur Morris, who had never shared the American disdain for monarchy, was so delighted with the election-for-life proposal that he switched sides once again.

Sherman, however, was unmoved. Monarchy is precisely what we would wind up with in an executive elected for life, he opined heatedly, and didn't we recently fight a war about that? Sherman had learned his lessons in the school of hard knocks and he could be as plain spoken as a drill sergeant. But the elegant Virginians agreed with him. The executive *needed* to be dependent on the legislature, Randolph and Mason both concurred, or yes, he might very well become a tyrant.

It was at this point that James Madison entered the fray. He hadn't thought much about executives before coming to the Convention, and his own plan had called for a weak one. But all the to-ing and fro-ing had gotten him into the library, as it were. He hauled out his copy of Montesquieu, and on its authority concluded that Wilson might well be right. Make the executive independent of the legislature and you might wind up controlling them both, he said. Each could be strong and vigorous. Each could do its own separate job. Each could be responsible to the people. And if either began stepping out of line, it would automatically behoove the other to resist. Madison even affixed a name to the idea: **separation of powers**.

The Convention voted McClurg's proposal down at this point but it still wasn't ready to adopt Wilson's. Indeed, the debate dragged on and on, and in the course of it the delegates changed their minds a dozen times more. How could they come up with a strong executive, they wondered, without making him too strong? The problem seemed unresolvable.

On July 26, George Mason took the Convention right back to the beginning with a proposal for election by the legislature, a seven-year term, and ineligibility for reelection. The weary delegates approved the plan 6 to 3 and sent it off to the Committee for Detail for drafting.

Another month went by. On August 24, the delegates turned to the Committee for Detail's write-up of what they had reluctantly agreed to back in July. It still didn't sound right. Election "by the legislature" meant what, exactly? they asked. Election by the House? That would give an advantage to the large states. Election by the Senate? That would give advantage to the small states. John Rutledge moved for the two houses to vote together in one body, which might balance the advantage out. His amendment carried.

But the dogged Wilson wasn't beaten yet. He and Gouverneur Morris made one more attempt to secure the election of the executive by the people. It was voted down yet again but this time by only a single vote, six states to five. The Pennsylvanians were making headway.

On August 31, the delegates, who were sick and tired of the whole business, voted to turn it over to an ad hoc committee. This was their last resort. The committee they chose happened to be unusually vigorous, however, with Sherman, Morris and Madison among its members. Morris and Madison must have had an earnest little talk with Sherman—who had left skid marks throughout the entire process—for quite without warning the committee voted to abandon election by the legislature and switch to Wilson's "electoral college." Further tinkering was required with the operating mechanism, but at last they had it. It was a delicate compromise, so delicate that the Convention might have rejected it out of hand had they not tried everything else. But, alas, they had.

So it was that the Constitutional Convention came up with a single executive, elected by "the people," with a four-year term of office, eligibility for reelection, and provision for impeachment. He would be stronger by far than any executive they had considered back in June, and with that in mind they now proceeded to limit his powers very carefully. Still, they wound up giving him entire charge of the military, entire charge of foreign affairs, extensive charge of appointments, and the conditional veto of legislative acts. That was a great deal.

Serendipity

In the course of its slow evolution, the idea of the executive had been turned inside out. Instead of a glorified servant who "executes" the will of Congress, the executive had become a kind of elective monarch, imparting energy and vigor to

"Remember, gentlemen, we aren't here just to draft a constitution. We're here to draft the best damn constitution in the world."

government. It was an extremely important innovation—one for which no single individual was responsible.

James Wilson had played the leading role, both as visionary and innovator. Sherman, Randolph, Rutledge, Mason, Charles Pinckney and Gouverneur Morris had hammered out the politics, backing and forthing many times. The delegates couldn't have reached a solution without all three roles being played—nor could they have reached it without an emerging theory, "separation of powers," expounded by Madison, to legitimize it. Many of them still suspected the theory and wondered whether it would actually work in practice. But somehow it felt right to them. A powerful executive might become a despot, true enough, but to do so he would have to cross swords with an equally powerful Congress.

This was the process. It was full of false starts and misdirection. But its tendency was to cut through layers of partisanship, prejudice and political myopia, and reach toward something like truth. For the American Presidency, despite its newfangledness, became one of the signal achievements of modern-day political science. None of the participants had gotten what he set out for in the beginning. Each had played off against the others, had backed into corners, had temporized, compromised, returned to the drawing board for new ideas, and together had groped for a deeper wisdom than any could call his own.

There was an odd kind of "synergy" in the outcome. A whole was created by the collective endeavor that was greater than the sum of its parts. And, as it turned out, there was an equally odd "serendipity." Things had just happily fallen into place.

CHAPTER 7 STUDY AID

1. Opposing Forces—Centripetal:

- Impotence of Confederation and the threat of anarchy
- A vision of American Nationhood
- Strong leadership and esprit de corps
- George Washington

2. Opposing Forces—Centrifugal:

- The Confederation
- The States
- The "convention breakers"—Large v. Small States
    a. Connecticut Compromise
- Slavery
    a. Three-fifths Compromise

3. Creative Chemistry:

- Visionaries
- Innovators
- Facilitators

4. Designing the Presidency:

- Innovation
- Role of the Executive
- A single, strong Executive
- Electoral College

5. Serendipity?

CHAPTER 8

The Constitutional Convention III
Structure and Dynamics

After Edmund Randolph's presentation of the Virginia Resolves, the delegates to the Grand Convention came to realize that they were not repairing the Articles of Confederation but designing a new government—a new *idea* of government—from the ground up. Thereafter, every motion, every agreement, every compromise, every tactical maneuver had to have some bearing on the development of "the plan"—a working constitution.

The Confederation experience had underscored the difficulty of creating successful republican governments. We might better understand the nature of that difficulty if we cast it in schematic form. The Republican Problem, let us call it, is to design a government able to accomplish three separate ends. It must:

- Express the sovereignty of the people.
- Protect the rights of individuals.
- Seek the public good.

Just why the Republican Problem was so difficult can be seen in a few quick comparisons. The classical republics often expressed the sovereignty of the people well enough but they ignored minority rights—with tragic consequences—and failed the public good. Liberal republics, such as Georgian England, did better with minority rights but not as well with popular sovereignty, and served the public good only intermittently. Neither the American states nor the Confederation had much better track records, some erring one way and some another. For all of them, it was a little like a carnival game where you can knock down one bottle or possibly two but never all three.

Madison referred to the Republican Problem in *The Federalist* when he spoke of the "diseases" of republican government. The Framers must now seek out a "republican cure" for the diseases in question.

Fundamental Ideas and Conceptions

In order to fully understand the structure and dynamics of the Constitution, we must begin with some fundamentals. These are the foundation on which the edifice of government must be built.

Fundamental Law

English Bill of Rights: http://wwlia.org/uk-billr.htm

> **English Bill of Rights:** Gave civil and political rights to the people and political supremacy to Parliament (1689).

> **Magna Carta:** English charter of liberties asserting fundamental rights and privileges (1215).

> **Habeas Corpus:** Writ which requires reason for arrest to be explained to court of law.

Throughout the controversy with Great Britain, the colonists had constantly invoked *fundamental law*—law that stood above the ordinary acts passed by a legislature—which they code named "the English constitution" or "the rights of Englishmen." Some scholars rejected the whole idea of fundamental law, as we will see in a later chapter, but Americans took it very seriously. Not many of them could have explained the concept very well. There was something in it of the law of nature—those basic ideas of right and wrong that most people accept—and something of historic agreements such as Magna Carta or the **English Bill of Rights**. There were even some common law precepts such as *habeas corpus*. Fundamental law came down to a belief that lawmakers couldn't do just any old thing they had a mind to—there were limits out there somewhere.

Over the long months of the summer, the Framers in Philadelphia thought a great deal about the document they were trying to work out. They referred to it as a "plan of government" but they realized it was more than just that. They came to understand that it was indeed a constitution—a set of fundamental laws created by the people themselves. A constitution exists prior to government and stands above the governing process. And government, in its turn, derives from the constitution and thus submits to the fundamental law.

These ideas had some important corollaries. First of all, constitutions could not be written by existing governments—that would be a case of the tail wagging the dog. And constitutions could not be adopted by governments either, only by the people themselves. Accordingly, it wasn't for the *states* to accept or reject the work of the Convention, nor was it for the Confederation to decide—it was for ordinary Americans. As the Framers finally worked it out, the people would call ratifying conventions within the various states and report their decision state by state, but the conventions themselves should have nothing at all to do with state governments.

This was truly uncharted water.

Written Constitutions

Writing out a constitution in black and white proved to be very different from the vague understandings of the English experience. With vague understandings there was always room for maneuver. With specific words on paper, by contrast, the wiggle room was much less.

> **Edward Coke:** A 17th-century English jurist instrumental in arguing that the king was subject to the law.

> **William Blackstone:** Author of the "bible" on English jurisprudence.

Constitutional vagueness was usually dealt with by legal experts like Edward Coke or William Blackstone. When the provisions were written out and published, by contrast, they became public property. Anyone could haul out a copy of the Constitution and ask questions of those in power. Unsurprisingly, we find that written constitutions soon had a track record of rallying and focusing public opinion. They held government responsible in ways undreamed before.

Finally, by writing it all out on paper, the people gained a means of correcting constitutional drift—the very thing that had "gone wrong" with the English constitution. It was harder for the fundamental law to wander off track.

Enumeration

Written constitutions changed the whole idea of authority. Heretofore, the powers exercised by a sovereign government were supposed to be unlimited. In ancient Rome, for example, there was a basic precept that *Caesar can do no wrong*—that literally any action willed by a Roman Emperor was backed by the authority of law. Constitutionalism rejected this notion out of hand, of course, for what else was a constitution supposed to do but subject governmental power to certain limits?

But how to accomplish this in black and white? How to give Congress or the President the authority they need without giving them too much? One might say, for example, that Congress shall have the power to pass "all needful laws," only to discover that such broad and general wording posed no limits whatsoever. On the other extreme, one might stipulate that Congress shall have power "to regulate wagon traffic on national roads"—using concrete and specific language instead—only to learn that the lawmakers had no authority, later on, to regulate canal traffic, rail traffic or air traffic. It is a knotty little problem.

After much discussion, the Framers hit upon the idea of *enumeration*. Instead of trying to frame a grant of authority in general terms, or trying to list out the multitude of possible particulars, they searched for a workable middle ground. They "enumerated" the powers of government, spelled them out, but did so in language that was neither too broad nor too specific. Enumeration reinforced the idea of limited government.

Republican Mechanisms

Republics did not necessarily require written constitutions or enumerated powers. But they did have essential requirements and the Framers had to think carefully about each of them.

Popular Sovereignty

The primary requirement was that the governance of a republic be for, if not always by, the people themselves. The power of government is always a benefit to *someone*. King, dictator, oligarch, plutocrat—there is always an interest to be served. In a republic the interest must be that of the *people as a whole*.

In order to achieve this benefit, the people as a whole must be "sovereign," which means they must consent to the laws that govern them. For John Locke (and many other political writers in the 1600s), consent was given tacitly rather than formally. People never sat around a table and negotiated an actual "social contract"; they agreed to the terms by remaining in a given political state, taking part in its activities, owning property and paying taxes.

With the American Constitution, the meaning of consent was dramatically altered. The social contract became overt and literal. Delegates to the Philadelphia Convention were sent by almost all the states, and the document they worked out had to be ratified by the voters in every state that joined the union. There was nothing tacit about it.

There was another point. In the older view, it could be argued that once the social contract was worked out, the people had little further part to play. The American Framers had a different idea. They saw the Constitution not so much as a signed and sealed agreement as an ongoing grant of power from the sovereign people—"power on loan."

For convenience, then, consent in the American system may be broken down into two varieties:

Original Consent. The people work out a plan of government formally and adopt it—reenacting the social contract. Original consent is renewed each time a constitutional amendment is considered. If the people reject the proffered amendment, they in effect consent to carry on with the existing constitution. And if they accept the amendment, they consent to a new constitution.

Periodic Consent. The people, by means of the electoral process, give their periodic approval of the government's political behavior.

Suffragettes demonstrating to obtain the right to vote. Can there be "consent" without an opportunity to participate? Is any form of exclusion acceptable under the rule of law? If so, under what circumstances?

The Framers' view of popular sovereignty only mirrored the facts. Americans were independent minded, jealous of their personal freedom, spread across a continent and armed to the teeth: it would be impossible to rule them without their consent. Since the adoption of the Constitution, American voters have often brought their government up short on election day. And no administration has been able to advance policies very far that the people themselves oppose.

Representation

Constituents: People being represented.

Representation was practiced in England, of course, where the Americans learned it firsthand, but it traced all the way back to ancient Greece. The idea was for a few elected individuals to represent the people as a whole, each acting in behalf of certain *constituents*. The process was two-way. The constituents, through their representatives, made their voice heard in government, and government, through those same representatives, got the constituents to approve unpleasant little matters like taxation.

Old Sarum is a notorious example of a rotten borough. It had no residents whatsoever.

With a bit of tinkering, the structure of representation could be designed to favor the rich, the poor, the intelligent, the powerful, and so on. So abuses were common. In the so-called rotten boroughs of England, the only "constituents" represented were a handful of local peers.

There were also at least two different ideas of how the process ought to work:

> *Attorneyship Representation.* The representative is conceived as a simple agent of his constituents, conveying their input to the legislative process without personal intervention. This was the kind of representation usually practiced in colonial governments. It was regarded as the more democratic form.
>
> *Virtual Representation.* Here the representative is more of a free agent, bringing his own ideas into the legislative process. He may not speak exclusively for the interests of his own constituents but instead may represent the larger public good as he understands it.

Virtual representation got a bad name during the controversy with Britain. For Parliament claimed that colonial interests, while not directly represented, were "virtually" represented in that body, and thus that someone would always speak up for them. To American ears, this rang false to human nature. Virtual representatives didn't have to pay the taxes they were levying against the colonies, nor did they have to account to colonial constituents at election time. The broadest-minded statesman in the world might never think twice about whom he was really "representing."

The Framers in Philadelphia discussed representation a great deal. George Mason favored attorneyship. "The representatives," he said, "should sympathize with their constituents, should think as they think and feel as they feel, and for these purposes should even be residents among them." Others opted for the virtual model. In the end, the delegates sought to incorporate both varieties into their constitutional architecture. They set up legislative districts in such a way as to compel the representation of specific interests—no rotten boroughs. Moreover, the representatives themselves would have to live under the rules—and pay the taxes—they voted for in Congress. Yet the Framers also expected representatives to be more than passive "attorneys" of their districts. They assumed that voters would choose representatives who were more virtuous and farsighted than the norm, and that the process would "enlarge and refine" public views.

REPRESENTATIVE PRINCIPLE

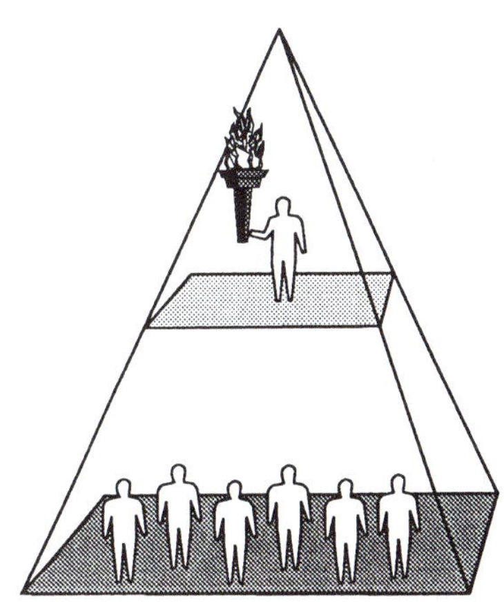

Representatives, having passed the election process, would generally be more virtuous and wise than the ordinary citizen. Consequently they were expected to be leaders, refining and enlarging the views of their constituents, rather than acting upon them mechanically. The large republic provided the necessary distance between the representative and the whims and passions of his constituency, to make virtuous leadership possible.

Virtue Mechanisms

In the instance described above, representation becomes a kind of "virtue mechanism." That is, solely by means of structural design, a political system could be engineered to place the most virtuous people in positions of greatest public trust—using human nature as a design principle. In the above

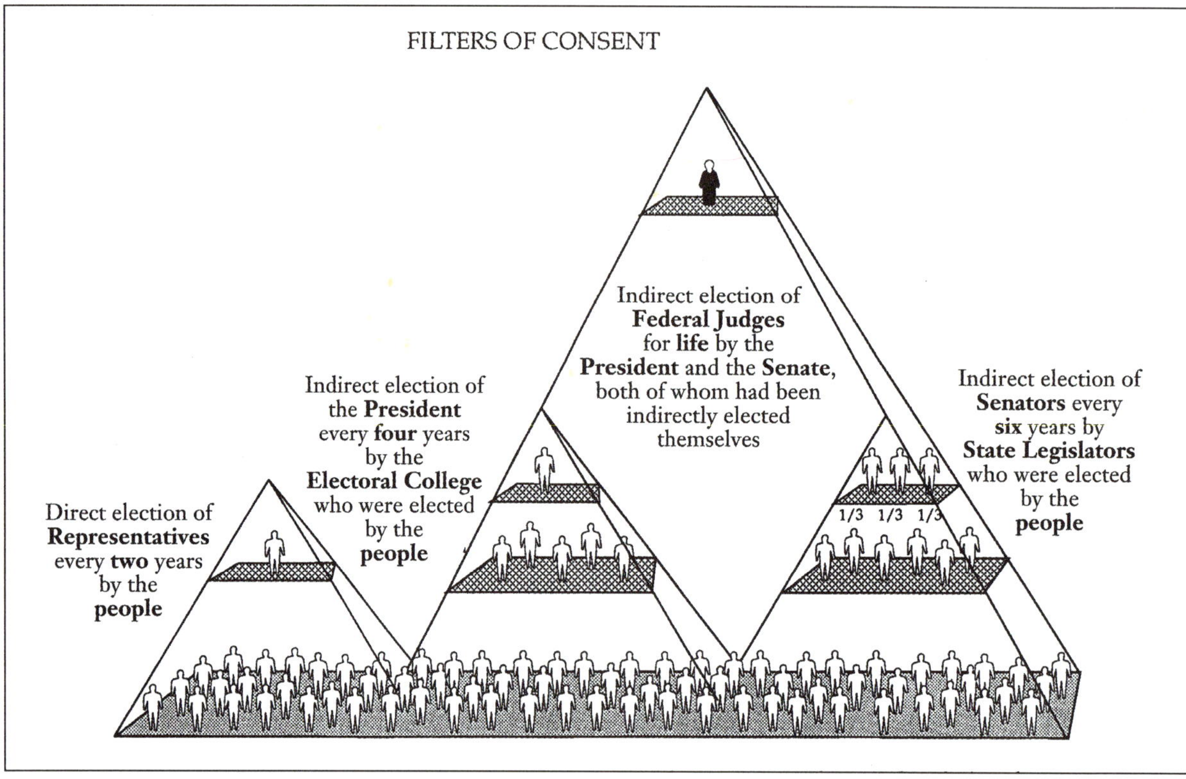

To maintain a popular government while ensuring against a democratic one, the Founders designed a complex system which would process the consent of the people through a set of filters. There were direct and indirect elections, staggered elections (in which one third of the Senate would be up for election every two years), and overlapping terms of office (two, four, six, life).

Those with few filters between them and the people would necessarily be closer and more accountable to the often capricious passions of the people. Those with more filters between them and the people would be further removed from the people, with greater opportunity for more cautious, reflective response to popular will.

case, representatives tend to be more virtuous than those who elect them because, as it happens, even the most rascally of voters dislike putting rascals in office. (For example, think about the outrage against President Clinton on the part of those for whom sexual transgression is commonplace.) The Scottish philosopher David Hume had called attention to such mechanisms, arguing that human nature was predictable enough to make them work. Indirect election (representatives choosing representatives), he said, could select virtue from the general populace the way a sluice refined gold from river sand.

The Framers found other uses for Hume's machinery. In the operation of the electoral college, for instance, electors chosen in the various states would meet together to select the President. The President, in turn, would nominate justices to the Supreme Court. These would have to be confirmed by the Senate, whose members were chosen by state legislatures. Such indirect methods were partly to insulate the electoral process from popular passions, true, but they were also to give Hume's "refining effect" fullest scope.

A second set of virtue mechanisms was provided by the principle of "fame." According to Aristotle, a public person's concern for his reputation would, under certain conditions, operate as a strong restraint on his conduct. What conditions? The conduct must be open and visible to the body politic, and the responsibility for it must always be clear.

The Framers focused responsibility in several ways. After extensively debating a plural executive, they decided on a single executive instead, reasoning that in the latter case "fame" would be a much stronger deterrent. The principle of enumeration also helped to pin the blame where it belonged by spelling out who was in charge of what. And the Framers required that all proceedings of Congress be open to the public and published for all to read: a congressman's floor speeches or voting record were not to be hid under a bushel.

A final set of virtue mechanisms was found in the last clause of Article VI, where the Constitution stipulated: "No religious Test shall ever be required as a Qualification to any Office or public Trust under the United States."

While a few of the Framers were religious skeptics, most were committed churchmen—and even the skeptics appreciated religion's power to promote civic virtue. The question was, how to bring such power to bear? Some delegates argued for a mild religious test, requiring officeholders to embrace mainstream Christianity.

Experience had shown, however, that religious tests, however mild, had the effect of cementing an official tie between church and state. Several of the Framers regarded such a tie as desirable. No republican society, they argued, could succeed without *some* official religion. But most of the delegates took the opposite view. Official religion worked two kinds of mischief, they said. It opened the way for the official church to corrupt the political system, and conversely opened the way for the political system to corrupt the official church. Those delegates representing states where a religious establishment still persisted could testify to the baneful effects.

By prohibiting a religious test, the Framers aimed not to destroy the influence of religion on politics but to purify it. The theory went something like this. All believers, now lacking any official capacity, were welcome to exert their influence on government in any way they saw fit—just as any citizens were. And since no denomination stood to gain by way of patronage, the motives of all must be relatively pure. The only real message a church might have under the Constitution would be to urge integrity, responsibility, and a sacralized civic virtue.

Most of the Framers had a strong confidence in the Constitution's virtue mechanisms. They believed that the national government would be more virtuous on the whole than the governments of the states, for it would draw upon the virtue of the entire nation.

Auxiliary Precautions

Yet the Framers were not willing to rely on republican mechanisms alone. Their view of human nature was rather skeptical, especially where politics was concerned, and they were not willing to assume the virtue of all public figures in

The Constitution:
http://www.britannica.com/eb/article?idxref=194987
http://www.jmu.edu/madison/gpos225-madison2/adopt.htm

all situations. "If men were angels," Madison wrote in an often quoted passage of *The Federalist*,

> no government would be necessary. If angels were to govern men, neither external nor internal controls on government would be necessary. In framing a government which is to be administered by men over men, the great difficulty lies in this: you must first enable the government to control the governed; and in the next place oblige it to control itself. A dependence on the people is, no doubt, the primary control on the government; but experience has taught mankind the necessity of auxiliary precautions.

The fundamental idea behind Madison's "auxiliary precautions" had emerged in the European Enlightenment. In simplest terms, it was that the power of self-interest could be utilized to control self-interest—control *itself,* in other words—the way the power of a steam engine could be utilized to brake a locomotive as well as drive it forward. The answer lay in structure once again.

The Framers termed their primary structural principle *counterpoise*. "Ambition must be made to counteract ambition," Madison explained. "The interest of the man must be connected with the constitutional rights of the place." What he meant was that through Hume-style structural engineering we saw above, it was possible to counterpose the self-interest of one player against that of another with the result that each would keep the other in line. "It may be a reflection on human nature that such devices should be necessary to control the abuses of government," Madison wrote, "But what is government itself but the greatest of all reflections on human nature?"

Runaway ambition was only half of the problem. The other half was what we might call group dynamics. Madison and several others had observed that a given group of individuals—say, an assembly of voters—could behave very differently in different circumstances. One "group personality," for example, might be hot and impetuous. The members could act with reckless abandon if their passions were stirred, if their condition was desperate, or if someone stood to gain from inciting them to panic. And the other personality might be cool and reflective. The selfsame voters could act with great deliberation, avoiding haste, even regretting and nullifying earlier misadventures, if the situation were just a little different. Think of the movie scene in which the hero dramatically halts the lynch mob, delivers a scolding, and sends everyone home feeling sheepish.

The Framers' constitutional engineering was thus twofold. First, they must make interest counter interest and ambition check ambition on the part of the political leadership. And second, they must incline toward the cool and deliberative side of popular judgment. Both ends could be accomplished through the careful design of structure.

Commonwealth of Oceana: Influential political work dealing with modern political ideas.

James Harrington had suggested a first step. In his influential *Commonwealth of Oceana* (1656), Harrington had described a good structure for creating counterpoise. Two children have discovered a small cake. They fall to squabbling about the division of the prize between them, each contending for more than a fair share. Without some means of reining in their self-interest, the conflict is bound to go on and on. But an adult arrives on the scene with a structural solution. One child shall be allowed to cut the cake while the other is given first choice of the pieces. Observe the magical transformation, Harrington wrote. Self-interest, which formerly promoted injustice (and led to anarchy), now moves both cutter and chooser to behave equitably. The Framers thought carefully about "cutters" and "choosers" in the constitutional process.

They also considered means of separating the warm and cool sides of popular judgment. Longer terms in office, for instance, worked for the latter. An officeholder who could remember back when a particular episode, or something like it, had occurred before was less likely to react precipitously. The size of a political assemblage also affected its psychology. Large bodies numbering into the hundreds would stampede more easily than small ones of twenty or thirty, for with the latter there was always a way for the voice of reason to be heard.

The Framers' main strategy, however, was to make it possible for—and even incumbent upon—certain officials in the structure to throw on the brakes when events began moving too quickly. In so doing, they necessarily forced others to stop and think. This was why the Framers came to put so much faith in "separation of powers."

Separation of Powers Mechanisms

Government embraces a wide variety of human activities. Passing laws, building roads, and pronouncing judgments in court are only three of the activities, but notice how dissimilar they are. In time, the various "powers," as they are called, were sorted into three general categories:

- *Legislative.* Enacting the laws that all must live by.
- *Executive.* Putting the laws into practical effect.
- *Judicial.* Applying the laws to specific cases and determining whether or not they have been violated.

In the classical republics, the legislative branch was held to be the most important of the three by far, for it was here that the sovereignty of the people became manifest. Stepping into the classical frame of reference for a moment, let us try to imagine how the executive and judicial branches might be viewed through the eyes of the legislature:

> *The executive.* This is the person charged with carrying out, or "executing," the legislature's will—a hired hand, so to speak. Suppose, for instance, that the legislative assembly decides to purchase new spears for the army. The

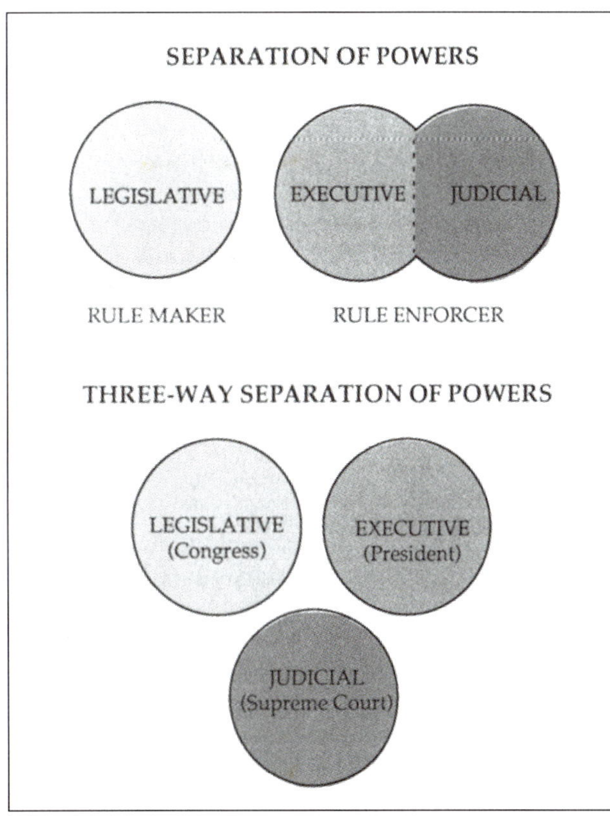

The traditional separation of Rule Maker and Rule Enforcer was modified by dividing the Rule Enforcer into two separate bodies (top) creating a three-way separation of powers (bottom).

> executive goes out and negotiates contracts with the smithies. He inspects the finished products, pays for them, and stows them away, returning to the legislature for further orders.
>
> *The judiciary*. This is the person charged with hearing and deciding cases. He too takes orders from the legislature. If, let us suppose, the legislature determines there is too much shoplifting in the marketplace, judges may be instructed to tighten up on the rules of evidence, throw out sob stories, and run up the sentences. They meekly obey.

Lest such thinking seem archaic, state governments commonly operated in this same manner before the Constitution was adopted, and some of them continued to do so afterward. It was considered to be very "republican."

It was this very system, however, that created much of the tyranny and anarchy of the ancient world—to say nothing of the American state governments—and for that reason Aristotle came up with an entirely different approach. We need to consider it too because it was equally influential when the Constitution was written. Aristotle associated the three departments of government with the three "estates" he found in most social hierarchies. Hence "the one," "the few" and "the many"—loosely connected with monarchic, aristocratic and democratic elements of society—could be equated with the executive, judicial and legislative branches respectively, and could be mixed in the structure of government in such a way as to check and balance one another. The one and the few, Aristotle supposed, would tug against the wishes of the many and vice versa, so that each would tend to hold the others in line. This was the origin of "counterpoise."

Aristotle's mixing and balancing of the estates became a central feature of English government. Crown and Lords checked one another, so the theory went, and in turn combined to check Commons. Several of the American Founders accepted this approach as a given. In some way, they believed, a republican government must be designed so as to play off class against class, and combine the psychology of the one, the few and the many into a dynamic working relationship.

The more modern idea of "separation of powers," which we encountered in chapters 5 and 7, seems outwardly similar to Aristotle's model, but it is actually quite different. It began as a critical insight. Observers of the English system noted that it did little good to mix and balance the various departments of government if one of them always proved to be stronger than the others and thus always succeeded in getting its way. One might as well go back to the all-powerful legislature ordering around its spear buyers and thief catchers, for in the end it came down to the same thing—as the American colonists had learned so well.

With this puzzle in mind, Montesquieu had begun to write about "separated powers." He supposed that if the various departments of government could somehow be separated from one another, such that none could directly control the others, the result might benefit the rule of law. But such separation would be tricky. How could a government coordinate its activities if each of its branches were a power unto itself? What would prevent the system from simply gridlocking? And how, in a practical sense, were the branches actually to be forced apart? Montesquieu had struggled with such questions. So did the Framers in Philadelphia.

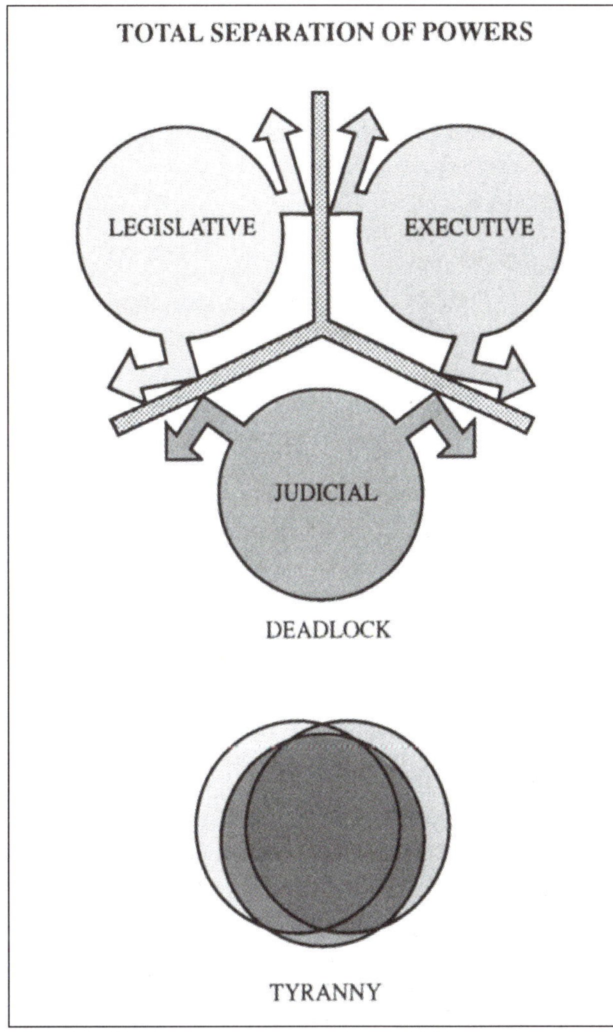

With total separation of powers, there is a risk of deadlock (top) or that one of the powers will obtain total control (bottom).

> *The accumulation of all powers, legislative, executive, and judiciary, in the same hands, whether of one, a few, or many, and whether hereditary, self-appointed, or elective, may justly be pronounced the very definition of tyranny.*
>
> —*James Madison*

Gradually, however, key delegates began to get hold of the idea. They had been trying to think of ways to make the government powerful enough to do the job required of it without making it too powerful for peace of mind. Separation might be the answer. If the legislative and executive branches could be pushed apart in some way without being driven to cross-purposes, each might become strong enough, and independent enough, to set up a working counterpoise. The executive would still carry out the laws passed by Congress, but he would not be just an errand boy. He would do his job blindly, as it were, without reference to any political purpose of the legislators. And vice versa. Once a law was passed, it would be out of Congress' hands entirely—so there was no point in shaping it so as to "get" someone.

But how to make the powers separate? State governments had struggled with the problem unsuccessfully, as we saw in chapter 5. Then the Massachusetts Model Constitution had come up with some answers. The key had proven to be the conditional veto, which John Adams had argued for in *Thoughts On Government*. Adams hadn't been keen on separation of powers. He had been thinking more along the lines of Aristotle's mixing and balancing approach. But strengthening the executive and giving him a negative on the legislature had indeed made separation possible.

The Framers were beginning to understand why. By enabling the executive to defend himself against aggressive legislators, the veto power allowed him to exercise an independent will. Yet it could never be *completely* independent because of the override provision. And, conversely, the legislature could protect its own integrity against an ambitious executive—simply by overriding his veto. So *its* will remained independent too.

Some Framers still weren't convinced. You *compromise* separation of powers by making the executive a kind of negative legislator—and that makes separation work? You hand an executive a timeworn instrument of blackmail and tyranny—and that creates freedom? It seemed like a parlor trick. At best, said the doubters, it would lead the executive and legislature to thumb their noses at each other and refuse to budge. But James Wilson didn't think so. With his portly bearing and cool self-assurance, Wilson always seemed to know how a baffling system would *really* work, and the inclination here, he thought, would be not toward deadlock but compromise. In the next chapter, we will work through a hypothetical example of what Wilson meant.

Beyond the conditional veto, there were other things the Framers could (and did) do to make the separation of powers effective. They could allow each branch to determine the qualifications of its own members. They could enable each to lay down the rules of its own proceedings. They could make it possible for each branch to provide for its own funding, its own officers and support services, its own physical and institutional location. Not least, they could enumerate a specific list of powers for each, so that each would have its own separate sphere of authority.

There was another odd kink. When the design of the national government was complete, a number of "checks and balances" were added to the structure that required the sharing, rather than separating, of power. These too violated the separation doctrine, it would seem, but they really didn't. For they enhanced the *purpose* of separation. That purpose was to enable the branches of government to exercise an independent will, to check potential incursions into each other's domains, and to provide a functional counterpoise—*without working at cross-purposes*. It was, in Richard Neustadt's phrase, "separated institutions sharing power."

Separation of powers created a whole new approach to republican government. Instead of mixing and balancing the various estates of a hierarchical society, it became possible to base *all* branches of government on the people as a whole. In time, this breakthrough would have far-reaching, even transforming consequences.

Separation of Powers Design

Separation of powers enabled the Framers to produce a coherent structure of government. Each functional department became a separate branch, and each branch could be strong, independent, and clearly defined. Given such a design, the government could be one of great "energy" without threatening personal liberty.

- Strong Legislature

Congress was granted the following enumerated powers by the new Constitution:

→ To lay and collect taxes, duties, imposts and excises.
→ To borrow money on the credit of the United States.
→ To regulate commerce with foreign nations and among the several states.
→ To establish uniform rules of naturalization and uniform laws of bankruptcy.
→ To coin money and regulate its value.
→ To fix the standard of weights and measures.
→ To provide for the punishment of counterfeiting.
→ To establish a postal system.
→ To provide patent and copyright protection.
→ To set up a federal court system inferior to the Supreme Court.
→ To define and punish felonies on the high seas.
→ To declare war.

- → To raise and support armies, and to provide and maintain a navy.
- → To make rules for the government and regulation of military forces.
- → To provide for militia forces when in service of the United States.
- → To govern the District of Columbia.

When reviewing this list there are two things to keep in mind. First, it represented an enormous increase in the authority granted to the old Confederation. And second, it is extremely limited when compared to the powers commonly exercised by nation-states today. As the Framers saw it, there was power to do what had to be done—and hopefully not a great deal to spare.

And there was a catch. In addition to the enumerated powers, Congress was given the authority "to make all Laws which shall be necessary and proper for carrying into Execution the foregoing Powers. . . ." This clause turned out to be the Constitution's wild card and was destined to be played in many a tense hand of politics. Since no one knew what might be judged "necessary and proper," the way was open to argue for just about anything.

- Strong Executive

Under the Constitution, the President was responsible for executing, or carrying out, the laws passed by Congress. This activity, on a continental scale, would require the efforts of a vast number of functionaries, whom the chief executive could hire or fire at will. Moreover, think of all the suppliers of goods and services to government projects, the management of government properties, the handling of government assets. There was a treasure house of personal power here in the form of *patronage*.

Furthermore, the President was made commander in chief of all military forces and given charge of foreign affairs. Both arenas were of crucial importance and both, historically speaking, had been the proving grounds of the Caesars. The President, in other words, was to be much more than a mere "executive." His office had a vast potential for developing charismatic leadership, for molding public opinion, for embodying the will of the nation. The President could be a kind of king.

- Strong Judiciary

The separation of powers mechanisms worked differently for the judicial branch of government. By its very nature, the judiciary exercised no independent will. It could not advance its own aims, urge its own program, reward its friends or punish its enemies. It could not even enforce its own decrees. Yet the judiciary was of utmost importance to the rule of law. If the legislative branch was the essential rule maker, the court system was the ultimate rule enforcer, for it applied the law to specific cases and determined whether or not violators would be punished.

The Framers' objective here was to insulate the judicial system from all means by which the other branches had tampered with it in the past. Supreme Court justices were provided a lifetime tenure of office and a guarantee that their salaries would not be reduced. The penal system was placed under their supervision, and they were given charge of their own bureaucracy. The Consti-

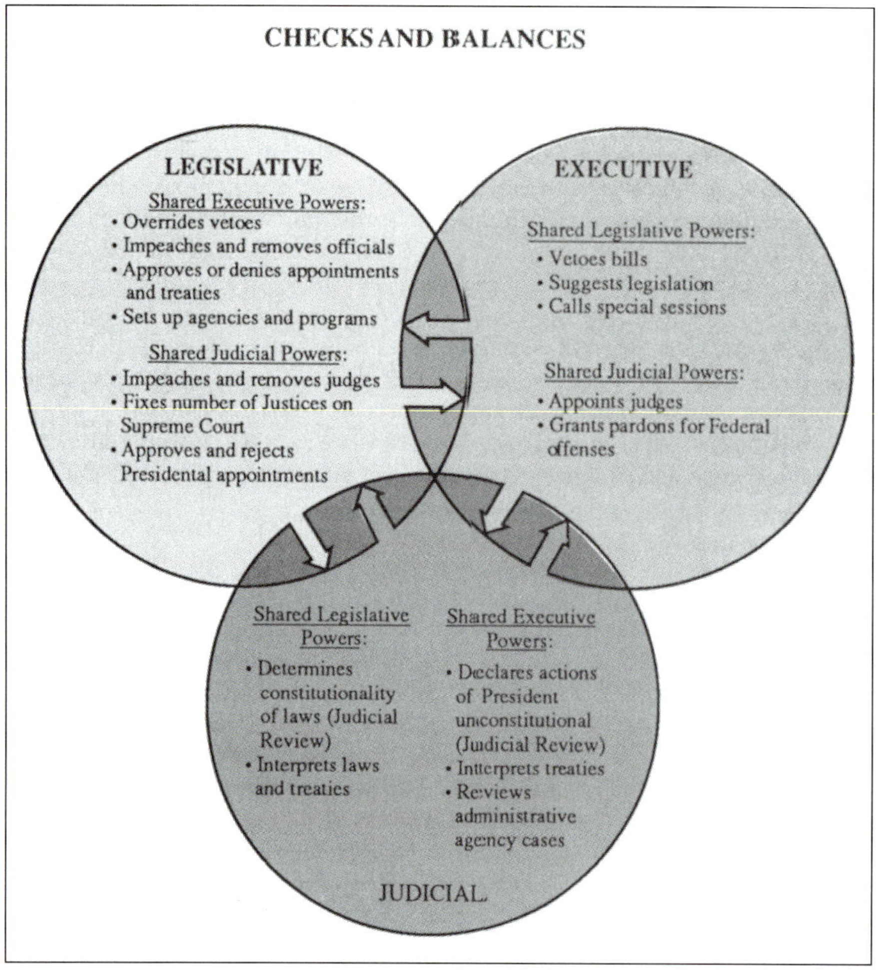

Interesting Paradox: The same kind of powers must be shared to maintain separation of powers.

tution made it very difficult for politics to subvert justice.

The federal court system has proven to be as strong as the Framers hoped it would—and for some a good deal stronger. Isolating the justices from the political fray and faithfully (for the most part) carrying out their decisions created a situation in which the judiciary could truly play the role of referee.

Check and Balance Mechanisms

As the Framers began laying out the structure of government, they repeatedly asked themselves how a given constitutional device was likely to play upon human nature. For instance, in *Federalist* 51, Madison argued the case for separation of powers in these words: "The great security against a gradual concentration of the several powers in the same department consists in giving to those who administer each department the necessary constitutional means and personal motives to resist the encroachments of the others." We will soon see how those "constitutional means" and "personal motives" would intersect when the Constitution was put into practice—almost precisely as the innovators anticipated.

In the political psychology of separated powers, means and motives often led to wariness between the branches. All the threats to veto or to override could make the governing process, in Woodrow Wilson's words, "at war against itself." The political psychology of *checks and balances* was quite different. In most of these situations, power was shared between two different branches or two different authorities in the same branch: in order to perform a given function, both would have to agree. To be sure, a bit of the veto psychology also applied to these situations, for either branch could essentially veto a contemplated action by refusing to perform its part. To a much greater extent, though, the branches were induced to cooperate with each other, as we will see in the following chapter.

The Constitution's principal check and balance mechanisms are as follows:

- Bicameralism

Congress was divided into two separate chambers, the House of Representatives and the Senate. Such a division would have taken place even without the Great Compromise, for the Framers were strongly in favor of its value. John Adams had staked his career on bicameralism, and it had proven its worth in the Model Constitution.

The Framers assumed that the House of Representatives would embody the "mind or sense of the people at large," as James Wilson put it, and "be the most exact transcript of the whole society." Accordingly, it must be kept close to the average voter and designed to reflect his momentary hopes and fears.

Original size of districts in Constitution.

Legislative districts of roughly 30,000 voters would each elect a single member to the House. Even though these districts lay within the boundaries of existing states, they were small enough with respect to most states (and would get progressively smaller as state populations increased) to focus upon discrete local interests. Thus, while a given member might represent a commercial constituency, his neighboring district might be industrial or agrarian. The interplay of these interests was expected to be lively. As various issues arose, the representatives would group together with little regard for state boundaries and have it out.

To keep the representatives in touch with public opinion, terms were kept short at two years—meaning that every congressman must account to the voters on a regular basis. Thus, as economic and political circumstances changed, the complexion of the House would change as well. There would always be something fleeting and transitory about the lower chamber, reflecting the emotional side of the body politic. In Madison's simile, it would be "hot."

The upper chamber, by contrast, had to be "cool"—the cream one poured into scalding coffee to make it drinkable. The psychology of the situation would be something like this. The people had their impulsive side, true, but they had their deliberative side as well. If the "cool" Senate caused them to pause and reconsider hasty measures, they might well change their minds. Put another way, the Senate might help the body politic to distinguish between apparent interest, short term, and genuine interest, long term.

But how could a constitution fill the upper chamber with such clear heads? It is easy to figure out ways of electing the tallest, the shortest, the smartest or whatever—but the *wisest*? The Framers struggled with this problem throughout the summer, and by September they were no nearer a solution than in May. In the end, they had to settle for an odd second best—wealth. By requiring senators to be elected by the various state legislatures, the Framers realized that wealth might well play a role in the outcome. And wealth wasn't wisdom. But at least the wealthy had a sizable stake in society, and for that reason would resist facile innovation. Furthermore, because the senators were one further tier from direct election, they would hopefully benefit from Hume's refining effect.

The lengthy (six-year) terms for the Senate had a similar effect, as did the comparatively small size of the body. Senators would know one another well and share a long institutional memory. If the system did not produce wisdom per se, it certainly did produce cool cream for the hot coffee. Historically, many measures of momentary popularity have sailed through the House of Representatives only to run aground in the Senate.

Bills could arise in either house, but to become law they had to be passed by both. The single exception was tax bills, which had to be initiated in the House. Since it was the people as a whole who had to pay the taxes, their representatives were given the right to settle on the amount required.

The two houses naturally checked and balanced one another. Sometimes they were thick as thieves and sometimes they were hardly on speaking terms. Whatever their relations, any piece of legislation that passed through both of them pretty well had the backing of the people.

- Power of Appointment

The appointment of many high-ranking officials—Supreme Court justices, ambassadors and ministers, Cabinet officers and the like—was to be made by the chief executive but subject to the "advice and consent" of the Senate. Advice and consent, especially the latter, altered the psychology of appointment considerably. The President would have a hard time installing mere cronies in high places if he must go before the Senate and explain each nomination.

Some of the Framers feared that extortion would result from this arrangement—that the Senate would blackmail the President into nominating people of its own choosing by refusing to confirm any others. This fear has proven as groundless as the obverse fear that the President would blackmail Congress with his veto power. In both cases, the players' personal motives are inclined by the structure toward a wary cooperation, for embarrassment afflicts both sides when a nomination fails—especially as a result of "political maneuvering."

- War Making and Foreign Policy

The Constitution gave the executive authority to send and receive ambassadors and ministers, and to conclude treaties with foreign powers. These provisions placed American foreign policy squarely within the executive domain.

Since foreign war had spelled the downfall of many republics, the Framers were deeply concerned about these powers. Some feared that the President might be tempted to take bribes from alien governments—or even from certain states—so checks and balances were used here as well. While the executive could negotiate a foreign treaty on his own, it had to be ratified by a two-thirds majority of the

Senate. As for war making, Congress alone was given the authority to declare war, but once that was done it was the President who took charge of the fighting.

Personal motives could become frightening here. Some Presidents have sought to avoid Congress (and hence the public) by conducting military operations without a declaration of war. Others have forced Congress' hand by placing US forces in harm's way. But Congress has fallen back on yet another check in the system—the power of the purse. By refusing to appropriate funds for a military operation, Congress can—and occasionally does—lodge a convincing negative.

- Powers of Investigation and Impeachment

While the Constitution did not specifically endow Congress with investigative power, such power inheres in the legislative process. For in order to frame a measure and debate its merits, the lawmakers must be able to consult experts, call in witnesses, and examine relevant data.

This was yet another check and balance, complete with its own set of personal motivations. Individual congressmen have made—and occasionally broken—their political careers by conducting investigations. Whether specific action follows or not, the mere existence of an investigation serves to capture public attention—and puts public officials on notice.

The impeachment power illustrates how delicate the check and balance mechanisms can be. Three grounds for impeachment were included in the Constitution: treason, bribery, and "other high crimes and misdemeanors." The first two were well understood. The third was intended as a catchall, and was added only after long discussion.

But what, exactly, did the phrase "high crimes and misdemeanors" mean? One could argue that an executive who failed to execute the laws faithfully (Article II, Section 3) might well be in violation, yet if the executive simply marched in step with the legislature he would not be exercising an independent will. If Congress violated the Bill of Rights by, say, passing a law to silence critics, would a President be guilty of "high crimes and misdemeanors" for refusing to carry it out?

All the Constitution's check and balance mechanisms were ambiguous in some way. They required historical experience, established precedents, and even adjudication in the courts before their operation would become settled. They also complicated political operations and encumbered the operators with all sorts of new calculations. Political novels such as *Advise and Consent* or *Clear and Present Danger* illustrate how entangled constitutional situations can become.

Critics have argued that so many "auxiliary precautions" make for a Constitution that is "all brakes and no engine." That may be true. It was not designed for efficiency, however, but effectiveness—against despotism. And in this purpose it has served us well.

Federal Mechanisms

The "federal" system of dividing government between sovereign states and an equally sovereign central authority was entirely a product of circumstances, as

Advise and Consent

For those who have not experienced it, the psychology of divided power is difficult to describe. It is a curious blend of hostility and friendliness, cooperation and obstruction. Allen Drury's 1959 novel, *Advise and Consent*, attempts to explore that psychology in detail. In the story, the US President, himself a rather doubtful character, has nominated the even more doubtful Robert Leffingwell to be Secretary of State. The novel goes on to probe the minds and hearts of the Senators who must now either confirm or oppose the nomination.

In the following excerpt, the President has just told Senator Orrin Knox that the Soviets have stolen a march on the United States and landed men on the moon:

> "It is against that background," [the president] said quietly, "that I must ask you to permit Bob Leffingwell to be confirmed."
>
> Senator Knox stared at him thoughtfully.
>
> "How do I know you're telling me the truth?" he asked calmly, and the President looked at him in some disbelief.
>
> "Do you think I would fabricate something as lurid as that?" he inquired dryly.
>
> "You might," Orrin said. "But," he added quickly, "I don't think you are. I fail to see, however, why it should change my attitude about your nominee. If anything, I should think it would make me even more determined that he not be confirmed."
>
> "I feel he is what we need in this situation," the President said with a quiet insistence. "It is inconceivable to me that you would stand in his way under the circumstances. I am asking you as a patriotic American to permit him to assume the office."
>
> "Well, by God," Senator Knox said sharply, "so now it's patriotism, is it? Well, let me tell you, Mr. President. To me it's patriotic to do what I deem best in my own judgment for the country; it isn't to give in to you and let you ride roughshod over everything decent just because you claim it's patriotic and imply that those who oppose you are unpatriotic. What kind of damned slippery argument is that?"
>
> "Senator," the President said, "do you have any conception of what I have just been telling you?"
>
> "I have a conception," Senator Knox said shortly. "I also have a conception of what is decent and honorable and best for the country which I think is just as good on the moon as it is on earth or Venus or Mars or any other place we're going to go to. And my conception had no room for Bob Leffingwell and his wishy-washy attitudes toward the mortal enemies of the United States. Why, good Lord. You know as well as I do exactly what they're going to start doing tomorrow. They're going to start pressuring us as they never have before. And you want an obliging stooge like Bob Leffingwell to deal with *that*? What kind of conception do *you* have?"

In the end, of course, all is well for the United States. With all its bumps and abrasions, the system really does work.

From America: A Study in Heritage, *Kendall/Hunt 7th Edition, p. 138.*

we have seen. Doubters were not slow to point out that federalism was unplanned, unprecedented and almost unimaginable.

But not undefensible. James Wilson, who had fought to the bitter end against the Great Compromise (and thus against federalism), later came out with brilliant justifications for it. Federalism, he argued, offered nothing less than a whole *new* approach to the control of tyranny.

- Federal Separation of Powers, Checks and Balances

Just as there was a separation between legislative and executive branches in the national government, there was another separation between the powers enumerated in the Constitution and the powers reserved to the states. The states could no longer conduct diplomacy or field their own armies, true; but the central government could not license taverns or arrest thieves either.

Furthermore, the two levels of government operated as a check and balance against each other, sometimes pulling together and sometimes tugging apart. The states checked one another as well, by means of political activity in the Senate. If a certain state or a certain region became too strong, there was a more or less automatic movement in that body to counter the buildup. In the curious architecture of the Senate, small states could hold their own against large ones, and isolated states could band together against regional concentrations.

- Federalism and the National Interest

The checking of regional power worked against the formation of dangerous majorities. In the House, regional majorities formed regularly. Wherever there was a concentration of population—usually in the Northeast—voting in the House was likely to become lopsided, arousing fears of sectional dominance.

But in the Senate such a concentration was almost impossible. For example, when the Constitution was adopted, four populous northeastern states (Massachusetts, Connecticut, New York, and Pennsylvania) could muster 27 votes out of the 65 possible in the House, perilously close to a majority, whereas in the Senate they could account for only 8 of 26. Any majority in the Senate had to be national rather than regional in character.

As soon as the various regions felt secure from a hostile majority, they could afford to relax. The protection of regional interest thus made it possible to articulate a purely national interest—something Americans had rarely felt under the Articles of Confederation. The people could rejoice in the peace of their greater republic and turn full attention to its promise.

Other Features of the Constitution

Beyond the Constitution's auxiliary precautions, there were other features of importance, and they too would exert a shaping influence. Before touching on a few of these, we should point out that the inclusion of a single phrase—even a single word in some cases—could (and did) have enormous consequences in the evolution of institutions.

Suffrage Qualifications

Virtually all of the states qualified the right to vote according to some standard of property ownership. For in the eighteenth century, "the people" didn't include *all* people by any stretch, only Aristotle's people of substance and account. In some states the requirements were quite restrictive while in others they were comparatively liberal. The Framers discussed voting qualifications extensively. Democrats among them argued for liberal standards in the name of a more inclusive democracy. Others hung back. In the end, they settled for the same qualifications as the several states. That is, anyone able to vote for an assemblyman in, say, Massachusetts could also vote for a US representative in the same state.

Advocates of democracy have seen the suffrage clause as restrictive and "aristocratic." Yet it might have been precisely the opposite. Many state governments were substantially democratic already and growing ever more so. If the Framers wanted to use the Constitution as a bulwark against this trend, they forfeited an easy opportunity to do so by not imposing an independent federal standard.

Commercial Regulation

The felt need for commercial regulation was one of the driving forces behind union. Thus, the Constitution's famous commerce clause. It was short, vaguely worded, and virtually without boundaries. It gave Congress the authority to "regulate Commerce with Foreign Nations, and among the several States, and with the Indian Tribes." Elsewhere, the states were enjoined from slapping their own tariffs on trade, as they had done in the Confederation.

With Congress taking charge of commercial regulation, the character of the regulation was bound be broad and general. It could not place a given state or group of states at a commercial disadvantage vis-à-vis their sisters, for to do so would run afoul of all that machinery outlined above. The effect was to create the largest—and eventually most prosperous—free-trade zone in the world.

Protection of Property

Historians have occasionally accused the Framers of devising a "backlash" plan of government expressly to protect property. Clearly the charge sticks and for reasons we ought to know well by now. For the Framers, *property* and *rights* were two sides of the same coin. A government unable to protect citizens' property could assuredly not protect their lives or liberties either, and come to that, one of life's greatest liberties was enjoying property to the full. Besides, a government based on ownership rested on people with something to lose.

The Constitution lodged the protection of property in several places. There were the implied limits on political participation described above. There was the protection of all minorities—including the wealthy—by means of the structural system. There was patent and copyright protection. There was even a specific prohibition against the states passing legal tender laws or impairing the obligation of contracts. And finally, there were the Constitution's notorious slavery clauses—which showed how far fifty-five property owners were willing to go.

Elasticity

One of the Framers' chief difficulties was that posed by the future. How to write a Constitution in 1787 that would still work in 1887 or 1987? While some regarded the task as impossible, others were more sanguine. If the Constitution's draftsmanship was kept reasonably brief and general, they thought, it might operate successfully for quite a while.

Accordingly, while the Framers pinned some things down very carefully, they left others vague and uncertain. Their strategy was to make the structure clear and concise while leaving powers and authority a little ambiguous.

They realized that sooner or later the Constitution would have to be amended. Just not frivolously. The process they set up required agreement by two-thirds of the Congress (both houses) *and* three-fourths of the states. Three-fourths is an exceptionally large fraction in politics. It required that the Constitution could only be altered by a substantial majority of the people nationally distributed. For every one of the twenty-seven amendments presently ratified, dozens have failed ignominiously, many for the lack of just one or two states.

At the same time, the Framers seem to have recognized that certain kinds of change—political, social, economic, technological—would probably require a subtler process of adaptation. The Constitution, in a word, would have to be stretchable.

The trick was to keep stretchability within limits. If the document proved to be too flexible, it wouldn't possess the detentive capacity necessary for constitutional government. And if it wasn't flexible enough, it would eventually have to be discarded. With such considerations in mind, the Framers appear to have included a discrete number of "**stretch-points**" in the constitutional text. These consisted of words or phrases so vague, general or ambiguous as to be subject to a variety of interpretations. What was "advice and consent?" "High crimes and misdemeanors?" "A republican form of government?"

Two of these stretch-points are particularly worth keeping in mind. The "necessary and proper clause," Article 1, Section 8, has already been mentioned. It would come up again and again in the course of American history. Similarly, the commerce clause, Article 1, Section 8, would fill the law books with adjudication as it was stretched first one way and then another by the Supreme Court.

Supremacy

The whole point of the Constitution, for the nationalists at least, was to provide a system of government able to counter the follies of the states. But how, exactly, to accomplish this? Within their own sphere of sovereignty, the states were still free to behave as they wished.

For James Madison, the only solution was for the national government to have an absolute (as opposed to conditional) veto over all state laws. But this concept was troublesome. What became of state sovereignty, Madison's colleagues asked, if the national government could strike down state laws at will? And what happened to those delicate balances of federalism?

Articles of Confederation versus Constitution

GOVERNMENT UNDER THE ARTICLES OF CONFEDERATION:	GOVERNMENT UNDER THE CONSTITUTION:
1. States are sovereign.	1. People of the whole nation are sovereign—federal union divides exercise of sovereignty between central government and states.
2. No independent executive.	2. Independent executive—chosen by electors.
3. No federal courts—national laws are enforced by state courts.	3. Separate federal court system, with power to consider Constitutional questions and resolve disputes between states.
4. No taxing power in Congress.	4. Congress may "lay and collect taxes, duties, imposts, and excises".
5. Congress has no power over interstate or foreign commerce.	5. Congress has power to regulate commerce with foreign nations, among the several states, and with Indian tribes.
6. Congress is an assembly of delegates chosen by state legislatures—delegates may be recalled at any time.	6. Congress is composed of representatives who have definite tenure—House is chosen by popular vote, Senate by state legislatures.
7. Articles may be amended only by the consent of all of the states.	7. Constitution may be amended with approval of three-fourths of the states.
8. Congress has only specific, delegated powers.	8. Congress has implied as well as specific delegated powers.
9. The central government cannot act directly upon the people.	9. The central government exercises its power directly upon the people, and concurrently with state governments.

From America: A Study in Heritage,
Kendall/Hunt 7th Edition, p. 158.

There were arguments on the other side as well. Of what possible use would the federal Constitution be if the states could simply ignore it? All the springs and pinions of the mechanism presupposed that national authority would have to be final.

The Framers struggled with this problem for months. Ultimately they had to settle for another compromise and more vague language. "This Constitution, and the Laws of the United States which shall be made in Pursuance thereof . . . shall be the supreme Law of the Land," it said, adding: "any Thing in the Constitution or Laws of any State to the Contrary notwithstanding."

As it turned out, the absolute and unqualified supremacy of the federal government was not established until the end of the Civil War—and then only on the battlefield.

The Constitution and the Republican Problem

We began this chapter with the Republican Problem. The first element of that problem, we recall, is to express the sovereignty of the people, as determined by numerical majorities. The Constitution clearly and consistently accomplished this end.

The second element is to protect the rights of minorities. Many of the Constitution's critics feared that it would never succeed at this decidedly more delicate task. Indeed, the movement for a bill of rights grew precisely out of that fear. It is worth noting, however, that the Framers, who were divided on so many issues, were virtually unanimous in believing that the constitutional mechanisms by themselves afforded ample protection to minorities. That was, after all, the whole point. And while it is true that the Constitution has not foiled every attempt at injustice, neither has the Bill of Rights. Together, they have foiled a very large number.

The third element of the Republican Problem is to act in behalf of the public good. Such good encompasses both majority *and* minority benefit, and thus can be extremely elusive. In short-run situations—where the public good is typically sacrificed—a republican government must be strong enough to resist the majority and wise enough to know when to do so. The American Constitution's rallying of virtue, together with its precautionary mechanisms, have often (though not always) made that strength-and-wisdom combination possible. As for long-run situations, exactly what the public good is and how it ought to be determined have provided the central text of our political drama from that time to this.

The Sun Rising

On September 8, the Grand Convention's last ad hoc committee was appointed. The Committee of Style and Arrangement, as it was called, was given final charge of drafting the Constitution. The Framers were commonly referring to their work by that name now, a "constitution," not just a "plan" or a "proposal." Gouverneur Morris was the committee's guiding spirit. He would later boast that he had written every word of the finished document.

Which wasn't quite true. For no sooner had the "finished" document been submitted to the General Assembly on September 12, then the debate broke out anew. This word was wrong. That phrase needed fixing. There was a clause over here that might be misunderstood in light of another clause over there. And a few of the Framers still had pet projects—a canal over the mountains, a national university—that deserved one last try for inclusion. The pulling and hauling went on right up to the end.

Saturday, September 15, was the Convention's last working day, and like so many before it, it was filled with trauma. The bitterest enemies of nationalism—with the exception of Luther Martin—had long since departed for home. Those left were the Constitution's friends, or so they supposed. There were some nasty surprises in store.

Elbridge Gerry was the first of them. He stood up and recited an eleven-point critique of the Constitution. His real difficulty came down to fear of a peacetime army, which was too great a temptation, he said, for a republic to withstand. Reluctantly, Gerry declined to affix his signature.

While Gerry's demurrer was not completely unexpected, Edmund Randolph's was. Randolph had submitted the Virginia Resolves in the first place, and had stood with the nationalists ever since. Now, with heavy heart, he said, he must withhold his own support of the final product. The federal government was too strong, he feared, its powers too vague and indefinite, and it posed too great a threat to the states. Randolph was governor of Virginia and a man finely tuned to his constituents. He believed the Old Dominion would never ratify the Constitution as it stood—and would never forgive those who did.

There was a palpable heaviness in the hall when Randolph spoke, and it grew heavier still when he was followed by George Mason. The source of Mason's complaint was depressingly familiar—too much bending and crimping, too many lapses from republican purity. The government they had sketched out "would end either in monarchy or a tyrannical aristocracy," he concluded. Mason, like Gerry, had wanted contradictory things in the Constitution. He had wanted a national government so weak that it could never endanger the people's liberties—and he had wanted to trim a few of those liberties himself.

George Mason favored laws to limit personal expenditures on fabrics, dress and furniture.

All three of these delegates had made strong contributions. They had served on the premier committees and taken part in the important debates. Madison was visibly shaken by their desertion. The Virginians had come to Philadelphia to create a strong national government, he supposed, and now, apart from the passive Judge Blair, only two of them would sign the final document—himself and George Washington.

Ben Franklin sought to place the reversals in a broader context. He had prepared a little speech of his own, to be read by James Wilson the following Monday. This was the last day of the Convention, and most of it was to be ceremonial. Franklin began with reference to that pall of gloom which had accompanied Saturday's defections—the fear that the Constitution had compromised too much. "Mr. President," he began:

> I confess that there are several parts of this constitution which I do not at present approve. But I am not sure I shall never approve them. For having lived long, I have experienced many instances of being obliged by better information or fuller consideration, to change opinions even on important subjects, which I once thought right, but found to be otherwise. It is therefore that the older I grow, the more apt I am to doubt my own judgment, and to pay more respect to the judgment of others. . . .
>
> In these sentiments, Sir, I agree to this Constitution with all its faults, if they are such. . . . I consent, Sir, to this Constitution because I expect no better and because I am not sure that it is not the best.

Though nearing the end of his astounding career as an American renaissance man and political savant, Benjamin Franklin contributed both his wit and wisdom to the unfolding of events at the Constitutional Convention.

Franklin was equally appropriate off the cuff. "Whilst the last members were signing [the Constitution]," Madison penned in his now book-length notes:

> Doctr. Franklin looking towards the Presidents chair, at the back of which a rising sun happened to be painted, observed to a few members near him, that painters had found it difficult to distinguish in their art a rising from a setting sun. I have, said he, often and often in the course of the session, and the vicissitudes of my hopes and fears as to its issue, looked at that behind the President without being able to tell whether it was rising or setting: But now at length I have the happiness to know that it is a rising and not a setting sun.

The sentiment was light and charming. All the same, the old man wept when he signed the Constitution.

CHAPTER 8 STUDY AID

Structures designed to solve the "Republican Problem"

1. Fundamental Law

2. Written Constitutions

- Fundamental Law→Constitution created by the People→Government
- Holds government responsible
- Corrects constitutional drift

3. Enumeration

- Holds government within limits
- Broad language
- Specific language

4. Republican Mechanisms

- Popular sovereignty: original and periodic consent
- Representation: attorneyship and virtual
- Virtue Mechanisms: representative principle, fame, and no religious test

5. Auxiliary Precautions

- Counterpoise

6. Separation of Powers

- Total separation of powers alone→either deadlock or tyranny

7. Checks and Balances

- Add checks and balances to separation of powers→incentive for cooperation and compromise while maintaining separate, independent branches
- Bicameralism
- Power of appointment
- War-making and foreign policy decisions
- Powers to investigate and impeach

8. Federal Mechanisms

- Federalism

9. Other features of the Constitution that address the "Republican Problem"

- Suffrage qualifications
- Commercial regulation
- Protection of property
- Elasticity: stretch-points
- Supremacy: national or state?

CHAPTER 9

The Meaning of the Constitution

When it appeared that the ratification of the Constitution might come down to the pivotal state of New York, Alexander Hamilton began to pen a series of newspaper essays for fellow New Yorkers. He would describe how the proposed system would work and then go on to explain why it would work, why it would mobilize virtue and contain self-interest, would empower the people while protecting their rights—would make possible the Good Society.

Hamilton was soon joined in this effort by friend and colleague **John Jay**, and then by the indefatigable James Madison, who had learned so many of the answers the hard way. Indeed, before ***The Federalist*** was completed, Madison would be their principal author. As far as readers were concerned, however, all eight-five essays were written by a single disembodied intellect named "Publius."

John Jay was a prominent New York statesman and first Chief Justice of the US Supreme Court. Wrote only 3 of the 85 essays in *The Federalist* due to ill health.

The Federalist was the deepest and most searching inquiry into republican government that had ever been made, and for many scholars still the best. They chartered a tradition of constitutional analysis that still flourishes today, for Publius would eventually be joined by a distinguished company of historians, political scientists, legal theorists and assorted *philosophes*, all of them probing the mysteries of a single parchment.

Federalist Papers:
http://www.yale.edu/lawweb/avalon/federal/fed.htm
http://www.leftjustified.com/leftjust/lib/sc/ht/fed/index.html

What follows are but a few of the insights of Publius and his friends.

The Constitution and Human Nature

Even though the Framers did not write the Constitution for mankind as a whole, they believed it must be "true" to human nature. Had they gotten that wrong, as one historian observed, their project would have undoubtedly been short lived. While they had no formal debates on human nature, their discussions touched on the subject over and over again. What kind of an animal was man, they asked, and how could his attributes be harnessed to the cause at hand?

Virtue or Self-Interest?

In chapter 1 we saw the ancients' answer to that question. Man, they held, was an animal in quest of virtue in all its forms, and political virtue most of all, for the *polis* was but the individual soul writ large. Thus, virtuous leaders were bound to seek justice, virtuous citizens must lose themselves in politics, and together they would honor the public good.

During the European Enlightenment, some thinkers—Machiavelli, Hobbes, the irrepressible David Hume—began to question the ancient wisdom, even though it had held sway for two millennia. These *realists,* as we might call them, questioned virtue from many angles. It was unstable, they argued, and could vanish overnight in trying circumstances. It was associated with visionary schemes, utopian projects, radical ideas, and monkish otherworldliness. It could foment a conquer-the-world spirit of aggression. And it could prompt a society to purge itself of "undesirables."

The most telling indictment was that virtue did not truly reflect human nature. To be sure, *Homo sapiens* could be forced, under pressure, to rise above themselves in the name of *areté*, but there was always a great deal of back drag. It was, to borrow a phrase from Madison, like asking men to be angels.

The realists sought the Good Society in a different way. After carefully observing actual (as opposed to theoretical) human behavior, they concluded that most people were driven by self-interest more than virtue. They sought ways to increase their wealth, enhance their comfort, and gratify their immediate desires. Self-interest did not necessarily incline them toward bad behavior even though it certainly might. It could also incline them toward creativity and cooperation, as in the marketplace. The trick was to observe self-interest and learn of its ways. Then, by designing an interest-driven—and interest-moderating—political structure, they could bring forth a Good Society that really worked.

Yet even with the realists in full cry, the older thinking still persisted. Advocates of virtue—we will call them the *idealists*—continued to insist that man could not abandon his moral sense, that self-interest could not answer human needs and that greed could never form the basis of a Good Society. They cited Montesquieu to this effect and the ever popular Rousseau.

Americans had a peculiar fascination with virtue, which they associated with their emerging identity. When they used the term, they imagined ideal characters, much like the action heroes of our own time, who fight on the side of right and never give a thought to reward. And the notion received a great deal of reinforcement. American founding myths—the Promised Land, the Chosen People—implied a special endowment of virtue. American churches, searching for common ground amid their differences, emphasized loving one's neighbor and living the Golden Rule. And not least, there was the commonwealth ideology with its creed of virtuous patriotism.

The Revolution proved that Americans were indeed willing to sacrifice for their country, and the virtue they spoke of was not empty prattle. But then came the postwar disillusionment. By the eve of the Constitutional Convention, it was unfashionable to speak of virtue in some quarters. Jefferson, for instance, who had once praised the ideals of his countrymen, now pedaled them softly. "The fantastical idea of virtue and the public good being a sufficient security of the state," he wrote to Edmund Pendleton, "I assure you was never mine."

As practical men, the Framers had to consider such realities. There was a sense among them that faith in American virtue must be amended. But *amended*, not *abandoned*. The centrality of virtue to a working republic was still accepted

The Problem of Sally's Two Dates

Sally Johnson would not see herself as a study in human nature. But Sally's dilemma presents some key issues of human nature in particularly convenient terms. Her difficulty is that she has two different dates for Saturday night.

It should be explained at the outset that Sally is a very nice person. She is radiant, wholesome, and attractive. Along with a career in marketing, she wants to marry, raise a family, live in the suburbs, and seek the Good Life. She is a good student at Wellesley College, is popular and well liked, is active in her church and an officer in her sorority. Toward other people she has been taught to behave with compassion and respect. She is the all-American girl.

But she has a problem. On Tuesday last, Fred Schultz called and asked her for a date on Saturday. Good old Fred. A date with him would unvaryingly include two lines of bowling at the student union, frozen yogurt, twenty minutes in the video arcade, and a walk back to the dorm. Fred was, bless his heart, something of a nerd. But Sally wasn't doing anything on Saturday night, and so she accepted.

Then on Wednesday, who should telephone but Lance—*the* Lance. Lance W. Rockefeller IV. He had seen her at the Black and White Formal, he said, and would like to get to know her. On Friday he and some friends were going to take the Lear jet down to the Bahamas for the weekend. There would be swimming and tennis, a short cruise on the family yacht, dinner at the Royal Poinciana, and perhaps a stroll along the beach at sunset. Mom and Dad were staying at the villa themselves, so everything would be chaperoned and aboveboard. Would she like to come along?

Would she ever! But there is that problem of the other date—with Fred. Breaking it would be a simple enough matter. She could call Fred and make up an excuse. Of course, that would mean telling a lie. It would also mean breaching a commitment and violating a trust. Sally has been taught that trusts and commitments are matters of no small moment, and that lies are unthinkable.

What should she do? Here are the thoughts that flit through her mind while the decision is being made.

In the real world there is such a thing as a permissible lie. Indeed, the real world is quite patently founded on permissible lies. And lies aimed at averting hurt feelings are obviously permissible.

Lance Rockefeller, with all his millions, is probably not the marrying kind. Courtship with him—allowing the imagination to soar—might well prove tantalizing, frustrating, and in the end, disastrous. Fred, bless his heart, is the marrying kind and would probably make someone a good husband.

Suppose word got around that Sally Johnson broke dates, jettisoned commitments, jumped at the best opportunity. This might not be good for her reputation in the long run.

Actually, it might be wrong *not* to break the date. She could never marry Fred. She was just leading him on. To go with Fred and be hating him every minute *would not be fair to Fred.* For Fred's own good she should be upfront with him and let him find someone his own speed.

Who was this Lance, anyway, that she should compromise her integrity for him? What could a single date, no matter how glorious, have to compare with her idea of herself? "This above all, to thine own self be true. . . ."

On the other hand, why not do it to Fred? Fred would probably do it to her. Others had. Bad things happened in the world. It was a bad world.

Suppose she turned Lance down. Maybe he would call again. Indeed, maybe he would respect her the more for her honesty. Maybe forthrightness would give her the winning edge over competitors.

Then, again, suppose he never called again. Suppose this was her only chance. Suppose he took some other—less upright—girl to the Bahamas and would wind up marrying her. "The saddest words of tongue and pen. . . ."

Mrs. Lance W. Rockefeller IV. . .

There they are, the sentiments of mankind. Enlightened self-interest. Corrupt self-interest. Virtue for its own sake. Virtue by calculation. All of them occur to Sally as indeed they would occur to most of us in her situation?

The question is, which of them will determine the outcome?

From America: A Study in Heritage, *Kendall/Hunt 7th Edition, p. 33–34.*

by most of them. The question was not whether to forget about it entirely, but whether it could somehow be made to serve.

Virtuous Self-Interest

To begin with, they must squarely face virtue's limitations. It was unreasonable, for instance, to expect virtuous behavior all the time. Experience proved otherwise. There were stories in the Revolution of men standing rocklike in their battle lines—only to flee in terror on a subsequent engagement. Jefferson himself, to his lifelong shame, abandoned the Virginia governor's office at the approach of Lord Dunmore's troops and scuttled off to Monticello.

It was equally unreasonable to expect virtue in everyone. In the Revolution, the conduct of a comparative few had inspired others to what might be called situational virtue. On the night of the famous attack at Princeton, two of the three assaulting forces had encountered obstacles and turned back. The third force, commanded personally by Washington and inspired by his battlefield courage, overcame the obstacles and fought its way through.

Finally, few in the modern world could be expected to live by a code that was two thousand years out of date. Renounce one's private life and lose oneself in politics? View the political community as one's own soul writ large? Sacrifice personal interest for the public good? Not in the eighteenth century.

The Framers, then, agreed to subject virtue to fewer demands. After all, a commercial society did not require citizens to live in barracks, as they had in ancient Sparta, or walk barefoot in the snow to prove their hardihood. It required them to settle accounts, abide by contracts, and be honest within reasonable limits. Nor were such behaviors altruistic, as *areté* might have been, or beyond the common range. Rather, they embodied what we might call *virtuous self-interest*, a milder and saner form of virtue that depended on horse sense to tame selfishness rather than zealotry to expunge it. With virtuous self-interest, citizens could lead ordinary lives and pursue their own happiness—subject mostly to the governance of the self.

Virtue and Structure

During the Confederation period, people like John Adams had thought a great deal about the possible use of structure to solve the virtue problem. It was an old concept, tracing all the way back to Aristotle, and it could be employed in a variety of ways. One of them was to provide a safety net in those instances when virtue—even among a virtuous people—might suddenly fail.

- Auxiliary Precautions

What Madison called "auxiliary precautions" were based on the realists' view of human nature rather than the idealists'. The guiding assumption was that in the best of societies some individuals would reach for power, some groups would ignore the public good, some voters would be easily manipulated, and some players would never play by the rules.

So the Framers designed a structure of government according to the principle of "counterpoise," to the end that interest would counter interest and ambition thwart ambition. Auxiliary precautions were not, as some political scientists have contended, the center ring of the circus disguised as a sideshow. They really were, in the Framers' minds, "auxiliary." The center ring was to be the American people peacefully living their lives.

- Structure to Mobilize Virtue

The Framers also realized that structure might have a lot to do with the "situational virtue" mentioned above. In chapter 8 we saw how they employed specific structural devices—many of them borrowed from David Hume—to insure that the most virtuous wound up in stations of greatest trust.

They employed even subtler devices. They sought to make certain, wherever possible, that the wise leadership they hoped for could bring virtue to bear in concrete, practical ways. How, they asked, could a President rally the forces of justice in a dicey situation—the way Washington had rallied his troops at Princeton? The following represents a rough scenario (based on experience) of how the question might be answered.

The US Congress, fearful that a small religious group known as the Druids has become a public nuisance, passes a law to deprive all Druids of their civil and political rights. This measure violates the rule of law, for it singles out a specific group for punishment. It also violates the Constitution, which gives Congress no authority whatsoever in religious matters.

The President doesn't like Druids any more than Congress does, but he has taken an oath to defend the Constitution. Apart from that, he is worried about political fallout. He won the last election only narrowly, and the loss of the religious fringe might make all the difference next time. Besides, he is becoming seriously concerned about Congress. Victory on such a high-profile issue may tempt some muscle-flexing congressmen into who knows what foolishness.

For reasons of both virtue and self-interest, then, the President decides to veto the measure. He writes out a veto message and marshals every argument he can think of to give it weight. Some of his arguments appeal to political self-interest, others to a sense of civic virtue. The message is available for the public to read, so it is bound to raise a dust. And since the President's leadership is now on the line, he begins calling his friends in Congress and twisting their arms to help out.

Some congressmen are persuaded by the President's dialectics and others by his personal appeals. Still others, who may reject both, are moved to stop and think. Defeating the President on this kind of issue may come back to haunt them at election time. After all, did *they* want to answer for constitutional trespass?

The President doesn't have to persuade *all* the congressmen. A third of them plus one will suffice to block the override, and since a quarter of them voted against the measure in the first place, the President only has to change *one-twelfth* of the votes. He probably will.

How many of the altered votes are motivated by virtue and how many by self-interest? It doesn't really matter. The point is that in a situation where the worst in human nature had a tactical advantage, the constitutional structure began altering personal motives and rallying the forces of justice. Put simply, the Constitution operated to make virtuous conduct a matter of self-interest.

Idea of the Extended Republic

James Madison:
http://www.jmu.edu/madison/
http://www.leftjustified.com/leftjust/lib/sc/ht/fed/mbio.html

We know something of the objections raised by many Americans to what Madison and the nationalists called the *extended republic.* To some ears the term sounded like an oxymoron. Republics were supposed to build a sense of togetherness—a feeling of "common wealth" and "public good." How could such a feeling be created across a sprawling continent? How could citizens maintain contact with one another and with the governing process? Worst of all, how could they withstand the turmoil of *so many* factions?

Madison himself had been bothered by these questions, but in the course of his deliberations with fellow nationalists (and the study that occupied most of his evenings) he had come up with an answer. It was to become Madison's great contribution to political theory—and a cornerstone of the American Founding.

In his reading of the ever pertinent David Hume, Madison became intrigued with Hume's concept of political representation as a "refining" process. In each ascending tier of representation, we recall, the persons chosen were held to be ever broader in outlook and ever better known for their virtue. As Madison pondered this notion, it struck him that its implications were exactly the reverse of the small republic idea—for it meant that bigger was *better* not *worse*. The larger the territory the more tiers of representation required, and the more tiers of representation the more virtuous the government. Madison coupled this insight with an observation of his own. Factional turmoil was not rarer in smaller states than larger ones. If anything, just the opposite. In tiny Rhode Island, for example, not only was factionalism alive and well, one faction had an absolute chokehold on the government.

Suddenly it occurred to Madison that the conventional wisdom on factions might be all backward. It wasn't a case of the smaller the state the fewer the factions, as the ancients had supposed—but the smaller the state *the more risk of a factional takeover*. For in a small state, a single large faction might find itself with little or no opposition. All at once Madison saw the classical republics through new eyes. Their governments had always been plagued by factional conflict and its dire consequences. The malady had usually been blamed on democracy—on the unrestrained passions of the people. Perhaps the blame really rested on the *polis* itself and its diminutive size.

What if, thought Madison, a republic were to seek peace through the opposite strategy? Instead of urging citizens to overcome human nature and rise above self-interest, as the ancients had done, suppose the republic simply acknowledged self-interest and welcomed all factions to the fray. Each faction would have many competitors in the scramble for influence—and in the extended republic many, many competitors—and the probability that any single one of them could dominate all the others would reduce to zero. The large republic, then, must be stronger and more stable, not weaker and more faction-ridden, than the small republic. Bigger really was better.

Such a recasting of the problem and rethinking of the solution was little short of astonishing. Madison had taken two thousand years of unchallenged assumption and turned it upside down.

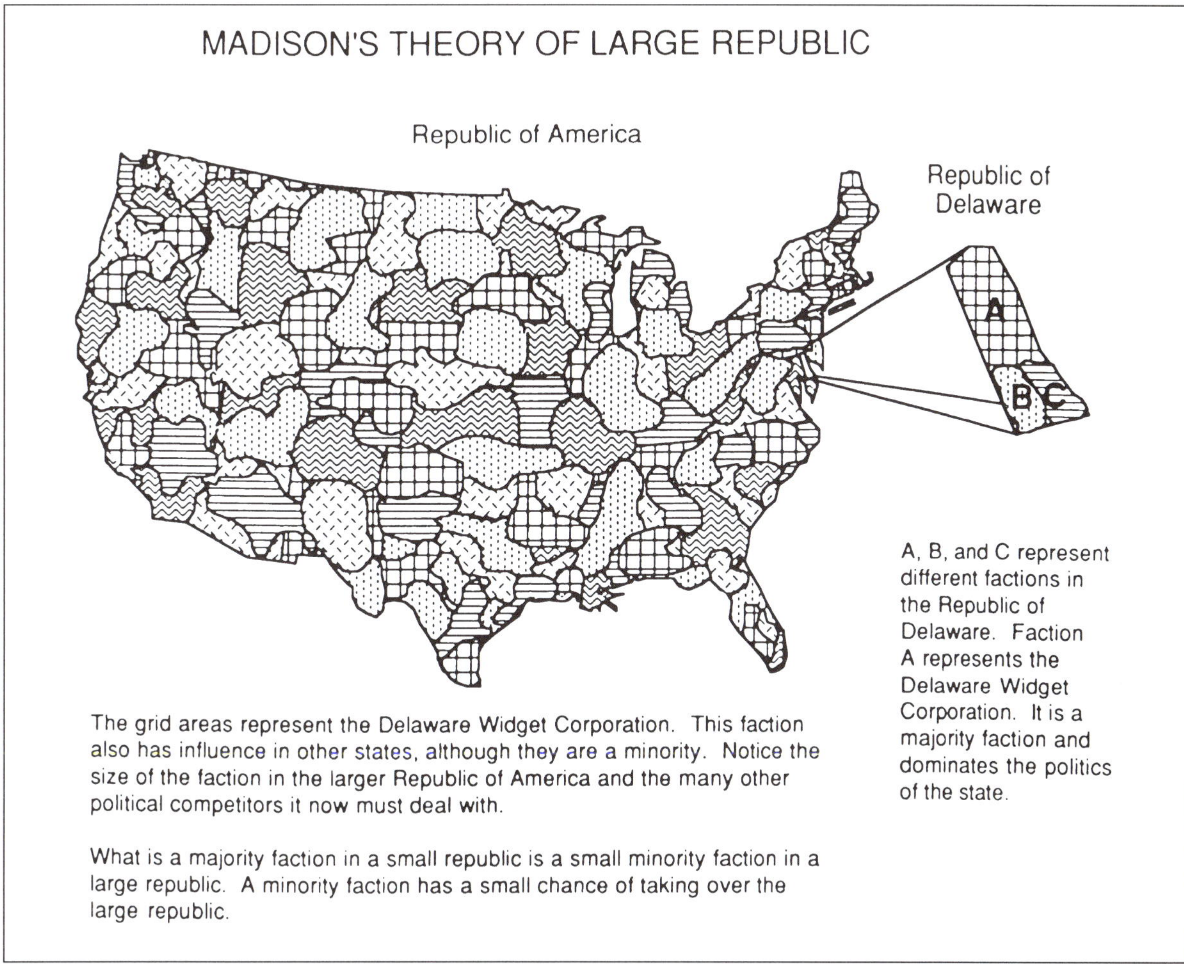

All at once Madison saw the classical republics through new eyes.

The Ambiguous Founding

The Founders' complex understanding of human nature led to a Founding which was ambiguous and multilayered. We have seen that all three kinds of republicanism—classical, liberal and democratic—derived meaning from the American Revolution. All three found fulfillment in the Constitution. As a result, the Constitution struck a balance between republican virtue and material self-interest—and redefined both. We have seen how virtue was redefined to reflect wisdom, prudence and common sense. Self-interest was redefined as well. It was the self-interest of a "moral and religious people," in John Adams' words—a self-interest of self-governance.

The Founding's ultimate reliance on virtue may point to a dark future, for our supply of moral energy might eventually run out. The Founders pinned their hopes on three possible sources of renewal. First, there was the religious

Factional Takeover: the Case of Grenada

We know it is difficult for single factions to take over large republics. Witness the United States. In small republics, by contrast, single factions can take charge much more easily. Witness Grenada.

In the spring of 1979, Grenada seemed anything but ripe for revolution. Life in the sleepy island had coursed along happily for more than twenty years under the uncontroversial Eric Matthew Gairy. True, there was some discontent. Youngsters unable to find work in the island's on-again, off-again economy idled about, filling their afternoons with fishing off the rust-colored reefs. When agents from Castro's Cuba came preaching an angry new gospel in the mid-1970s, some of these kids were ready to listen. So began the New Jewel Movement.

The movement's tactics were peaceful at first. Its presidential candidate, Maurice Bishop, sermonized about socialism and wrote blistering editorials in the *Torchlight*. Bishop was not content with constitutional politics, however, and his two lieutenants, Kendrick Radix and Bernard Coard, were even less so. With tactical and logistic support from Havana, they set about planning a coup.

The takeover proved to be absurdly easy. The NJM commandos merely waited until Gairy left the country, gathered up units of the People's Revolutionary Army, and began battering their way into the country's clapboard police stations. By the end of the day, March 13, 1979, Grenada was their very own.

Of course, consolidating the NJM's rule, especially among a non-ideological and often unwilling populace, proved another matter. Gairy's entire government was jailed. His small defense force was rounded up and neutralized. Radio Free Granada was set up to portray the former President as a fascist dictator, while the *Free West Indian* began inundating the island with Castroite propaganda. As for the *Torchlight*, which had earlier befriended the NJ, it was shut down completely. And when twenty-six leading citizens tried to launch a publication of their own, they were arrested and locked away in the Richmond Hill prison.

Meanwhile, Soviet "cruise liners" began putting into St. George's, and one of them unloaded three thousand AK47 automatic rifles for the People's Revolutionary Army. Popular uprisings were well and good, as Grenada's new masters understood—but they had to be kept under control.

From America: A Study in Heritage, *Kendall/Hunt 7th Edition, p. 152.*

influence. Once separated from the political process, churches would hopefully focus their attention on the making of citizens as well as the saving of souls. Second, there was education. So highly did the Founders regard this influence that several of them sought to provide for a national university in the Constitution. And finally there was the old platonic hope that self-renewal might be generated by the Good Society itself—that good politics could forge good people. Of course all three propositions were iffy.

The Constitution and the Rule of Law

In chapter 5 we saw how the rule of law helped to explain shortcomings of both the Confederation and state governments. The Confederation was abandoned with the Constitution's adoption. Many of the state governments, learning from experience, reformed or completely rewrote their constitutions with the rule of law in mind.

The US Constitution was also written with the rule of law in mind. Using our earlier five principles as a guide, we can see some of the specific ways the Framers expressed rule of law concerns:

> *Generality*. Congress was forbidden to create titles of nobility (which would impose different categories of citizenship) or to single out any person in a **bill of attainder**.
>
> *Prospectivity*. Congress was prohibited from passing any *ex post facto* law.
>
> *Publicity*. Congress was required to keep a journal of its proceedings and publish it to the people.
>
> *Consent*. Suffrage was guaranteed to all who enjoyed that right in their respective states.
>
> *Due Process*. Trial by jury was required in all federal cases, and Congress was forbidden to suspend *habeas corpus* under all but the most extreme circumstances.

Bill of Attainder: Legislative act in which an individual is simultaneously charged with a crime and found guilty.

But the way in which the rule of law figured into the Constitution was both broader and more subtle than these features indicate. The whole structure of the federal government was designed so that rule making and rule enforcement would be placed in different hands.

Congress was made the essential rule maker. The executive plays some role in the process—in his use of the veto, for example—but the *essential* rule maker is Congress. And the judicial system is the essential rule enforcer. Once again, the picture is muddied a little by the executive, who has certain enforcement functions as well—but the *essential* rule enforcer is the court system. For in the end, judges and juries must decide how, when, where and against whom the laws will take effect.

The executive is in charge of arresting lawbreakers and bringing them to trial.

We have seen what happens when rule making and rule enforcement are not separated. In republican governments based on an all-powerful legislature, the outcome was often simple tyranny. For as with any Nero or Caligula, the unchecked majority sought to have its way at all cost, victimizing outsiders, minorities, eccentrics, advocates of unpopular opinion, or those who seemed ripe for plucking. In the notorious case of postwar Rhode Island, the debtor interest pushed laws through the legislature requiring creditors to accept depreciated paper currency in payment of loans—or go to jail. In contrast to the rule of law, this was the *rule of will.*

The Framers of the Constitution wanted no such tyranny. They took utmost pains to separate the judicial branch from both the others and shelter it away from politics.

Rule of Law Politics

With this separation in place, the psychology of governance was all different. The rule maker, lacking enforcement ability, was obliged to operate blindly. He could no longer single out specific targets, for he couldn't be sure how his rules would be applied. Think about the "Druid" case once again. It would be useless for Congress to pass laws against the Druids, even without a presidential veto, for in the end, conscience-stricken judges could render such laws unenforceable.

So rule makers were obliged to change their ways. If they wanted to draft enforceable legislation, they must now try to frame their objectives as generally as possible. Instead of singling out the Druids by name, they might try legislating against a particular Druid practice such as, oh, jaywalking. The law now becomes more enforceable as far as the courts are concerned, but it also enjoins *everyone* from jaywalking, not just Druids. Judges like this kind of law. It is precise, morally defensible and amenable to adjudication. Unlike being a Druid, which is hard to prove and even harder to get juries riled up about, it is something the law can really sort out. Druids, of course, like anyone else, can avoid the law's sanctions simply by waiting for the light to change.

The Constitution not only protects Druids from Congress but also from the courts. Suppose the judicial branch loathed Druids and wished to lock them all up. They might nail the ones who jaywalk, true enough, but lacking the means to legislate, that is all the judges can do. They are as blind in their function as the lawmakers are. They can only address the general behavior of jaywalking.

Separating rule making and rule enforcement, then, brings the process of governance closer to the rule of law. The laws have an automatic tendency to become more general, more prospective, and so on than they might otherwise.

What about checks and balances? These too can promote the rule of law, as yet another elaboration of our Druid story might illustrate. Suppose the original Druid measure passes through Congress and is vetoed by the President, as we saw above, and suppose the override fails. Congress is still out to get the Druids, but it has learned a hard lesson. And the jaywalking approach won't help either. At this point, congressional leaders may actually sit down with the President and say, *What sort of measure would you accept?*

This is an important question, one often asked in American politics. The President is likely to answer: *I would accept a measure closer in keeping with the rule of law*. Why? Because, being a canny politician, he has a sharp eye on the people. He knows they distrust laws that single out one group for punishment. He knows they realize that persecution of Druids today might lead to persecution of Mormons tomorrow and Libertarians the day after. And he knows the jaywalking subterfuge won't fool them. The people's virtue, knowledge and sophistication will weigh into the checks and balances of the system in large ways and small. So either the Druids (along with everyone else) must be held accountable to real laws or they must be let off the hook. If Druids rob banks or kidnap children, the laws can really put them out of business. Otherwise they (along with everyone else) are scot-free.

How the Rule of Law Creates Freedom

How, precisely, does the rule of law create freedom? For some people, freedom and law seem antithetical. How can anyone be free, they ask, if there are so many rules to obey? Others point out that some of the world's worst tyrannies have had the same machinery as democracies—i.e. legislatures, legal codes, court systems—so why didn't the law create freedom there? A third objection arises from the fact that majorities, often very slender ones, speak for everyone in the legislative process, and often pass laws of dubious popularity. What "freedom" is there for the hapless minority?

The best way to see how rules create freedom is to visit a hot dog stand on a busy day. The hungry patrons automatically queue up into a line. The head of the line is served first. No crowding in. Places can be held by friends under prescribed restrictions. *No one has to have any of this explained.* Or think of a group of eight-year-olds hunkering down to a game of marbles. Before the play begins, they call out a set of rules for the match at hand. There isn't much "legislation"—just quick agreement. In both cases we see the almost instinctive

Cars lined up during a fuel shortage in California. There were no rules posted or people directing the traffic into an orderly fashion, it happened automatically.

creation of rules to head off potential conflict. These rules don't constrain freedom—they make it possible.

Such unconscious rule making illustrates a central truth about the rule of law. For when the laws of society adhere to the rule of law, they function precisely like the examples given above—as facilitators of freedom. Put another way, the rule of law makes the laws of society work like the laws of nature. They become as steady and predictable as gravity, and living by them enables people to interact with mutual benefit. This fact underscores the difference between regimes of liberty and regimes of tyranny. The latter may have all the visible machinery of law, but because they lack the *rule* of law they don't resemble nature in the least. They are full of nasty surprises.

Consider the case of car theft. I see an automobile I have always wanted and discover an easy chance to steal it. Knowing the laws of society, I can assume that if I do steal the car, I will be caught and punished. Accordingly, under the rule of law, I alone determine what my fate will be—for I alone can decide whether or not to bring on the consequences. In a rule of will society, none of this may apply. I might decide to steal the car and, depending on my connections, may never go to jail at all. Conversely, I might decide not to steal the car and yet *still* wind up in the slammer—just because someone in power doesn't like me. Without the rule of law, I cannot govern my own destiny. With it, I can.

The rule of law also provides an answer to the slender majority problem. As long as the laws meet the tests outlined above, it doesn't really matter whether every single person approves of them. The important quality of law is predictability not popularity. Any law is bound to offend *someone*—work through some examples and see for yourself. So long as the laws satisfy the majority and are aimed at "bank robbers" rather than "Druids," they cannot pose a threat to freedom.

Law Versus "Administration"

There is an inherent tendency for all governments, not just tyrannical ones, to fudge here and there on the rule of law. Governments have to administer things—precisely in the way large corporations do. They must build roads, operate schools, collect garbage, care for the poor, maintain armies and navies, and so on. They acquire large and complex bureaucracies to perform these tasks. And for most of the tasks, government need pay no attention whatsoever to the rule of law. *Administration* does not require it.

It is only *governance* that requires it. It is only when laying down the rules governing people's lives that the government must be careful about the rule of law. Consider any administrative situation and the difference soon becomes apparent. In building a road, let us say, a government can favor certain contractors, back out of agreements, ignore the public, conduct shoddy inspections and keep vital information secret. While these practices violate every principle of the rule of law (and certainly do the government no credit), they are not tyrannical. *For they don't govern people's lives.*

Whitehall: Administrative center for British Empire.

Some governments never distinguish between the two cases. They suppose they are *governing* their citizens when in fact they are *administering* them as they might administer highway construction. This is precisely what brought on the American Revolution. In order to reform their ramshackle empire, the bureaucrats in Whitehall began overhauling everything—forgetting that they were *altering the laws of society* in the process. What the American colonists reacted to was not the improvement of customs collection or the untangling of imperial authority—but the "administration" of their lives.

Behind the Constitution, then, there was a strong commitment to the rule of law. For as the Framers realized, that and that alone could create freedom.

The Constitution as a Moral Document

The American Constitution is sometimes said to lack a soul. It is not about truth or justice—those passions of the ancient republics—nor does it embody what philosophers call a moral theory. It is merely a design of government. So goes the critique anyway.

Those who hold this view go on to argue that the American people, faced with such emptiness, have had to turn to the Declaration of Independence or the Bill of Rights for the moral inspiration that a founding requires. Those documents bolster our liberal individualism, to be sure, and give us claims against society in the form of rights. But that is not quite a moral theory either. What of our claims *as* a society? What about our sense of nationhood and common purpose, our place in the world and the meaning of our national undertaking? What is there to say of the American people *as* a people?

For some, these questions are irrelevant. Americans, they say, are primarily understood as individuals, autonomous and independent, not as *a people*. The Founders seem not to have taken this view. They appear to have regarded the founding of the United States as just that—a *founding*, a full-fledged announcement of nationhood. And they regarded the Constitution not just as a plan for living together, but as a call to live by certain ideals. Why else would they be so concerned about virtue or the rule of law?

If this is true, why didn't the Founders come out and say so? They had their reasons. One of them was to get the Constitution adopted. They didn't want to antagonize the states (any more than they already had) with heady nationalist rhetoric. And they generally shied away from declamatory prose, which was often vague, usually sententious, and always a source of confusion. For them, the real truths were the "self-evident" ones.

The Dignity of Mankind

Let us consider the following case. Suppose that for the Founders, the moral theory of the Constitution was among those self-evident truths. It wasn't so much a point to argue as a given to assume. We should keep the Scottish Enlightenment in mind while exploring such a possibility, for thinkers like Francis Hutchinson had held that truth operated on the level of common sense and didn't require elaboration. And several of the Founders were among his disciples.

In order to see the world as the Founders did, we must glimpse their view of the past. They had recently fought a great war, which they saw as a war for all mankind, and against daunting odds they had won. They attributed their victory to civic virtue, and belief in that virtue both set them apart and bound them together. Moreover—and here is another crucial point—they widely understood the outcome in terms of divine favor.

At the same time Americans had proven the success of republicanism. They hadn't just *proclaimed* republican principles, they had *applied* them. They had come together as "the people" and created their own governments, and defects notwithstanding, the governments had actually worked.

Finally, and most importantly, the Founders identified both the Revolution and republican government with "the Laws of Nature and of Nature's God." In their minds, the reason for their double success was that they had acted in conformity with universal truth. For them there really *was* a moral law in the universe—and it made all the difference.

This aspect of the Founding is often glossed over. Yet it is difficult to make sense of the Founders' actions, and still less of their explanations, without weighing in these assumptions. Natural law, as we have seen, was understood to represent the higher law, the moral law, the law by which all human actions must be judged. Belief in such a code had a long and respectable history. The American Founders did not give lengthy disquisitions about it precisely because they more or less took it for granted.

It was according to natural law, as Charles Kesler points out, that the Declaration of Independence and its arguments for human equality, natural rights, and government by consent held together and made sense. The Revolution worked, one might say, because the Laws of Nature were believed to operate for rather than against it.

And the same went for republican governance. The Founders recognized that most republican governments—indeed, practically all of them—had dismally failed. So why bother with republicanism at all? Why not just take the best features of the British system, elect a Parliament, find a worthy person to be crowned king and call the job done? This might well have been a useful course under the circumstances.

For an answer, we must read the Founders broadly, gauging assumption as well as opinion, and searching for a coherent center. Republicanism offered something that constitutional monarchy did not—a fundamental principle, embedded in the Laws of Nature and accepted by the Founders as true. To give it a name, we might try *the dignity of man*.

More a commonsense assumption than philosophical precept, the dignity of man meant that there was something distinctive about the human personality—something that no other life-form possessed. Mankind alone enjoyed the faculty of reason. Mankind alone spoke of right and wrong. Mankind alone had the ability to make free choices. Man wasn't just another animal.

The Moral Founding

Government had rarely acknowledged human dignity. Most government, in fact, was based on the opposite assumption, that some *Homo sapiens* were far superior to others, and it was these exalted beings—king, conqueror, dictator, true believer—who deserved to rule. As a result, most people on the planet had never thought about equality or rights or a voice in government. They were regarded as something like clever draft animals, suited only for exploitation.

Republican societies had attempted to swim against this tide. They had searched for every means imaginable to enable human beings as ordinary citizens to realize their full potential. In the process, they had taken every sort of wrong turn and wound up with tyranny, anarchy, dissolution or war, for the animal in question was extremely complex and its hopes seemed ever self-defeating.

The republican project, then, wasn't simply to devise regimes of justice but to figure out how human beings could live together in dignity. In this quest, the American Founders enjoyed a curious advantage. The thirteen states were like so many laboratories, each of them trying out various approaches to the common goal. Some results were pretty good while others were all but disastrous. Anyone could see, for example, that Pennsylvania was drifting into dangerous waters or that Massachusetts might go right over the falls. What was right or wrong with various constitutional strategies was being energetically debated all over the country.

The point wasn't that Americans had failed to get it right the first time—the point was that *they could get it wrong without giving up*. Such was their confidence in the republican idea that they could dare to tinker with the mechanisms for realizing it. We should think about what that meant. Thirteen independent sovereignties spread across a continent, and all of them convinced that republicanism was True. None had an entirely successful republican regime, nor could they find one *anywhere* they might use as a model. Yet they tinkered on. How on earth had such certainty made its way into so many minds?

This question points to the true Founding of America. We can think of it as a sea change, a "paradigm shift," in the consciousness of an entire populace, one that despite all particulars of locale and vagaries of circumstance rendered them into *a people*. It was the intuitive understanding that government had to be accountable to the dignity of mankind—and that there was a way to make it so.

This conception of the Founding helps to explain some of the "mysteries" of the Constitutional Convention. We asked how it was that key players could lose on crucial issues and still press forward. Or how the Framers could turn compromises into strokes of creativity. Or why they could forsake time-honored doctrines with such nonchalance. The answer was that they were held on task by a faith that transcended mere politics, and this faith gave them both strength and flexibility. They believed they could make all sorts of design adjustments and theory modifications without compromising the end in view—so long as they understood what that end truly was. They could avoid being doctrinaire because they were so certain of being right.

Faith in human dignity had some important corollaries for the Founding. One of these, as Lance Banning points out, was a belief that the people as a whole could be trusted. If the Constitution was skeptical about human nature, it was strangely sanguine about "the people." The entire citizenry, properly assembled, was not going to subvert the law or pervert democracy, the Framers thought. Individuals might try such things, yes, and factions surely would—that was human nature. But the people *as a whole*, acting constitutionally, must always be on the side of right.

A second corollary held that a constitution honoring the people's dignity could not fail to win their support. Thus, the Framers could sign their document, however unorthodox, and defend it with serene confidence. For when all was said and done, the people were not going to reject that which would give their lives fullest meaning.

The third and final corollary held that happy endings were embedded in the Constitution. Truth and justice may not win out in the short run—they hardly ever did—but they may well win out in the long. For once the people were empowered, truth and justice must surely be in their own interest. Given a sovereign choice, they would eventually, if not immediately, mend their errors and repent of their wrongs. They could muddle their way through the sorriest of difficulties if they believed that the cards were not stacked against them.

Much of the Constitution's structure was based on these assumptions. The principle of counterpoise, for example, was to insure that the people's considered wisdom, not their hasty impulses, would hold sway. And those fire walls between power and the populace were put in place not to rob the latter of decision making but to enlarge and refine the way they went about it. To borrow a phrase from Lincoln, the Constitution was written for "the better angels of our nature."

Faith in the dignity of man also shaped the way the Framers justified their work later on. In *Federalist* 43, for instance, Madison explained "good government" in precisely such terms. Good government must have energy, he said, must promote stability, must provide for the public safety and promote public happiness. It must comprehend the "character" of the people and speak for "the honor of mankind." It must "make possible a government informed by and fostering intelligence, virtue, and a dedication to greatness." The Constitution, in other words, emphatically *did* give an account of justice.

And it did something else. By seeking to embody the Laws of Nature—which Madison specifically mentioned in *Federalist* 43—the Constitution set up a supralegal standard for judging legislation and even for judging the popular will. Charles Kesler has argued that if the people themselves sought to demean human dignity, the Constitution in this larger moral sense would stand against *them*. Thus, even though the text of the document recognized slavery and the people themselves practiced it, the Founding as a whole condemned it, a fact which Lincoln saw clearly. For the dignity of man could never abide human servitude.

The principles behind the Constitution, such as the dignity of mankind, would eventually result in a rejection of slavery, even though the text of that document protected it.

The idea of a moral founding puts a lot of things back together. It tells us that the Revolution wasn't one sort of phenomenon while events in Philadelphia were quite another. It links up the Declaration of Independence, the Constitution and the Bill of Rights, so that each becomes part of a larger whole. The law of Nature and Nature's God mentioned in the Declaration's first paragraph is reconnected to the doctrines of equality, natural rights, and government by consent mentioned in the second paragraph, and these in turn are embodied in a government capable of putting such principles into practice.

Thus, when we say the American Founding "worked," we don't mean that Americans became rich and powerful, or that they were able to keep their government reined in, or that they came to enjoy unprecedented personal freedom, though all of this may be true. We mean that the American Founding realized the republican project. It created a system of government that was accountable to the dignity of mankind.

The Constitutional Polity

The Constitution was more than an outline of government. It was also a blueprint for nationhood. Partly, of course, the nation in question was already in existence, but mostly it lay off in the future as a set of possibilities.

In examining such possibilities, we should keep in mind the ancient Greek connection between "polity" and "soul." The polity was thought by the ancients to be more than a collection of individuals—it was a higher embodiment of personhood. Thus, the capacity of the individual for moral thought or virtuous action was realized in the polity as justice. And, conversely, the just society was supposed to shape and guide the development of the individual soul.

Empire of Liberty

The Framers thought a great deal about the size and shape of that future nation, to say nothing of its character. It was principally the West that troubled them. There were not a few Americans who suspected the West of faltering in the march of human progress. How, they asked, could the institutions of civilization be left to the care of men wearing fringed buckskin and smelling of bear grease? Even scarier, maps showed that the West stood to become much larger and more populous than the East.

The problem of the West for America, then, was like the problem of the colonies for England. If its development got out of hand, there might be a terrible price to pay down the road. One could easily imagine a vast expanse of bailiwicks, fiefdoms and other jerrybuilt entities, most of them half-wild and semi-governed, quarreling and fighting with one another in a manner to shame the Balkans—and absolutely out of control.

Greed came into the picture too. Seven states had western land claims, and speculators had been busy figuring out how the various tracts could be turned into profitable ventures. However, in a remote backcountry, at the mercy of Indian attack and foreign intrigue, orderly settlement wasn't feasible without political institutions. But precisely *what* institutions? States? Territories? Protectorates? And by whom should these be governed? By existing states? By the national government? By themselves? Just how these questions were answered would determine the ultimate character of the United States.

Various statesmen suggested various solutions to the problem. John Jay had submitted a proposal to Congress that would have effectively closed the West to further settlement. Gouverneur Morris wouldn't go quite that far but he would make sure that the West always remained politically inferior to the East—a kind of imperial dominion. Elbridge Gerry would allow for the development of states in the West, but he would stipulate in the Constitution that those states could never outnumber the original thirteen. How else to hold them in line?

An artistic rendering of agrarian republicanism.

Thomas Jefferson disagreed with them all. He envisioned the West not as a barbarous outback but as an Arcadian idyll, a land of prosperous farms and self-reliant yeomen tilling their own soil. That vision would find expression in the lithographs of Currier & Ives a half century hence, and in the prose of modern writers like Willa Cather. If America was the chosen land, the West was to become the *heart*land.

Jefferson's solution to the western problem was an outgrowth of that famous passage he had written in the Declaration of Independence. Let the West create itself, he proposed. Let settlers build their own political societies, with a bit of supervision from the general government, and when those societies have reached a certain stage, let them take their place in the Union. Grant them full equality with the original thirteen, and never think twice about their mounting numbers. The time may come when the entire Mississippi Valley would be divided into states, and who knew but what even Canada, Mexico and Cuba might eventually come aboard. Let America spread across the hemisphere, Jefferson rhapsodized, and become an "empire of liberty."

A scary thought? It was at the time. In retrospect, the scarier thought would have been the Morris or Gerry alternatives—holding the West in bondage. For that would have defined the United States in a far different way. And the likely result would have been not one secession movement (as in 1861) but a host of them, along with insurrections, filibustering adventures, plots, intrigues, tumults and, yes, civil wars. It would have been a case of the East trying desperately to hold on, as the British had tried before them, and in the end losing their grip.

Northwest Ordinance: A 1787 Congressional ordinance that established the coequal status of states created out of the Northwest Territory with the original states.

These proposals were embodied in the famous Northwest Ordinance, a piece of legislation drafted by Jefferson himself and passed by Congress in New York while the Constitutional Convention was meeting in Philadelphia. The Framers discussed the measure at length. By committing all US territories to the supervision of the new federal Congress, they knew they were effectively making the Northwest Ordinance a part of the Constitution—and that the United States would grow to resemble Jefferson's "empire of liberty."

The Federal Republic

Just as no one had ever thought about "**federalism**" before, so no one had imagined what sort of polity a "federal republic" might be.

Federalism enabled the United States to realize Madison's extended republic with all its promised advantages. At the same time, it retained America's small "classical" republics—with all of *their* advantages. And it accomplished this double benefit while neutralizing the dangers of both. The large republic did not morph into a new Rome, nor did the small ones dissolve into anarchy and war. It brought the peace, stability and harmony of an Old World empire, together with the blessings of an imperial marketplace, and at the same time it brought the creative pluralism of the ancient Greek city-states, each with its own mark of distinction. And citizens were at once identified with a proud nation of consequence, and with close-to-home polities in which they could make a difference.

There was both toughness and flexibility in the combined system. It was tough because it was unified and strong. It was flexible because the constituent parts were free within limits to go their own way. The toughness never hardened into despotism, nor did the flexibility soften into chaos. Unseen balances seemed to play off one extreme against the other.

The Commercial Republic

Diversity was frowned upon in the classical republics. All members of a given class were expected to dress alike, act alike, and pretty much think alike. Commerce was disparaged too. For all the buying and selling of the marketplace was thought to augment private interest at the expense of public. Together, in fact, commerce and diversity exerted a baneful effect on the ancient *polis*, or so it was believed. Goods for sale, especially luxury goods, implied that one citizen would have possessions different from another's, and this would strengthen the sense that both lived apart. And who knew but what the private life enjoyed by each might not eclipse the public life they shared.

Commerce led to an even deeper malady—"desire." A person who hankered after fine clothing or objects of art displayed a disturbing side of the personality, for desire was thought to be erotic in character and therefore ignoble. One reason for distrusting democracy was that the ordinary people, those lacking noble birth and high learning, were the ones most likely to be ensnared by desire. Let them have their "shiny things," sniffed the philosopher Zeno, while the virtuous meditate upon justice.

In the polity foreshadowed by the American Constitution, all this ancient wisdom would be turned inside out. For in the "commercial republic," as some were calling it, all interests, including the ignoble ones, would be placed on a par and given a political voice. No way of life would have a favored status and none would be disreputable. Public matters would be confined to a few discrete areas, carefully listed out, and the protection of property would leave individuals in control of their own lives.

What sort of world was implied here? First of all, it would be a world of diversity, even divergence. It would lack a single lifestyle or value system, just as it would lack a single religious truth. People would live differently in various parts of the country, and there would be all sorts of pathways to success. The idea of *society* as an organic whole would eventually be replaced by *societies*, plural. These would exist not only as separate communities across the land but as separate groups, identities, interests, and points of view within each.

Second, commerce and the things attending it would become as important as the traditional symbols of value, such as land. There would be an increased emphasis on business activity and the restless, mobile, open-ended way of life accompanying it. Old World standards based on heredity would give way to New World standards based on personal achievement. Money would talk.

There would be a new stress on innovation. Inventors would become folk heroes. Learning would emphasize science and technology, and business would emphasize entrepreneurship. Americans would come to prize change, all kinds of change, and they would cheerfully call it progress.

Finally, the "pursuit of happiness" would expand in every direction, and it would be fueled wholly by "desire." As commercial activity delivered an ever greater abundance of goods and variety of choices, "happiness" would come to be defined as acquiring things.

The Inclusive Republic

Because the American Republic was so diverse, a traveler moving across it might seem to be moving through different worlds. Where communities were so unalike, there wasn't much chance for any one of them to become normative, so there would be no "right" America, just lots of variations on a general theme. And there would always be room for another community, a different idea, or a new variation.

The Founding documents not only accepted such inclusiveness, they sanctioned it by their very wording. All men are created equal. No person shall be deprived. Representatives shall choose. There was nothing here about insiders and outsiders or "old settlers" and "new arrivals." The Founding was based on universals. No wonder it ushered forth a land of immigrants.

Inclusion would require the building of bridges. If you have to live with those who are different from yourself, you can't simply dismiss them as aliens. Sooner or later you must try to understand them and search out common ground. Toleration lay not solely in the national temper but in the national charter as well.

These were only a few attributes of the constitutional polity. History would reveal many more. What they all held in common was a taproot in that single sheet of parchment.

Inclusion would require the building of bridges "across cultural differences."

The Inspired Constitution

At the turn of the twenty-first century, it is no longer common to assert that the US Constitution was divinely inspired. The notion savors too strongly of national chauvinism. It was popular a century ago, however, and not regarded as especially boastful. For the argument was that God had inspired the Constitution as a model for other regimes of liberty—much as the British Constitution had been a model before it.

Inspired or not, the Constitution certainly did become such an exemplar. It was copied all over the world, with suitable variations, and whether by its own influence or some other, there are many regimes of liberty today. The belief in an inspired Constitution has not been without benefit.

Legitimacy

The government of the Confederation was discarded with the Constitution's adoption and is today all but forgotten. Nevertheless, it too was once regarded as an inspired document. There is a clear need, in other words, to look upon the fundamental law as beyond mere human agency. For if such law is of purely terrestrial origin, can it be completely "legitimate?" Will the people honor and venerate it sufficiently?

James Madison was concerned about such questions. The Constitution, he realized, could be no better than the public faith it inspired. If it seemed a frail and finite thing, it might lose its aura of mysterium and cease to hold sway. Having lost much of our innocence in the twentieth century, Americans have approached that very pass with the Constitution, as a number of recent commentaries suggest. Still, some good reasons may exist for the earlier view of an inspired Constitution. We might pause and consider a couple of them.

A Remarkable Generation

The success of the Constitutional Convention had a lot to do with human chemistry. It wasn't that the Framers were all brilliant, learned and forceful—although some of them certainly exhibited each quality. Fifty-five brilliant, learned and forceful men would never have succeeded as well as these did, for they would have recognized no leadership, brooked no compromises, and braved no departures from the settled wisdom. Thus, "chemistry," in the sense used here, suggests that a peculiar mix of traits—brilliance *and* wisdom, forcefulness *and* flexibility—somehow fell into place in Philadelphia. But how did it fall into place? And why?

As a thought question, imagine trying to replicate such a mix today. It has been attempted, of course. Some of the best minds of the Western World came together to work out the League of Nations Covenant in 1919. And the League fizzled. Human chemistry can be very elusive.

Still more elusive is the character of an entire people. Yet it required a certain people to send those particular individuals to Philadelphia in the first place, and to fathom the significance of what they accomplished. The Convention delegates were in certain ways typical of their generation. In a James Madison, a James Wilson, a Roger Sherman, a Gouverneur Morris, we can see patterns of thought and approaches to life that appeared in some form all over the country.

One need only browse through *The Federalist* to feel the weight of this point. What is most remarkable about these essays is not their insight or sophistication but their intended readership. Who, one wonders, was supposed to curl up with such writings, assess their implications, appreciate their wisdom, and be persuaded by their logic? And yet Americans did so, by the tens of thousands. Accordingly, whatever else we think about the Founding, it was the product of an extraordinary generation. We search in vain for another quite like it.

Appeal for Divine Aid

Benjamin Franklin's appeal for prayers at the Convention soon became famous. The important part of the story is that Franklin, like many others, literally did not know where else to turn. The "centrifugal" forces rending that assembly seemed all but irresistible, and time and again the outcome was in doubt.

It was mostly a fear of failure that drove the Framers on—that and a pretty good idea of failure's end result. They bolstered one another's flagging faith by alluding to the eyes of the world, which they imagined to be riveted on Philadelphia, and to the eyes of posterity—which would praise or blame them in epic measure according to how it all turned out. Such references might have been a kind of prayer in their own right. For in the Founders' world, meaning worked itself out in the long run, not the short, and one looked to history and to futurity for the shape of God's purposes.

Then, too, the vision before their eyes was an essentially spiritual one. It was Winthrop's city upon a hill—the New Jerusalem, the Promised Land. It had to be seen in large historical terms to be fully apprehended. Americans had come this far and with this sense of purpose, they told one another. They had uncovered the old within the new—rediscovered the secrets of the ages. It was unthinkable that they should falter now.

A Lucky Roll?

The breakup of the Convention was one possibility. Another might have been a constitution that didn't quite work—like the Articles of Confederation. Given the swirl of politics among them, the Framers might easily have created a system in which forces were out of balance, controls ineffective, and powers dangerously askew.

If this seems beyond imagining, consider the following. The electoral college did not operate remotely as the Founders had anticipated. The election of President and Vice President quickly proved to be unworkable, even hazardous, and required a short-order constitutional amendment. The document was not without its surprises.

With surprises in mind, here is one way to think about the outcome. If two dice can be rolled to produce one of eleven numbers (that is, numbers 2 through 12), we might imagine ten of the eleven as representing the possibilities for a constitution with fatal, debilitating flaws, of the kind we have seen so often. Only one of the eleven numbers represents that rare combination of qualities necessary for long-term success. The Framers had only a single roll of the dice—and they *had* to come up with a "seven."

In Retrospect

More than two centuries later, the US Constitution has come to seem inevitable, even foreordained. Moreover, it has become the admiration of that world of futurity the Framers imagined.

Yet when the Convention came to a close, the draft document wasn't admired by anyone. The Framers had debated every line of its text, not once but repeatedly, and had wound up compromising even the compromises. In the process, there was no philosophy, no authority, no conventional wisdom they hadn't more or less hashed up. To a few of them, it seemed best to confess failure and call for another convention.

If the Constitution had a prayer of ratification, however, the Framers must return to their states and represent it as the "the most brilliant document struck off by the hand of man"—to anticipate Gladstone's praise. That didn't look easy. Yet as they rose to speak in the various ratifying conventions, the Framers began to see their work in a different light. The most accomplished of them led the way—Wilson at the Pennsylvania convention, Hamilton at the New York, and Madison at the Virginia. As it would have been impolitic to speak of compromises, they instead subtly reshaped political theory to account for the way things had come out—as though the compromises had been bold strokes of creativity. And in the process, oddly enough, they actually *became* bold strokes of creativity.

Take federalism, for example. While the delegates had been working it out at swords' points, federalism appeared to be a hopeless mishmash of concessions. In retrospect, however, it took on an aura that was almost cosmic. "Let our government be like the solar system," John Dickinson rhapsodized. "Let the general government be like that of the sun and the states the planets, repelled yet attracted, and the whole moving regularly and harmoniously in their several orbits."

The Constitution was filled with such serendipity. From small features opened worlds of fresh meaning. And in action it was not just a new form of government but a new way of governing. Congress scuffled with itself and bickered with the President. Each made up its mind and changed its mind and had its mind changed by the other. The people watched them both and talked incessantly of approving or disapproving their actions. Every once in a while they became mad as hornets and turned the rascals out. The process seems so normal today that we forget how astonishing it is—and how it seemed to pop up out of nowhere.

So it was that the Constitution began to look more like a revelation than a workpiece. And so it was that Americans began alluding to it as the hand of God in history. For in operation, it seemed to be bigger than any of its authors—or all of them combined.

Ultimately it is the people who accept or reject the Constitution, then as well as now.

CHAPTER 9 STUDY AID

The Meaning of the Constitution

1. Virtue, self-interest, and the good society:

- Human nature
- Virtue and self-interest revisited
- Virtuous self-interest
- The uses of structure: auxiliary precautions; mobilization of virtue

2. Extended Republic:

- Madison
- Factions
- Bigger is better

3. Ambiguous Founding:

- Blended republicanism: classical, liberal, and democratic
- Rule of law: generality, prospectivity, publicity, consent, and due process
- Separation of powers: congressional, judicial, and the executive branches
- Politics
- Checks and balances
- Freedom and the rule of law
- Administration

4. Moral Document:

- Implicit ideals: i.e., dignity of mankind
- Moral founding: the "people"

5. Constitutional Polity:

- Nationhood
- Empire of Liberty: the West
- Federal republic
- Commercial republic
- Inclusive republic

6. Inspired Constitution:

- Legitimacy
- Human element: a remarkable generation
- Divine aid
- Luck and serendipity
- Inevitability?

CHAPTER 10

The Ratification Campaign

Many Americans believe that the adoption of the proposal written in Philadelphia was all but a foregone conclusion. Actually, the ratification of the Constitution was the product of a lengthy and highly contentious debate, and the outcome was anything but certain.

At the debate's beginning, most Americans clearly opposed the new Constitution. After all, they knew nothing about it, had not participated in its creation, and at the very least could suspect it on grounds of "innovation." This meant that the "**Federalists**"—those who favored adoption—had a colossal selling job to perform. And the buyers had a job of their own. They must examine the Constitution in minute detail and ask every tough question they could think of.

It is in this task that the people themselves assumed a role in the founding process, and a demanding role at that. Public debate in America had never been on such a high level—and might never be again. It was, observed John Adams, "the greatest single effort of national deliberation that the world has ever seen."

The fact that the ratification battle was vigorously and closely fought alters the nature of the Founding. For one thing, we can dismiss allegations (once quite common) that the Constitution resulted from an "elitist conspiracy," a "counterrevolution" or any other such coup-by-cabal. There is simply no way that a small group could have accomplished what had to be in order to get the Constitution adopted.

The ratification battle not only broadened participation in the Founding, it altered the terms. Where the Philadelphia Convention had been a deliberation among a small homogeneous group, the adoption campaign involved a very large number, and their backgrounds were as diverse as their thinking. There would be a winning side and a losing side, and thousands of Americans would have to make an up-or-down choice between alternatives. And no compromise was possible, only the most diligent kind of soulsearching. Americans had to ask themselves the tough questions:

What was the true meaning of the American Revolution?

What was the true essence of republican government?

What was the true destiny of the United States?

What had come out of the Philadelphia Convention was a mere proposal. What came out of the ratification debate was a binding commitment, pledging both present and future. And it was that commitment alone that rendered the Constitution legitimate.

The Ratification Process

Like everything else, the Framers in Philadelphia had debated every facet of the ratification process. Some had assumed that the existing state governments would ratify. Edmund Randolph, however, pointed out that if the question were put to state governments, many local politicos fearful of losing their places would either vote against it or else find some subtle way to bring about its defeat.

And there was a deeper issue. Both Madison and Wilson strongly argued for ratification by the people rather than the states, for the people alone were the source of political authority. Gouverneur Morris certainly agreed. In the Constitution's preamble, he opened with words that were certain to raise the ire of state governments: "**We the People** of the United States . . . "

Not everyone went along, however. Elbridge Gerry, still nervous about Shays Rebellion, submitted that the people had "the wildest ideas of government in the world." And several other delegates concurred. Would the people be able to follow the nuances and subtleties of constitutional deliberation? they asked. Did "the people" even exist? Nevertheless, the democrats had their way. Ratification would take place in the several states but by means of special conventions called independently of state government—and representing *the people*.

The Framers also hotly debated the question of how many states must ratify before the Constitution would go into effect. Any number less than thirteen would underscore the proposal's revolutionary nature. After more to-ing and fro-ing, they settled on nine as the magic number. But could they get that many? Rhode Island must be written off immediately. Maryland was almost as doubtful and so was New York. North Carolina was becoming questionable and would in fact be among the last to ratify. Subtracting those four, *only nine states were left*.

The document itself was sent off to Congress in New York. A bare eight days after receiving it, Congress dispatched printed copies to the several states and recommended that they call ratifying conventions.

The Federalists

Federalists:
http://members.tripod.com/~dzierba/index-50.html

The "Federalists" were not just the thirty-nine signers of the parchment drafted in Philadelphia. They were a kind of political party that gusted into being as soon as the Constitution was distributed.

Political parties are complex organisms compounded of ideology and interest. To be successful, they must find a way to recruit dynamic leaders and industrious workers. Even more importantly, they must find a way to engage the hearts and minds of ordinary citizens at the "grass roots." All this was accomplished at fast-forward in the fall of 1787—without telephones, telegraphs, or television. The sheer energy of the task was enormous.

It is one thing to ask why an Alexander Hamilton or James Wilson would put aside business affairs and attend to high matters of state. It is quite another to ask why an ordinary citizen would interrupt his farming or shipbuilding and begin writing letters to old comrades in the regiment. What did *he* expect to gain from constitutional politics?

We should keep the Founders' analysis of human nature in mind as we consider the possible answers. Many kinds of interest—often local in character—were clearly at play in the ratification campaign. But such interests were at play on *both* sides. If a certain farmer in upstate New York supported the Constitution in the hope of gaining a better market for his corn, we can be sure that at least one of his neighbors was ready to oppose it for some equally self-interested reason. We can't explain the ratification of the Constitution, nor the opposition to it, solely in terms of bread and butter.

Ideas can be powerful motivators as well, and it is on this level that the ratification debates fall into patterns that make sense. Federalists and Anti-Federalists alike were urged along by a set of fundamental assumptions, or "founding ideas," and it is these that need to bear the weight of our scrutiny. We already know something of the Federalists' ideas, for these had dominated the Philadelphia Convention. They included the vision of a strong and united nation-state, the promise of a dazzling national prosperity, and a belief that America was destined somehow to become that City upon a Hill.

These were compelling ideas. They appealed to younger men, to optimists, and to those of high personal vitality. Indeed, where the Federalists distinguished themselves was in the sheer assertiveness of their campaign. They seized the initiative early on and refused to let go. They were well organized, capably led, and determined to the point of doggedness. The campaign, some said, called back the glory days of the Revolution.

But knowing the Federalists tells us only half of the story. Voters had to consider the arguments of the *Anti*-Federalists as well, and lacking our hindsight, they didn't know but what the latter had the stronger case.

Federalists envisioned an America of unlimited economic possibilities and prosperity.

A few Anti-Federalists had attended the Philadelphia Convention themselves and could attest to the bargains and compromises behind the scenes—and to the personal motives of some delegates. Of course they too had personal motives. Some were men of prominence in the state governments, and their fortunes were bound up with their circumstances. Others feared the interference of a strong national government in various local enterprises. Yet once again, most Anti-Federalists were men of principle and were every bit as high-minded as their adversaries. The story of the ratification struggle is not a tale of personal pique or thwarted ambition.

The Anti-Federalists

Anti-Federalists:
http://users.erols.com/wmwalker/anti.html

Anti-Federalists have been identified with both classical republicanism and democracy. Accordingly, their opposition has been taken to imply that the Constitution was dangerously deficient in one of two respects: either it violated ancient republican principles or else it placed power in the hands of an aristocratic elite. Anti-Federalists have also been identified with the "true" spirit of the Revolution. They were the Revolution's authentic heirs, so the argument goes, while the Federalists were its interloping stepchildren.

Whatever may be said for such arguments, it is certainly true that many Anti-Federalists *saw* themselves in this way. Thus, understanding the Anti-Federalists helps us to better understand the Founding. To be precise, why did the Founding take one fork of the road and not the other?

For most present-day Americans, the Anti-Federalists have suffered the fate of all historical "losers"—to be ignored, misunderstood, or regarded as irrelevant. They are often associated with the Confederation, a government with few friends today, or with the corruptions of errant states. They ought to have known, we say, that theirs was a lost cause.

In the fall of 1787, however, the Anti-Federalists had a great deal going for them. They better reflected American public opinion than their opponents did. They held the upper hand in crucial states. Their spokesmen were among the famous orators of the country. Their objections to the Constitution were both real and substantive—and wherever they went toe-to-toe with the Federalists, they landed heavy punches. As late as May of 1788, only four states out of thirteen had ratified the Constitution while two others had considered and rejected it. In Massachusetts, one of the four, the vote had been nail-biting close. And in both Virginia and New York, which promised to be crucial, the tide seemed to be running the other way.

Why, then, did the Anti-Federalists lose? They were not, on the whole, as well organized as the Federalists or as capably led. They took too long in mustering their forces to full strength. And they allowed themselves to be outmaneuvered in two semantic particulars. First, they became the "anti's"—those who were against something rather than for it. This made them seem carping and negative. And second, they lost control of the key term in the debate—"federal." What they stood for was the old meaning of the word *federal*—to disperse power and authority—

while their opponents stood for a new meaning which oddly enough signified just the opposite. What the Constitution's supporters wanted was consolidation. But how do you argue against consolidation when it is called "federation" instead?

Below these difficulties lay something deeper. The Anti-Federalists' arguments, however meritorious, began to ring hollow upon the patient, steady thumping of Federalist persuasion. In critical numbers, Americans came to believe that the Anti-Federalists were not just ineffective but actually wrong. The strange story of the Anti-Federalists is that being as prestigious, as knowledgeable and as strategically well placed as they were, they still managed to lose. Or, put another way, if you really do stand for republicanism, democracy and the American Revolution, how can you possibly fail?

Our task, then, is to figure out why the Anti-Federalists' idea of the Founding wasn't the "right" one.

The Anti-Federalist Indictment

The Anti-Federalists spun a lengthy bill of particulars against the proposed Constitution. The President had no council to stay his hand. A Vice President seemed unnecessary and dangerous. The West was sure to be sold out. There would be a peacetime national army at the beck and call of any tyrant. There would be a "federal city" off in the boondocks somewhere, a cross between Rome and Babylon, where power would multiply and debauchery fester. "The greatness of the powers given, and the multitude of places to be created," declaimed Richard Henry Lee, "produce a coalition of monarchy men, military men, aristocrats and drones, whose noise, impudence and zeal exceed all belief." And don't forget, it was all hatched in secrecy. "The evil genius of darkness presided at the Constitution's birth. It came forth under a veil of mystery."

Analyzing Anti-Federalisms' most effective arguments, we find a five-point indictment:

1. *The Constitution would create a government violating most if not all of republicanism's classical precepts.* Republics require a cohesive, homogeneous citizenry. They require a polity small enough to insure the citizen's direct participation. They require the denial of self-interest and the creation of civic virtue, and so they must look askance at commerce with its wealth and luxury. They must eschew concentrations of power, especially executive power, for such is the fodder of tyranny. And finally, republics must depend on the sovereignty of the people, and therefore must trust the will of the majority. The proposed plan of government would accomplish none of the above—on the contrary, their very opposites.

2. *The Constitution would reestablish the hated British forms and thus nullify the Revolution.* The powerful executive would become a "fetus of monarchy"—the phrase still ringing from Independence Hall. The representatives, so small in number, would open the way to aristocracy. The bicameral legislature would

The image of the simple yeoman farmer as the bedrock of democratic republicanism is one of the most enduring images to emerge from the dawn of the American nation.

look suspiciously like Parliament with its upper and lower chambers, and the elitist Senate would transmogrify into an American House of Lords.

3. *The Constitution would separate the people from the governing process and subject them to arbitrary restrictions.* Look at the long terms of office. Look at the blocks against direct election. Look at the smug and aloof judiciary with its lifetime tenure. And look at Madison's "counterpoise"—what was it but the defeat of popular sovereignty?

4. *Federalism in practice would amount to "splendid empire," for the residual sovereignty of the states would prove but a dream.* Note all those loosely worded clauses in the Constitution—they would never bind anyone. The powers enumerated in Article I, Section 8, while seeming to be limited, were in fact made infinitely stretchable by the "necessary and proper" clause—which Alexander Hamilton had already interpreted to mean "anything." And if one didn't like Hamilton's interpretation, there was always that standing army to back it up.

5. *There was no bill of rights.* Nor was the omission of such guarantees a mere oversight. Delegates to the Philadelphia Convention had thoroughly discussed the possibility of a bill of rights—and unanimously rejected it.

This was an impressive list. But the Anti-Federalists were not just "anti." They had their own vision of America, and it was quite as compelling as that of the Federalists. It emphasized, in the first instance, a healthy diversity. Anti-Federalist America would be a patchwork of local cultures, each of them colorful and distinctive, and none of them answerable to national norms. It emphasized, in the second instance, the old republican idea of virtue. Americans would live a pastoral life close to the soil and close to the primary verities, a life of republican plainness and simplicity. Finally, in the third instance, Americans would exercise power themselves, not pass it along to some distant capital. Theirs would be an energetic, town-meeting style of governance, and as a result, government would remain under the guidance of decent, ordinary, God-fearing citizens. We need only think of the murals of John Steuart Curry or the movies of Frank Capra to appreciate how "American" this idyll was. If the Federalists' vision of America was a City upon a Hill, that of the Anti-Federalists was something more like *Our Town*.

***Our Town*:** Play by Thornton Wilder dealing with everyday life in small New England town.

Anti-Federalist Strategy

Ratification of the Constitution: http://www.nara.gov/exhall/constitution/conhist.html

Not only did the Anti-Federalists have an alluring vision and solid arguments, they had a promising political strategy. They would press hard on the bill of rights issue, where the voters might feel most vulnerable, and use this to force the calling of a new convention. The Anti-Federalists well knew that the Constitution was compounded of delicate compromises and hairspring understandings. If a new convention were called, everything would undoubtedly come unstuck.

But why indeed was there no bill of rights? Surely if Americans felt vulnerable without one, the Framers would have been savvy enough to include it. Yet they hadn't. And they had thought carefully about their reasons. In the first place, under the Constitution the people themselves were sovereign, so there was no authority beyond their own which must be hedged about with restrictions. In the second place, by limiting powers which were not granted in the constitutional text, a way might be opened to claim the existence of such powers—a boomerang if there ever was one. And finally, with no means of enforcing the rights in question, such rights might go the way of all parchment barriers, to be ignored and forgotten. "No bill of rights," wrote Roger Sherman, "ever yet bound the supreme power longer than the honeymoon of a new married couple, unless the rulers were interested in preserving the rights."

These were worthy arguments. But the Anti-Federalists had equally worthy replies to them. Take the first one, for instance, that the people did not need a bill of rights against themselves. Was that really true? Anti-Federalists asked. Were the people any better guarantors of rights than kings were? In particular, were factional majorities apt to respect the rights of their foes? Madison himself had stewed about that.

Or take the second argument, that there was no need to limit powers not given. If there was no such need, the Anti-Federalists countered, then why had the Framers included clauses guaranteeing trial by jury and *habeas corpus*, or prohibiting bills of attainder? In all of its enumerated powers did the Constitution even hint at such authority?

Or the third argument, that bills of rights were unenforceable. The point granted, the Anti-Federalists said, that had never stopped anyone in the past. A bill of rights was not supposed to be *enforced*—it was supposed to be a rallying point for public opinion. Whoever "enforced" the *English* Bill of Rights?

These were sobering responses, especially the last one, and not without effect. Score one for the Anti-Federalists. What they didn't count on, however, was the rapid deployment of a counterstrategy, and precisely that soon followed. Rather than continue denying the need for a bill of rights, Federalists pulled themselves up, conceded the point gracefully, and then urged that the Constitution be ratified anyway since it included all the machinery for adding a bill of rights by amendment. Such a reply was typical of the Federalists. It promised, in effect, that voters could have their cake and eat it too. They could have the Constitution with all its advantages *and* their bill of rights.

Federalists knew the American people well.

The Federalist Defense

The Anti-Federalists realized that the game would be won or lost in the big states: Pennsylvania, Massachusetts and Virginia. Without those states, the Union couldn't possibly succeed—nor could the Confederation possibly survive. And in any case, nine states had to ratify before the Constitution took effect.

As the battle took shape, it became clear that the Framers themselves would be the frontline defenders of their work. They were well equipped for the task. And so we see them outside the stuffy hall with its heat and flies, where they could speak their minds openly at last. There were the Pinckneys down in Charleston, drawling softly as they outlined the Constitution's advantages to fellow planters. There was Hamilton in New York, intense, obsessive, fairly bursting with energy, anxious to face down his nemesis Melancton Smith. There was Oliver Ellsworth in Connecticut, his clothing dusted with snuff, citing scripture to the state delegates—eighty of whom had names out of the Old Testament.

Clearly too, all stops would be pulled out and all authorities pulled in. The two most important authorities were John Adams, serving as minister in London, and Thomas Jefferson, similarly stationed in Paris. Adams was pleased with the Constitution—he pronounced it "the greatest exertion of human understanding"—though he couldn't help noting that it simply copied his own thinking. Jefferson, on the other hand, had his doubts. He didn't like the idea of the executive succeeding himself over and over, and he wondered about that missing bill of rights. But after an urgent letter or two from his friend Madison, he was ready to say: "The Constitution . . . is unquestionably the wisest ever presented to men."

Thus primed, Federalists took the first hand. Delaware ratified unanimously on December 6. New Jersey followed on December 16, also unanimously. Georgia ratified on January 2, 1788, and Connecticut came through a few days later with a heavily lopsided vote. Four states were in the bag—but they were all small.

Pennsylvania was the first large state and the Constitution's first major test. Three of the Core Framers—Wilson, Morris, and Franklin—were Pennsylvanians of high standing. Still, the state's politics was a donnybrook under the best of circumstances, with partisan bushwhacking, mass demonstrations, even the occasional riot, and the constitutional question promised to be an especially lively one.

Federalists had a comfortable majority in the Assembly. Their foes, however, had the ear of the democratic electorate and were prepared to make the most of it. If they could somehow delay the vote for a ratifying convention, they had every chance of electing a friendly majority in the Assembly and Anti-Federalist delegates to the convention later on. It soon came down to bare-knuckle tactics. The Federalists pushed for an immediate vote. The Anti-Federalists sought to block the vote by keeping their own members away and preventing a quorum. Mob action ensued. Bully boys broke through a door, roughed up two of the Anti-Federalist truants, and bodily carried them to the State House, where they were held in their seats for the roll call. The resolution for a ratifying convention passed 45 to 2.

Nor was the convention itself much quieter. Many of the delegates knew little of the proposed Constitution and there were some who didn't want them to learn. As Davie of North Carolina said, "It is much easier to alarm the people than to inform them." But James Wilson delivered a masterful performance. His head held high to keep the thick glasses balanced on his nose, Wilson patiently answered objection after objection, his Scottish burr trilling out the "r's." It proved to be an experience in democratic politics to cheer the most jaded skeptic.

The convention sat for a respectable five weeks and then ratified the Constitution by a vote of 46 to 23. Wilson's reward was to be mugged by a band of ruffians while he was out celebrating. He was severely beaten and might well have been killed.

The next test came with Massachusetts. This was the largest of the conventions, with 355 delegates. They gathered at the Brattle Street Church in Boston, amid a storm of controversy in the press. As in Pennsylvania, local politics played a manifest role in the proceeding as backcountry farmers rattled into town to express themselves on the subject of seaboard merchants and their political shenanigans. Massachusetts, said the pundits, would never ratify.

But once again an odd chemistry took effect. Elbridge Gerry, the Massachusetts Framer who had refused to sign in Philadelphia, expected to dominate the proceeding with his eleven-point critique, but the mayhem in Brattle Street was too much for him and he departed in a huff. After that, the accomplished trio of Rufus King, Nathaniel Gorham and Josiah Strong—the remaining Massachusetts delegates—held sway. Following Wilson's path in Pennsylvania, they explained the Constitution's fine points with a calming self-possession and fielded question after question from the floor.

The other distinguishing feature of the Massachusetts battle was the role played by ordinary citizens. One of them, a rough-hewn and self-taught farmer named Amos Singletry, discoursed at length on the Constitution's evils:

> Does not this constitution . . . take away all we have—all our property? Does it not lay all taxes, duties, imposts, and excises . . . ? These lawyers and men of learning, and moneyed men that talk so finely, and gloss over matters so smoothly, to make us poor illiterate people swallow down the pill, expect to get into Congress themselves. They expect to be the managers of this Constitution, and get all the power and all the money into their own hands. And then they will swallow up us little fellows. . . .

Federalists replied in kind. One of them was also a farmer, also unlearned, and also down-to-earth eloquent. "Mr. President," said Jonathan Smith,

> when I saw this Constitution, I . . . got a copy of it, and read it over and over. I had been a member of the convention to form our own state constitution, and had learnt something of the checks and balances of power, and I found them all there. . . . My honorable old daddy there [referring to Singletry] won't think that I expect to be a Congress-man, and swallow up the liberties of the people. I never had any post, nor do I want one. But I don't think worse of the Constitution because lawyers, and men of learning, and moneyed men, are fond of it. . . . These lawyers, these moneyed men, these men of learning, are all embarked in the same cause with us, and we must all swim or sink together. . . . We sowed our seed when we sent men to the Federal Convention. Now is the harvest. Now is the time to reap the fruit of our labor. And if we don't do it now, I am afraid we shall never have another opportunity.

It was a telling performance. Equally telling was Sam Adams' dramatic swing from opposition to support for the Constitution, and Gov. John Hancock's own bellwether endorsement. Anti-Federalist ranks broke on February 5 and Massachusetts ratified 187 to 168. Had ten votes gone the other way, the Union would have undoubtedly died stillborn then and there.

Anti-Federalists took heart from this fact. If they could block ratification in New York and Virginia, both of which were due to convene in June, the Constitution might be defeated yet. But Maryland, one of the doubtfuls, ratified in April and South Carolina followed in May. That made eight states, one short of the nine necessary, and New Hampshire was leaning toward a scrape-by affirmative vote. Still, without Virginia and New York on board, it was hard to see an American nation setting sail.

If Massachusetts' had been the largest of the state conventions, Virginia's was the ablest. The Assembly Hall of the House of Burgesses was packed with members of the tidewater elite. Madison was there of course, and so was George Washington, but white-haired and blue-blooded Hugh Grigsby, a shrewd observer of the Old Dominion aristocracy, submitted that neither of those two Founding Fathers "stood in the estimation of Virginia on the same platform with Patrick Henry and George Mason as statesmen." Henry had boycotted the Philadelphia Convention when he "smelt a rat." And Mason had sat through every hour of it and denounced the outcome.

Henry showcased the Anti-Federalist approach to constitutional debate at its most passionate. "Whither is the spirit of America gone?" he asked in lamentation:

> Whither is the genius of America fled? . . . We drew the spirit of liberty from our British ancestors. But now, Sir, the American spirit, assisted by the ropes and chains of consolidation, is about to convert this country into a powerful and mighty empire. . . . There will be no checks, no real balances, in this government. What can avail your specious, imaginary balances, your rope-dancing, chain-rattling, ridiculous ideal of checks and contrivances?"

Against such a performance, James Madison must have felt puny indeed. He stood before the delegates beggar-like, his hat in his hand (and his notes in his hat), his voice so frail that those in the rear had to strain to hear him. And in the way of most Federalists, he simply plodded along with the dull, prosaic facts of the matter, answering questions, allaying concerns, parrying Henry's oratorical thrusts.

The real star of the show was Edmund Randolph. He too had been at the Philadelphia Convention and had refused to sign the Constitution. However, unlike his friend George Mason, who now hurled thunderbolts against the document, the stage-handsome governor had undergone a change of heart worthy of Damascus Road. Not only was he now in favor of ratification, he was in favor of it heart, soul and pocketbook. When Madison became so exhausted that he had to take to his bed, Randolph rallied the flagging Federalists and took command. It was, said Catherine Drinker Bowen, the political performance of his life.

The Virginia convention made its decision on June 25. The Constitution won by a slender ten votes. When news of the victory reached Poughkeepsie on July 2, the New York delegates had been convened there for two weeks. It was another case of tenpins wobbling precariously before knocking each other down. For if ratification had failed in the Old Dominion, it would surely have failed in New York as well, but as matters now stood, Virginia was in the Union (along with dithering New Hampshire) and the Constitution had sprung into life. If New Yorkers opted out now they would be literally on their own.

All the same, the New York convention brought forth its own Federalist hero. Alexander Hamilton, who had been abandoned in Philadelphia by his New

York colleagues, battled for the Constitution like a Roman gladiator and won almost single-handedly. "He is," observed Jefferson in another context, "a host within himself." The vote, taken on July 25, saw ratification squeak through 30 to 27.

As the New York debates had heated up after the Federal Convention, Hamilton and his friend John Jay, supposing that their badly splintered state might prove pivotal in the ratification contest, had teamed up with James Madison in what would prove to be the most famous collaboration in American political writing. *The Federalist* was a series of eighty-five newspaper essays about the Constitution. Under the pen name Publius, they appeared first in New York papers and were then reprinted throughout the country. They were written in great haste, of course, and many of them made for a ponderous read. But in terms of sheer insight, there was nothing to compare with *The Federalist*, then or since.

Hamilton and Madison (Jay, whose health was poor, participated in only three of the papers) undertook a searching inquiry into the nature of republics, and the effort paid off handsomely. For whatever else *The Federalist* accomplished, it pointed up the Federalists' cardinal strength. Having debated every line of the Constitution in Philadelphia and having learned principles through painstaking research, the authors didn't need rhetorical fireworks because they could use real dynamite.

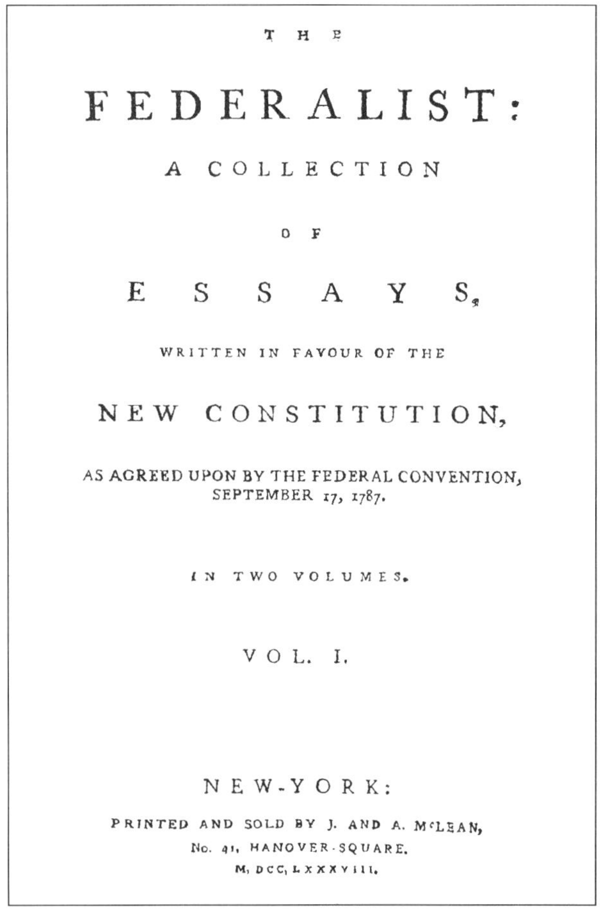
THE

FEDERALIST:

A COLLECTION

OF

ESSAYS,

WRITTEN IN FAVOUR OF THE

NEW CONSTITUTION,

AS AGREED UPON BY THE FEDERAL CONVENTION, SEPTEMBER 17, 1787.

IN TWO VOLUMES.

VOL. I.

NEW-YORK:

PRINTED AND SOLD BY J. AND A. M'LEAN,
No. 41, HANOVER-SQUARE.
M, DCC, LXXXVIII.

The Federalists Papers *were and remain one of the most influential documents in American history. Representative of the spirited public debate surrounding the ratification of the Constitution, the Papers offered cogent arguments favoring its passage.*

"Publius" was particularly adept at justifying the Constitution's many compromises. Accordingly, what might be depicted by Anti-Federalists as a sellout (or even a washout) took on brilliance and sagacity in *The Federalist* essays. There was Madison's theory of factional behavior in a large republic. There was Wilson's theory of sovereignty as "power on loan." There was the idea of the President as representative at large. There was separation of powers. These weren't just clever apologies—they were breakthroughs in political thought.

It has been pointed out that *The Federalist*'s profundity was largely wasted. For the essays were not widely read by their intended audience and it can't be shown that they changed a single vote in Poughkeepsie. This critique misses a salient point. *The Federalist* was but an example, though an inspired one, of the Federalists' entire approach. What the essays accomplished in formal prose was also taking place on street corners, in city taverns, and in public meetinghouses throughout the country. The Federalist style of calm, steady, rational argument, emphasizing cognitive depth rather than rhetorical display, triumphed in the ratification campaign. It remains a force in US politics to this day.

There were two areas in particular where the Federalists could go for the jugular. The first was the order and stability argument. With the Confederation clearly failing, the only alternative to a stronger union seemed to be no union at all, and that would mean continuing trade wars, political frictions, boundary disputes, contested land claims and all the other baleful offspring of anarchy. Americans didn't want that.

Equally troublesome for Anti-Federalists was the human nature argument. Most Americans realized that human beings respond to their selfish desires rather than some abstract public good, and that this fact explained the failure of ancient republicanism. Madison's spread-eagle nation-state with its candid avowal of self-interest brilliantly solved this problem. Interests could conflict and collide, Madison promised, and the outcome would automatically redound to the public good. People could serve the public interest by serving their own.

In both of these assertions, the Federalists were once again promising it both ways. State sovereignty would preserve all the vitality of localism—while union would bring the blessings of empire. And Americans could gain all the benefits of virtue without repressing their human appetites. Such arguments were simply unbeatable.

Capitulation

In the end, the ratification campaign turned into a rout. For as soon as key states had ratified, the Confederation lost all viability. At that point, the only alternative for the holdouts was going it alone, and even feisty Rhode Island wasn't up

With the ratification of the Constitution and the creation of the office of President, George Washington was once more called upon to serve the American cause.

to that. Together with equally reluctant North Carolina, the tiny polity eventually threw in the towel.

Another kind of capitulation occurred on the floors of state conventions after the deciding vote. There were a few diehards who stormed back to their domiciles in disgust, but for the most part, winners and losers alike offered apologies and reconciliation.

Celebrations in the various states were something to behold. In one of them, held in Philadelphia, there was a "Grand Procession" a mile and a half long winding through the city, its centerpiece a float of the "Grand Federal Edifice," drawn by ten white horses. Harking back to the ancient world, this temple of American nationhood was supported by thirteen Corinthian columns, and upon its dome, bearing a cornucopia, stood the figure of *Plenty*. The symbolism seemed right.

Even grander celebrations were held in April of the following year when President Washington took the oath of office. He had performed his own quiet heroism in the ratification campaign, writing to Federalists and Anti-Federalists alike, urging, chiding, cajoling and dolefully lamenting as the skirmishes were fought. We don't think of Washington as a politician, but he truly was one of the best.

Why Federalism Won

The politics of the ratification campaign and the swinging of the balances were only visible manifestations of a much deeper phenomenon. The American people as a whole were making a decision, irrespective of local situations, and that decision was clearly *for* the Constitution.

Every once in a while, one of them sat down and wrote out his thinking on the matter, as in the case of Jonathan Smith described above. For thousands of others, we can only make guesses based on fragmentary evidence. But the general picture seems reasonably clear.

To begin with, Americans were more apt to side with faith and hope than with doubt and fear. Thus, where Anti-Federalists decried "commercialism," most Americans tended to think of prosperity instead, and where Anti-Federalists denounced "empire," their countrymen envisioned that City upon a Hill. Their optimism was bolstered by the Constitution's amendability. The document may not be perfect, they knew, but its flaws could eventually be ferreted out and repaired.

Then, too, as children of the New World, Americans truly believed in having it both ways. They were less interested in theoretical purity—as in, say, classical republican doctrine—than they were in practical outcome. So when the Federalists promised them all the benefits of the ancient republics and none of the costs, they replied: "Sounds good."

As a third point, Americans began to see that the Federalists were every bit as devoted to republicanism as their adversaries—they just weren't shackled to it. For the Anti-Federalists, republicanism was a kind of civil religion, and one had to accept its doctrines as a confession of faith. Federalists were more pragmatic. For them, classical republicanism with its rigid moral demands was no more valid or compelling than the liberal or democratic strains of republican thought, and it was far less relevant to American life. In consequence, they were

Anti-Federalist Fears Founded?

One of the main concerns of the Anti-Federalists was the specter of empire that the Constitution raised. In their arguments, they cried over and over again that the concept of parallel sovereignty was no more than a pipe dream and would lead to a large, tyrannical national government. While we all like to laugh at the seemingly unfounded worries of the Anti-Federalists today, some of their ideas have played out in American history. Ever since the signing of the Constitution, states' rights have been continually eroded. The tension between national and state sovereignty was played out most dramatically in the Civil War, but the issue still has not been fully resolved. Even in the twentieth century the fears of the Anti-Federalists have been recalled by event after event.

Consider, for example, the made-for-a-textbook case of Little Rock, Arkansas, in 1957. In the fall of that year, a number of the community's citizens, angered by the prospect of school desegregation, decided to defy a federal court order and simply refuse to integrate. They knew, of course, about the supremacy clause of the Constitution, but they supposed that states had their rights too, and that somehow, such rights could be enforced independently of the federal judiciary.

On the first day of school, nine black children appeared at Central High ready for class. There was trouble. A venomous crowd was milling about on the front steps and growling angrily about a lynching. Hesitantly, school authorities appealed to the Little Rock police department, and from there to the mayor's office, where it was decided to close the school. But the mayor of Little Rock was not sovereign, and so the decision was appealed still higher. The governor of the state, Orval Faubus, also decided against integration and called out the state militia to support the local police. However, since a federal court order was also at stake, the issue had to go higher still. Ultimately, it wound up on the desk of Dwight Eisenhower, President of the United States, who decided that the court order should be upheld. This is exactly the kind of scenario the Anti-Federalists predicted would occur. And what happened next was also what the Anti-Federalists knew would happen.

Eisenhower won out. To prove his point and establish the supremacy of the federal government, he nationalized the Arkansas state militia—meaning that Faubus lost command of his own troops—and sent in the 101st Airborne for good measure. The division's crack 327th Battle Group lined up outside Central High with fixed bayonets. Then the soldiers slowly advanced on the shouting throng of obstructionists. One stout soul had to approach within six inches of a gleaming bayonet before fully comprehending where ultimate sovereignty was now vested.

Still the question of whether or not sovereignty can reside in parallel entities continues to haunt the nation. Consider the outcome of the 2000 presidential election. Who finally determined the final vote count in Florida? True, the federal courts became involved on several levels, but in the end the state of Florida was still responsible for certifying its election results. And the nation was bound by that outcome.

Revision of vignette found in
America: A Study in Heritage,
Kendall/Hunt, 7th Edition, p. 146.

willing to jury-rig a system from component parts, some classical, some liberal and some democratic, mixed and matched to their own situation. What was important to them was not adhering to a particular doctrine but pursuing a general spirit—one familiar to all Americans. *This thing doesn't work*, they said in effect, *so let's figure out how to fix it.*

Finally, Americans became persuaded that union was necessary to complete the American Revolution and endow their national experience with a true "founding." The Constitution's ultimate promise was that Americans, as a people, would be able to live a stable, secure and happy republican life. Nothing short of that promise could make their history add up. If all those terms—freedom, justice, opportunity, virtue—were to have real and practical value, then Americans, who had risked so much already, had to take that final historic chance.

In sum, each side had offered a solution to the Republican Problem. The Anti-Federalists' solution was familiar but problematic. The Federalists' solution was bold and risky but not unreasonable. There was something decidedly "American" in leaning toward the second alternative.

From Ratification to Politics

Anti-Federalism collapsed so quickly that the ratification controversy came to seem chimerical. It was not. Anti-Federalists had raised a number of issues that were not only not chimerical, they would never go away. We could rummage through the debate and pick out some questions that were destined to cast long shadows into the American future:

> ***Was the grant of power to the federal government indeed too broad and loosely worded?***

This question would dog the heels of the Washington and Adams administrations, and culminate in the dramatic election campaign of 1800. Themes of big government, busy government, distant government and unresponsive government have run through American politics ever since.

> ***Could sovereignty truly be vested in parallel entities, or was it unitary by nature, as the Anti-Federalists claimed?***

Questions of states' rights have been around as long as questions of curbing the government. Even in the early days, when the states seemed of much greater consequence than the Union, there was a steady erosion of power from the periphery to the center. The American Civil War resulted from the inability of states to assert themselves against a determined national government. When push came to shove—as the Anti-Federalists had prophesied—one kind of sovereignty or the other had had to be decided on the battlefield.

> ***Was republican society compatible with "imperial" nationhood, as the Federalists argued, or were the two hopelessly at odds?***

America lived up to every detail of Hamilton's dream, becoming a nation to reckon with, even to fear, in the eyes of the world. For the most part, its international influence has been salutary, as republican idealism would require. But there have been some gray areas. The US wrested large chunks of the continent from neighbors to the north and south, then became a player in the global power game. As for its relations with native inhabitants, there is little or no defense for them. Anti-Federalists had their point about the choice between republic and empire.

Was a real "public good" to be found among contending private interests, as Madison claimed, or was that promise simply too good to be true?

Pork Barrel: Government projects or appropriations allocated on the basis of patronage.

Unquestionably, the extended republic delivered the strength and stability that Madison promised. The "public good," on the other hand, has proven more elusive. The way the federal system operates, representatives frequently seek measures of benefit to the folks back home without considering anyone else. The result has often been government by pork barrel. The Anti-Federalists may have been right when they argued that public good can only be achieved through civic virtue.

Finally, was republican society compatible with "commercial" values, as the Federalists maintained, or were the two destined to be forever at odds?

PACs: Political action committees formed to lobby for specific issues.

Commerce run amok has become yet another American theme. From the "trusts" of Rockefeller's day, to the "malefactors of great wealth" denounced by Franklin Roosevelt, to the corporate-sponsored PACs of the present day, "big business" has repeatedly been cast as the bogeyman. And in a different way, so has bourgeois materialism. What good is it to gain the world and lose one's own soul? the poets have asked of our pleasure- and gadget-loving society. And the main thrust of the counterculture has been against our enslavement to possessions.

These were only a few of the questions tumbling out of that first face-off, between supporters and detractors of the Constitution, and every one of them was to wind its way through US history. All of which suggests that the ratification debate was but a rough draft of American politics.

In the end, it was the American people who cast their lot with an untried form of government and brought the United States of America into being.

The fact that the choices were not simple or the outcome foregone establishes the adoption of the Constitution as the central event of the Founding. The American people decided to cast their lot with a strange form of government, man-made, oddly shaped and without historical precedent. For all they knew, that decision truly was the great mistake of their time—as the Anti-Federalists might have put it—and the downfall of the American Revolution. Whatever happened from that point forward, for good or ill, it would trace back to the moment the ninth state voted to ratify.

CHAPTER 10 STUDY AID

1. The Constitution and its accompanying issues:

- Innovation
- Behind closed doors
- Was it elitist and self-serving?
- How was it to be debated?
- What did the Revolution mean?
- What was the essence of republican government?
- What was the United States?

2. The process of ratification:

- Options for ratification
- Solution: by the people of each state; special convention; by representatives chosen by the people
- Majority ratification

3. The Federalists:

- Almost a party: interest, ideology, leadership, rank & file, grass roots appeal
- Endorsed strong national government to enforce rule of law
- Prosperity through commercial republic
- Republican government (liberal republicanism to an extent)
- Vision of a city upon a hill

4. The Anti-Federalists:

- Not a quite a party: reflected initial public opinion; dominated key states; made strong arguments
- Feared interference from a strong national government
- Feared loss of autonomy and influence
- Republican government (classical and democratic to an extent)
- Constitution too liberal and elitist
- Saw themselves as true heirs of the Revolution

5. Anti-Federalist arguments:

- Constitution violates classical republican precepts
- Constitution mimicked corrupt British forms
- Constitution separated people from the government

- Constitution meant empire not republic
- No bill of rights
- Anti-Federalists' America: diversity, virtue, self-government

6. Anti-Federalist strategies:

- Demand bill of rights: protect people from own excesses
- Demand new convention: enumerated powers insufficient

7. Federalist strategies:

- Yield on bill of rights
- Focus on three most dominant states
- Gain endorsements of Adams and Jefferson
- Argue for order, stability, and security in union of states
- Argue structure needed as precaution against human nature

8. The voting:

- Delaware, New Jersey, Georgia, Connecticut
- Pennsylvania (dirty tricks), Massachusetts (close)
- Maryland, South Carolina make eight
- Virginia (close) is ninth, New Hampshire
- New York: The Federalist Papers

9. Why?:

- Bandwagon
- Will of the people
- Vision of city upon a hill
- Having it both ways: theory and practice
- The spirit, rather than the letter, of republicanism
- Necessity: promise of stable, secure, happy, republican life seemed to fulfill the promise of the Revolution
- Then and now Americans tend to be Anti-Federalists at heart but Federalists in their mind and purse; the tendency is to vote the latter

10. The long term outcome:

- Big government: fears not unfounded
- States' rights: diminished, everything becomes a federal case
- Imperial nationalism: imperialism; imperial presidency
- Public good v. pork barrel: virtue v. self-interest at work
- Commercialization: the business of America is business?
- Where do we stand on the issue of the Constitution today: would we undo it or would we keep it?

CHAPTER 10 STUDY GUIDE

NAME: ______________________ **SECTION:** ______________

Key Terms

Ratification

Federalists

Anti-Federalists

The Federalist Papers

Publius

Multiple Choice Questions

1. All of the following were aspects of the ratification process except:
   a. Citizens within the states must call for the ratifying convention.
   b. The ratifying vote need only encompass a simple majority.
   c. The Constitution had to be approved by all state legislatures.
   d. Once ratified by nine states, the Constitution became the law of the land.
2. Which of the following was not one of the Anti-Federalist indictments against the Constitution:
   a. It lacked a bill of rights.
   b. The national government would be too far removed from the people.
   c. States would lose sovereignty.
   d. The ratification process was not legitimate.
3. Which of the following Founders was not an author of the Federalist Papers:
   a. Alexander Hamilton
   b. Thomas Jefferson
   c. John Jay
   d. James Madison
4. Which was the fourth state to ratify the Constitution:
   a. Delaware
   b. Connecticut
   c. New Jersey
   d. Georgia

Review Questions

1. According to the text, how did the ratification process alter the meaning of the Founding?

2. What did the Anti-Federalists have going for them? The Federalists?

3. Why were the Anti-Federalists so interested in including a bill of rights in the Constitution?

4. Why did the Anti-Federalists feel that the Constitution violated republican principles?

5. What is meant by the notion that public good can only be achieved through civic virtue? Did the Anti-Federalists violate this principle? Explain.

Review Exercises

1. Briefly describe the Anti-Federalist vision of America.

2. Summarize the arguments made in Massachusetts by Amos Singletry and Jonathan Smith. What were their main points of disagreement?

3. Describe the Federalist approach to endorsing the Constitution, as exemplified by the Federalist Papers.

Big Essay Question

In the previous question you described the Federalist approach to endorsing the Constitution. How might the Federalists alter their tactics if the ratification campaign was going on right now? What would they do differently, or the same, to "sell" the Constitution? Be thoughtful and creative.

CHAPTER 11

The Bill of Rights

When the Bill of Rights became part of the Constitution in 1791, it was added by way of the amending process spelled out in Article V. Its enumerated items, arranged in clusters, became Amendments One through Ten.

Placing the Bill of Rights at the Constitution's end underscored its importance. As an addendum, it has occasionally threatened to overshadow the document it was meant to complement.

The Constitution and the Bill of Rights

Where the Constitution (like most blueprints) made for dry reading, the Bill of Rights had a ringing, rhetorical quality—as if asserting the cause of all mankind. There was another contrast. For those who saw the Constitution solely in terms of power, the Bill of Rights changed the subject, as it were, to freedom and dignity.

Bill of Rights:

http://www.nara.gov/exhall/charters/billrights/billmain.html
http://www.aclu.org/library/pbp9.html

Modern-day political culture has sometimes gone even further and imagined a contradiction between the Constitution and the Bill of Rights—as if they were written by two different groups bound in opposite directions. Yet James Madison, the Constitution's primary Framer, also drafted the Bill of Rights. And he saw the two not in contradiction but in consonance—twin cornerstones of the Founding. He even suggested placing the Bill of Rights within the body of the Constitution rather than at the tail as a series of amendments.

For us to appreciate Madison's point of view, we must retrace the history of the Bill of Rights and take a closer look at its key ideas.

The Demand for a Bill of Rights

When Washington had invited fellow delegates to sign the completed Constitution, George Mason had stunned them all by refusing to comply. The Constitution, he pronounced, was sadly and unforgivably wanting a bill of rights.

The term was not unfamiliar. After all, the English Bill of Rights (1689) was a landmark of America's constitutional heritage. And the Revolution itself had been propelled by rights-conscious slogans like: "No taxation without representation!" "From the beginning," write Kurland and Lerner, "the language of America has been the language of rights." More broadly, bills of rights were commonly discussed throughout the western world in the eighteenth century and here and there they actually materialized. In the US, no fewer than eight state constitutions had bills of rights attached to them. That of Virginia, which was often a model for the others, had been drafted by George Mason himself.

In order to understand what a bill of rights was and what it was intended to accomplish, we must recall that most governments at the time were monarchies and their powers were the prerogative powers of kings. Limiting such powers had proven to be a difficult task. A monarch might grant concessions of one kind or another in a time of crisis, but little by little he usually gained back the lost ground—and even added a few acres. Accordingly, the function of a bill of rights was twofold. First, the king's subjects got a chance to assert their rights in black and white. And second, the words on paper served to remind both parties, king and subjects, that the assertion had been made. As John Dickinson argued in an influential newspaper essay, a bill of rights would establish "in the minds of the people truths and principles which they might never otherwise have thought of, or soon forgot."

The rights asserted in such bills were almost always derived from ***natural rights***, which like natural law were held to be universal in character and comprehensible to everyone. But because natural rights were vague and open-ended—the right to live, the right to be free—it was thought necessary to draw up a list of particulars to bring the rights down to earth. Some bill of right documents were rather lengthy, listing out scores of specific items. Most, however, were fairly short, numbering between ten and twenty. Brevity added rhetorical punch.

Rights may be asserted, yet their enforcement may be another matter altogether. Witness the events at Tiananmen Square in China in 1989. The Chinese constitution offers an extensive list of rights that were asserted by student demonstrators. However, the government interpreted these rights quite differently as it systematic suppressed pro democracy political dissent.

At their most effective, bills of rights took the general concept of rights and shaped it into what might be called ***Great Oughts***. Whenever you have suffered an injustice, a voice within often cries something like: *A person ought to be allowed to* ______________ (fill in the blank). Think of the Great Oughts as the most profound and fundamental utterances of that voice. And when other people hear them, they can become banners to rally around.

The language of rights is not the language of supplication. Indeed, asserting a right is in effect a claim coupled with a threat. And yet, paradoxically, rights by their very nature are virtually unenforceable, which may be one reason why we get so upset when they are violated. Still, in the general way of eighteenth-century politics, bills of rights did serve to remind the king and his ministers that they might be treading on shaky ground.

At the time of the Founding, some Americans were downright skeptical about bills of rights. For one thing, the whole concept was easily misunderstood. "Rights," if they existed at all, were necessarily limited by the rule of reason, by political and economic circumstances, and by the logic of governance. Those who failed to grasp such limits could set up a clamor for their rights that might easily lead to anarchy. Nor did rights enhance the spirit of community. The rights of the people *as a people* were one thing; the rights of individuals were quite another. For individuals must always assert rights *against* society.

Then, too, listing rights out always posed a problem. Why was this right to be included in a given catalogue and not that one? What happened to the rights that were excluded—did they cease to exist? And who got to decide?

James Iredell, an early Supreme Court Justice, pointed out that the language of a bill of rights could easily be reduced to mere babble, and Noah Webster provided a handy illustration. On his list of rights, he said, he would include stipulations "that Congress shall never restrain any inhabitant of America from eating and drinking, at seasonable times, or prevent his lying on his left side, in a long winter's night, or even on his back, when he is fatigued by lying on his right."

Tacking a bill of rights onto a *republican* constitution was especially troublesome. Unlike royal prerogative, the powers of republican government were those of the people themselves, and it seemed absurd for the people to curtail *their own* sovereignty. On the other hand, the *difficulties* posed by bills of rights could afflict a republican government quite as much as any other.

Even more so. The Framers in Philadelphia were particularly troubled by the notion of mixing bills of rights with enumerated powers. For the strategy of enumeration to work, there had to be a binding assumption that the government could act only in the enumerated areas and no other, yet any mention of rights was bound to suggest otherwise. For example, stating that the government could not abridge religious freedom implied that it had such authority in the first place, even though that authority was never enumerated.

Most importantly, for the Framers in Philadelphia, the Constitution was its own bill of rights. By enabling the people to live republican lives under a government they themselves controlled, it secured for them all the blessings that a bill of rights could ever hope to. Accordingly, as George Mason well knew, the Framers had carefully considered his arguments for a bill of rights and rejected them 10 to 0.

The Bill of Rights in the Ratification Debate

Anti-Federalists, as we know, saw an important opportunity in this situation. If opposition to the Constitution could be mustered on the grounds of its lacking a bill of rights, they might be able to force the calling of another constitutional convention—which would presumably deal with their *real* objections. Many Federalists were angered by what they saw as a cynical use of the bill of rights issue. It was a political red herring, they steamed, and deserved to go unheeded.

Red Herring: Something that distracts from the real issues.

Yet as the ratification debate unfolded, thoughtful Anti-Federalists made penetrating forays into the bill of rights issue. Suppose, they submitted, that the

Constitution's "necessary and proper" clause were interpreted the way that Alexander Hamilton had advocated. The government had the enumerated power to coin money and regulate currency, which meant that it also had the power to punish counterfeiting. What was to stop a go-for-broke Congress from declaring counterfeiting subject to general warrants, secret trials, and punishment by drawing and quartering? In other words, a bill of rights clearly *could* apply to republican governance.

Federalists responded to this challenge just as thoughtfully—and with even more cunning. Perhaps there *was* merit in the Anti-Federalists' complaint, they conceded. But since the draft Constitution had been provided with an amending process, why not ratify it first and then, if the American people truly wanted a bill of rights, well, they knew how to get one. Few Federalists were sincere in making this offer. After all, once the Constitution was ratified, why would they have to bother? The Anti-Federalists had used a red herring—they could use one too.

James Madison, as usual, was out ahead of everyone. He considered the arguments for a bill of rights, and concluded that whatever the motives behind them, they might well be valid. Besides, such a mechanism could be made part of the Constitution's structural interplay, checking the government against the people themselves. And, as a newspaper commentator added, a bill of rights could be made "to secure the minority against the usurpation and tyranny of the majority," and defend "the sober and industrious part of the community . . . from the rapacity and violence of the vicious and idle."

But there must be some provisos. First of all, the rights listed in any bill would have to be framed with great care so as to not conflict with the Constitution proper or interfere with its other mechanisms. Second, they must be framed as *civil* rights, not *natural* rights, meaning that they must be concretely described, carefully limited, and made to address real-life situations. Finally, they must be enforceable. The only sense it made for including a bill of rights in a republican constitution was to allow individuals, when their rights were violated, to seek actual redress in the courts.

Madison, in other words, was not primarily interested in Great Oughts. He wanted a bill of rights that was less a matter of ringing declarations and more a matter of nuts-and-bolts constitutional protections. But those protections, he believed, could be of inestimable value.

And so it was that James Madison almost single-handedly kept the bill of rights issue alive in Congress after it was politically dead. When the Constitution was ratified, Federalists and Anti-Federalists alike were willing to forget about the promised amendments. Not Madison. He had become a believer.

Drafting the Bill of Rights

It was easy for Madison—now *Congressman* Madison—to secure appointment to the appropriate congressional committee, which then solicited suggestions from the states. (Send us your favorite right.) These arrived in a farrago of natural rights, civil rights, historic privileges, and pious declarations. Madison sifted through hundreds of items and winnowed them down to a discrete few.

DRAFTSMANSHIP OF THE CONSTITUTIONAL CHAIN

The Founders drafted some clauses with the rigor of a steel link (left); and other clauses with the flexibility of a rubber link (right).

Then he turned to the problem of draftsmanship.

Legal draftsmanship must always choose between two general approaches to the casting of an idea into prose:

> *Broad and abstract language.* Here the terms are vague and open-ended: "All people have a right to justice." In attempting to apply such language to specific cases—as a court would have to do—a variety of possible meanings and innumerable exceptions are always likely to come up, and the courts wind up with carte blanche in sorting them out. Yet this type of language has pronounced rhetorical value. It makes the right being stated sound like a Great Ought.
>
> *Narrow and concrete language.* Here the terms are precise, legalistic and freighted with qualifiers. "All members of the XYZ Club, in good standing, with dues paid up, shall have the privilege of using the club swimming pool during the hours of 9:00 A.M. to 5:00 P.M. on Tuesdays and Thursdays, holidays excepted." Courts are far more restricted in the interpretation of such language, so it is far easier to apply it to specific cases. On the other hand, it has little or no rhetorical value.

Any draftsman wishing to assert a natural right would undoubtedly use the first alternative, or wishing to grant a legal privilege would use the second. But what about the sort of thing Madison had in mind—civil rights, subject to adjudication? There were arguments for leaning either way. Yet for the most part, Madison leaned toward the narrow alternative, for if you really mean the rights in your bill to appear in court, you must frame them as specifically as possible.

Accordingly, we can see a number of places where Madison consciously employed the narrow and concrete drafting strategy. Take the Third Amendment, for example:

> No Soldier shall, in time of peace be quartered in any house, without the consent of the Owner, nor in time of war, but in a manner to be prescribed by law.

The wording is full of concrete terms such as *soldier*, *house* and *quartered*, and replete with qualifiers like *in time of peace*, *in time of war*, and *without the consent of the Owner*. "Shorthand" words and phrases are also used. While these may sound vague—*in a manner to be prescribed by law*—they actually refer to well-established procedures.

Clearly, then, Madison intended the Third Amendment to appear in court. True, attorneys might still quibble over the meaning of some terms. They might argue that the alleged quartering was actually in a "barn" rather than a "house," or that the "soldier" was really a "sailor." Yet the courts would have little difficulty applying this kind of language to real-life situations.

Madison could have drafted the Third Amendment the other way, in which case it would go something like: "Government shall not interfere with private property." But think of the implications. Such broad and abstract wording would prohibit quartering, all right, but it might also prohibit much else. And the courts would have a field day deciding what the words *interfere*, *private* and *property* did or did not include. In the end, the legitimate work of government, which must always include *some* interference with property, could easily be undermined. Madison understood this irony. Simply by stating a right that most Americans strongly believed in, the Constitution could seriously be crippled—while the right itself remained unprotected.

This explains why many of the rights listed in the proposed amendments were drafted narrowly and concretely, and the situations they addressed were not general woes of humanity but difficulties Americans had encountered before:

Second Amendment:	The right to keep and bear arms.
Third Amendment:	The conditional right to be free from quartering.
Fourth Amendment:	Security from unreasonable searches and seizures.
Fifth Amendment:	Security from the expropriation of property without just compensation.

In the Fifth through the Eighth Amendments Madison included a list of procedural guarantees for persons accused of crimes. Since virtually all of these protections were found in the English common law, they could be described by "shorthand" references:

Double Jeopardy: Putting a person on trial for an offense for which he or she has previously been tried.

Fifth Amendment:	Grand jury indictment required for serious crimes.
	No double jeopardy.
	No self-incrimination.
Sixth Amendment:	Speedy and public trial.
	Trial by jury.
	The accused to be informed of the nature and cause of the accusation.
	The accused to be confronted with witnesses.
	Compulsory process for obtaining favorable witnesses.
	Assistance of counsel.
Seventh Amendment:	Jury trial for civil suits.
	Common law to govern appeals.
Eighth Amendment:	Reasonable bail.
	Reasonable fines.
	No cruel or unusual punishment.

Even though the meanings of these terms were well established, some of them would be subject to the same difficulties outlined above. For example, the meaning of the phrase "cruel and unusual punishment" was rather different in the eighteenth century than it would be in the twenty-first. Some present-day Americans regard capital punishment as "cruel and unusual."

More problematic still was the inclusion of the following words in the Fifth Amendment: ". . . nor shall any person . . . be deprived of life, liberty, or property, without due process of law. . . ." The phrase *due process of law* was yet another shorthand expression familiar to the legal fraternity. It meant that accused persons should not be punished unless all the safeguards of the common law had scrupulously been applied to their case. By connecting this phrase with *life, liberty, or property*, however, an odd situation was created. The words *could* be interpreted to mean that legislative acts must never deprive people of life, liberty or property *at all*—that such ends can only be accomplished by courts. Once again, consider the effect on government. Most government regulations have the effect of curtailing liberty. If I am commanded by the law to provide brakes for my automobile, I lose a part of my "natural" freedom to drive any way I choose. So, does the Fifth Amendment mean that government may not require me to have brakes? And if so, is it still a real government?

The due process clause illustrated what dangers could befall a bill of rights. Meanings that were clear and unambiguous in the minds of the authors could develop all sorts of ambiguity when considered by subsequent generations. The words *due process of law*, once divorced from their traditionally accepted meaning, were so vague that they could be taken to mean almost anything.

The Great Oughts

There are three items in the Bill of Rights that were not drafted narrowly and concretely. They are often taken to be the most important rights of all—the ones that come to mind when the Bill of Rights is mentioned. Clearly Madison did not approach them in the way he approached the others. They are much more in the character of natural rights and thus fully qualify as Great Oughts. They are:

Freedom of conscience. Found in two separate clauses of the First Amendment.

Freedom of expression. Found in four clauses of the First Amendment.

Right to privacy. Implied by language of the First, Third and Fourth Amendments.

An obvious first question is why, when Madison took pains to draft most of the rights so narrowly, did he leave these three vague and general? One part of the answer is that there is simply no way to define religion, expression, or privacy in narrow and concrete terms—words literally fail. Another part is that the subjects themselves are of overarching importance. Freedom of conscience, freedom of expression and freedom to live one's own life may be considered the fundamental ends of republican government—the "happiness" that free people seek. In other words, these truly are *Great Oughts*.

Religious Test: See page 213 for discussion.

Freedom of Conscience

The First Amendment begins with these words: "Congress shall make no law respecting an establishment of religion, or prohibiting the free exercise thereof. . . ." Matters of worship, in other words, are beyond the reach of the federal government.

In drafting the Constitution, the Framers had forbidden religious tests for office holding. This was considered a bold step toward religious freedom, and it was taken only after agonized debate. But the First Amendment went further still. Not only must there be no religious test in politics, there must be no religious anything where the state was concerned.

In framing these clauses, Madison drew upon his own experience in Virginia. At the time of the Revolution, he and Jefferson had authored the bill that disestablished the Anglican Church as the "official" religion of the colony and proclaimed religious liberty. That too had been a momentous step, for it had cut a tie that many in the Western World took to be essential.

Take the English case, for example. England itself enjoyed a kind of religious freedom. All churches were at liberty to preach their doctrines and practice their faith as long as they obeyed the law. There remained the established Anglican Church, however, and it was supported by public tithes. Granted, the church had an inordinate and occasionally untoward influence in politics, as its critics tirelessly maintained, but it also made an important contribution, something we might call "divine legitimacy." If an Anglican Archbishop was present at occasions of state, giving his benediction with grand ceremony, it must mean—and for many did mean—that God backed the government.

In America, as we have seen, the presence of an even greater number of contending faiths pointed toward a certain logic: that all be allowed to preach their own gospels and that none be accorded special favor. Some of the Founders believed that this approach would enhance civic virtue, as we saw in chapter 8. But others believed that a degree of religious influence still needed to be officially recognized, as it was in England, in order for the state to enjoy that legitimizing sense of mysterium.

The First Amendment cut through that debate like a butcher knife. Religion was a private affair, it said. Citizens of the Republic ought to be free to accept or reject whatever faith they chose, subject only to their own conscience. Church and state must remain separate and apart.

Freedom of Expression

Just as it was thought daring to separate church and state, it was thought daring to allow freedom of expression in politics. England had pointed the way there as well—and it was a way fraught with obstacles.

In *Leviathan* (1651), Thomas Hobbes' famous dissertation on political philosophy, Englishmen were presented with an argument for the complete annihilation of free speech. It came down to the simple fact that speech could be extremely dangerous. Hobbes had his supporters too, many of them in high places. In one notable case, a critic of the Stuart monarchy had had his books burned, his property confiscated, his tongue cut out, and his person thrown into a dungeon for life.

Part of the problem was conceptual. When the head of the state actually embodies the state in some mystical way and serves at the pleasure of God—which is precisely what the Stuart monarchs believed—there can be no concept of "loyal opposition" such as we have today, for dissenters cannot possibly be loyal: they are quarreling with the state and with God. The English skirted this difficulty by aiming their complaints at the king's "evil counselors" rather than the king himself. In separating king from counselors, they were trying to separate state from government so that opposition to the one would not mean disloyalty to the other. (After all, the king could always get different counselors.) Counselors, however, were powerful men in their own right, and they took criticism no better than most kings. On the contrary, they hid behind the king's robes and labeled all critics as rebels.

Yet in time even the most hidebound came to see that there must be *some* political expression. Here and there, under varying restrictions, members of Parliament were allowed a degree of freedom to take issue with government policies. But only a degree. There was always the occasional backlash and offending MPs were sent to the tower. Free speech was in no way guaranteed.

And there was another problem. In alleging certain kinds of wrongdoing, such as bribery or corruption, a critic could also be charged with libel, for impugning the character of a public official was viewed no differently from slandering one's neighbor down the street. Here again, though, common sense eventually came to the rescue. It became plain to see that making public officials immune to corruption charges only invited worse corruption, and slapping vigilant citizens with lawsuits only killed off public virtue.

Freedom of speech involved both of these—opposing government policies and pointing out official corruption—and both took root in the American colonies. (One of the landmark legal cases involving free speech took place in Boston in 1734–35.) Both helped bring on the Revolution too. Those who opposed a given policy—such as the Stamp Act—alleged personal corruption on the part of the officials connected with it, and personal corruption—as in, say, the taking of bribes by customs officials—often led to a broader kind of political dissent. Both types of criticism were damaging to the royal government, and both were squelched whenever possible.

In the trial of editor John Peter Zenger for libel, lawyer Andrew Hamilton argued that the truth of the allegation constituted a sufficient defense. The jury agreed.

As a result, republican fervor became strongly identified with free speech. Virtue must be at liberty to speak its mind. Moreover, with the king and his evil counselors out of the picture, the concept of free speech became clearer. Since republican government served no one but the people (there were no references to "divine right" or God's pleasure), the people had a right to scrutinize it down to the smallest detail. A citizen could oppose any or all of its policies *without being disloyal.*

Freedom of speech, then, was often prominently listed in bills of rights. And it was expanded to include freedom of the press, freedom of assembly, and freedom to petition the government for redress of grievances. Free expression was held to be nothing less than a republic's lifeblood.

The Right to Privacy

Nowhere in the Bill of Rights is there mention of the right to privacy. Moreover, the courts began to use that phrase only quite recently in US constitutional history. Some critics have denied that a "right to privacy" even exists.

Which brings us to the doctrine of ***implied rights***. Those who contend that the right to privacy truly does exist must argue by implication. Their case goes like this. Freedom of religion implies that the private conscience ought not be intruded upon by government. Freedom of expression implies that people ought to make up their own minds and not be told what to think. The quartering amendment implies that the government ought to leave people alone in their own homes, and so does the search and seizure amendment. From all four it is possible to abstract an implied right to privacy. Furthermore, since the Ninth Amendment specifically states "The enumeration in the Constitution, of certain rights, shall not be construed to deny or disparage others retained by the people," the case for implied rights is arguably quite strong.

This still leaves the question of exactly what "privacy" is. Of course that question would be no different if an amendment actually stated: "Congress shall make no law abridging the privacy of individuals." The root idea is a very old one, found in some of the earliest common law cases. It is simply that there ought to be a sphere of life in which government cannot tread. For example, in medieval England there were instances where a fugitive was declared "home free" once he crossed his own threshold, and the pursuing sheriff had to give up the chase—a man's home being quite literally his castle.

It is possible to formulate a conception of privacy that is in keeping with other ideals of the Founding. It would go something like this. The rule of law is itself a means of securing privacy. For if government cannot use the laws to manage the lives of citizens, the laws create a private sphere in which citizens act solely on their own. Privacy, in other words, is only another word for freedom.

Described in this way, the Great Oughts form a conceptual whole, one that reinforces the American Founding. We have seen that the moral text of the Constitution was all about individual dignity and responsibility. The Great Oughts underscore this point. They affirm that one's mind, one's conscience, one's understanding and values, one's volition, one's expression and one's personal life belong to the individual alone—and that the individual must hold himself accountable for them.

So it was that the Constitution acquired its famous amendments. The articles submitted by Madison's committee were approved by Congress in the fall of 1790 and submitted to the states for ratification. It is of passing interest to note that Madison drafted several amendments that were not ratified. One of these—his own personal favorite—would have applied the Bill of Rights to the states. *They* needed it, he thought, more than the federal government did.

The Bill of Rights and the Supreme Court

After the advent of **judicial review**, it was up to the Supreme Court to interpret the Bill of Rights and apply it to specific cases. While the details of judicial review will be covered in a later chapter, one aspect of the interpretive process concerns us here.

Constitutional interpretation came to focus on those areas of the text where the broadest and most elastic wording was found, such as the commerce clause or the necessary and proper clause. The Bill of Rights amendments increased the Constitution's elasticity by perhaps a factor of three. This, of course, was largely due to the Great Oughts, which had to be drafted in the vaguest of language.

Even before the Court began to practice judicial review, it became obvious that the Bill of Rights would alter the way the constitutional system worked. Consider a couple of examples. With reporters poking into government operations, it became clear that every contemplated action must now be weighed according to how it would play in the press—and as a result governance by gentlemen behind closed doors would soon be a thing of the past. Or this. With that same freedom of the press, scurrilous attacks on public officials were bound to multiply and respect for government was bound to diminish. Where the imagery of republicanism always supposed government by virtuous statesmen, America's began to look like government by politicians.

The US Supreme Court has firmly asserted authority under the Constitution to interpret the meaning of the Bill of Rights and other provisions of the Constitution.

With the Bill of Rights assuming an ever more important role, it was only natural that the Supreme Court would pay ever greater attention to puzzling out its meaning. In countless specific cases, the Justices found themselves obliged to define "speech," characterize "religion," stake boundaries around the terms "abridge" or "deprive"—translate the Bill of Rights into constitutional law.

In the process, the Bill of Rights began to change. The reason why had to do with the way courts work. In theory, the Supreme Court, like any Anglo-American court, was bound by its own prior decisions, its *precedents*, in adjudicating a case at hand. If in an early decision the Court defined the phrase "freedom of speech" as referring only to *political* speech, then all subsequent decisions must respect this definition. In fact, it was impossible to hold the Justices to this rule, for their decisions could not be appealed, and thus they often departed from precedent to lasso new ideas. As a result, there was a built-in tendency for constitutional terms to become ever broader, more abstract, and more inclusive each time a given word or phrase was adjudicated.

Judicial Review: http://www.constitutioncenter.org/sections/basics/basic_id.asa

Take freedom of expression, for instance. What was clearly limited to *political* expression in the minds of the Framers gradually grew to include all kinds of expression. Everything from pornography in magazines to vulgarity in the movies to obscenity on the stage came to claim the protection of the First Amendment—and this was but a single avenue of development.

How Free Should Free Speech Be? The Daniel Schorr Case

At sixty, CBS news correspondent Daniel Schorr, silver haired and distinguished looking, cut a figure of journalistic responsibility second only to Walter Cronkite. In truth, however, he was not quite the same benign uncle. For, in the fall of 1976, amid the passions of the presidential election, Schorr suddenly stunned the country by releasing to a New York newspaper the text of an extremely sensitive government report. The Pike Report, resulting from an extensive investigation of US intelligence activities, was so damaging to the pride and prestige of the CIA—which was faulted (among other things) for plotting the murder of foreign dignitaries—that the House had voted to suppress it. Schorr, moreover, was wholly unrepentant. He would not answer questions of the House Ethics Committee; he would not produce a copy of the report; and he would most certainly not reveal the source by which he obtained it.

There had been pirated documents before, the most famous of being the Pentagon Papers of the Vietnam era. But there was a difference in the two situations. Because the Pentagon Papers had shed light on a war that was already intensely unpopular, there was a clear sense that the public interest had been served by their publication. Here, by contrast, many Americans were uneasy. Some of Schorr's most ardent defenders worried about the impact of the Pike Report on the intelligence community's morale. National interests were at stake on both sides.

For his part, Schorr saw only one of those interests: free speech. At issue, he complained, "was not my right to report, but the public's right to know." The national security interest Schorr waved aside. Others in the journalistic community weren't so sure. An editorial in the *New York Times* condemned Schorr for "selling secrets." At the annual Gridiron Dinner, the correspondent found himself not a roast*er* but a roast*ee*. And CBS, after agonizing over the situation, finally decided to suspend him from its news staff. He subsequently resigned.

Where did the public come down? Ditheringly in the middle. When the House Ethics Committee moved toward a confrontation with Schorr (with the possibility of jail for contempt), the public rallied to his defense. Yet polls showed that Americans also feared the power of the press, power that in this case might have caused real injury. In the end, it came down to a question of whom was mistrusted most, the government or the media.

Found in America: A Study in Heritage, *Kendall/Hunt, 7th Edition, p. 176–77*

Daniel Schorr: Freedom of speech or the abuse of? (UPI/Bettmann Newsphotos)

Many Americans came to deplore what they saw as the perversion of the Bill of Rights. The words of the first ten amendments ought to mean only what the Framers originally intended, they complained. But there was another side to consider. When the Bill of Rights was drafted, electronic surveillance or mind-altering drugs had not yet been invented. Should the protection we hope for at the turn of the twenty-first century apply only to the quartering of soldiers?

The double-sidedness of the question illuminates an important fact of republican life. Where fundamentals are concerned, the people cannot afford to crouch behind parchment barricades. No constitutional wording, however carefully composed or well intentioned, can prevent the deterioration of rights or the proliferation of license if the public itself lacks vigilance. Virtuous citizens, playing by the rules, are still the ultimate custodians of truth.

The Bill of Rights and American Freedom

There is a certain conception of freedom that emphasizes forward motion. It holds that democratic institutions are on a historical march from past to future, forging ever onward to mark the progress of mankind. This idea, rooted in the philosophy of Wilhelm Friedrich Hegel and others, has become popular among Americans. We have come to expect that each generation will be happier, more affluent and above all freer than its predecessor.

Against the backdrop of such thinking, the US Bill of Rights has come to play a role that few of its Framers would have imagined for it. By redefining and continually expanding rights, so the argument goes, a democratic people can continue to realize that promise of the future.

Viewed from this perspective, rights become identified with "justice" and "equality." After all, we never think of them as being attached to race or gender, nor to wealth or background or social class—but only to human beings *as* human beings. As a consequence, the scouting up of rights becomes a way of redressing whatever is wrong. When a group is found to be at a disadvantage, the problem can be corrected by giving that group its "rights." In this way the language of American politics has largely become the language of expanding rights. We speak of new rights for this group, that gender, some other ethnic or cultural identity, even for the disabled, the elderly, the young, the poor, and those of varying sexual preference. If everyone has more and better rights, the argument goes, outsiders will eventually become insiders, inequality will vanish, and justice will prevail.

In such a world, the rights that James Madison carefully narrowed down and "civilized" are often conceived in absolute terms—much like natural rights. And no matter how much they grant, it can never be quite enough. Thus, a wide latitude of self-expression in the arts needs to be widened still further, special facilities for the handicapped must be constantly improved, and "gender equality" in the workplace runs to nuances and subtleties beyond fathoming. Furthermore, by substituting the contemporary term *human* for the eighteenth-century term *natural*, this rights-oriented conception of life can be projected all over the

Technology Versus Rights

One argument for the existence of implied rights in the US Bill of Rights is that the Founders couldn't have imagined all the ways that later technology might be used to intrude upon rights. Electronic surveillance, for example. In James Madison's day there was no such thing, of course, but we have since raised it to a high art. In a world of satellite communications, eavesdropping antennae and computerized decoders, it is possible for us to listen in on telephone conversations around the world. Not coincidentally, the issue of privacy, which may not have seemed important in Madison's day, has suddenly become so.

Or consider some recent advances in chemistry. The perfection of sodium pentothal and other "truth serums" has made it possible to abridge rights in ways hitherto unimagined, for a single injection can make a victim spill everything. Other kinds of chemicals might conceivably be added to, say, a municipal water supply to create various dispositions—languor, hostility, euphoria—in the populace. Such dispositions might also be produced, perhaps more easily, by skillful manipulation of the mass media.

The list goes on and on. Nuclear power and nuclear weapons testing have created unimagined civil dangers, as has research into chemical and biological warfare. Genetic engineering raises the apocalyptic specter of man-made monsters. Recent forms of pollution—such as the toxic waste dumped into the Love Canal—give a whole new meaning to phrases in the Fifth Amendment.

If all of this sounds Orwellian, it was George Orwell who first called our attention to it. In *1984*, we recall, truth was continually victimized by the Thought Police, and Big Brother's TV cameras snooped everywhere. Science-fiction writers have taken it from there, inventing every imaginable invasion of rights from mass produced embryos (*Brave New World*), to emotional repression (*Fahrenheit 451*), to robotic tyranny (*Blade Runner*). Perhaps the ultimate derogation of rights was imagined in a story by Orson Scott Card, in which a victim was sentenced to die "by every means known to man." Each time he was executed, his existing consciousness—including the pain and horror of execution—would be taped and programmed into another mind, and the whole thing would be repeated.

There is, then, something truly to be said for a concept of human rights that transcends the Founders' own listing of them. After all, the worst thing they could imagine was an illegal tax on tea.

Found in America: A Study in Heritage, *Kendall/Hunt, 7th Edition, p. 164* world, with the result that much of the world's tyranny, anarchy, poverty and exploitation can be explained simply as a want of rights.

Because this idea has important implications for republican governance, it might be worthwhile to make a few commonsense observations about it. First, rights are only "expandable" up to a point; beyond that point they rapidly come into conflict with each other. A customer's right to dine wherever he chooses necessarily conflicts with a restaurant owner's right to select his patrons. A woman's right to an abortion is at odds with her child's right to life. Tom's right to express himself through obscenity runs up against Dick's right to shape community standards and Harry's right to shield his children from smut. All rights cannot be respected for all people in all situations.

Second, dynamically expanding rights tend to become divorced from their accompanying responsibilities. In republican theory, rights and responsibilities were inseparable. The right to participate in the political process, for example, was also the responsibility to participate honestly and well. The right of free speech was also the responsibility of careful and thoughtful speech. When the expansion of rights is taken as a cure-all for social ills, responsibility is often left in the dust.

Finally, rights in conflict with and divorced from responsibility do not provide a recipe for social peace—or for the progress those rights are meant to facilitate. Quite the contrary. Groups see the expansion of other groups' rights as a threat to their own, fostering high anxiety. Casualties of the ensuing free-for-all tend to be, first, a sense of virtue in public affairs, and second, a conception of the public good as embracing everyone.

There is a subtler difficulty. If we lean too heavily on rights as a remedy, we may risk the "double jeopardy" described by political scientist Ralph A. Rossum. We discover that rights are not really enforceable by the courts at all, and thus that we have sought security behind a parchment barrier. At the same time, we ignore the fact that the Constitution is the *real* guarantor of our rights and we further weaken constitutional government.

These difficulties conceded, the Bill of Rights has not been without enormous good effect. No reasonable observer looking back over American history would say that those ten amendments were superfluous and unnecessary. Nor could one compare the fortunes of minorities in 1900 with those in 2000 and conclude that the Bill of Rights hasn't hastened the "march of freedom." If it has done nothing else, it has prodded Americans incessantly to live up to their founding ideals.

The Bill of Rights as a Moral Charter

What changed Madison's mind about the Bill of Rights was thinking about the Republican Problem. He had asked himself again and again how a republic could be designed so as to embody the will of people while at the same time protecting individuals and pursuing the public good. The Constitution's structure had been designed for that purpose. That it would operate successfully in most circumstances Madison and the other Framers were reasonably confident. They were less confident about unusual circumstances, and they had reason to be.

Virtually every system failure in republican history has occurred at a time of war, domestic upheaval, economic catastrophe, or rapid social change. For in these situations, the perception of minorities is subject to drastic revision. Consider a few episodes. In the 1830s, during a time of rapid social change, Mormons and Masons were suddenly identified as bogeymen and targeted for persecution. In the 1890s, during a time of economic breakdown, "Jewish bankers" were accused of plotting to manipulate the money supply. In the 1940s, during a time of war, Japanese Americans were condemned as potential spies and locked up in prison camps. And in the 1950s, at the height of the Cold War, suspected Communists were investigated, deported, fired from their jobs, blacklisted, and made to suffer appalling indignities. With every feature of the constitutional system in place and operating, majority opinion turned violently against these groups, and the results were tragic.

In the last analysis, the Bill of Rights was written for them. That Congress "shall make no law . . . prohibiting the free exercise of religion" does not concern Catholics, Lutherans, Anglicans, Baptists, Methodists or Presbyterians. It concerns small sects with strange doctrines and disconcerting practices. Similarly, the First Amendment's freedom of speech is not for Republicans or Democrats. It is for Anarchists, Syndicalists, Communists, Radical Socialists, Skinhead Nazis, and even less-popular political faiths. And when the Bill of Rights promises privacy from intermeddling, it is not to protect the magazine-cover family in suburbia but motorcycle gangs, hippie communards, and ragtag Bohemians. The overwhelming majority of Americans neither embrace nor sympathize with such alternatives—so why bother protecting them?

Part of the answer has to do with freedom. If one is free only to adhere to the norms, such freedom has little meaning. Not only do the unpopular minorities lose out, so does the majority itself, for its members act not out of choice but compulsion.

A second part of the answer has to do with justice. The justice delivered by real-life republics often falls short of perfection, as we have seen. David Hume argued that justice is not a fundamental passion. We seek it mostly for its utilitarian value, not its moral necessity, and that value is limited by all sorts of considerations. With the Bill of Rights, there is a kind of justice by fiat, moving

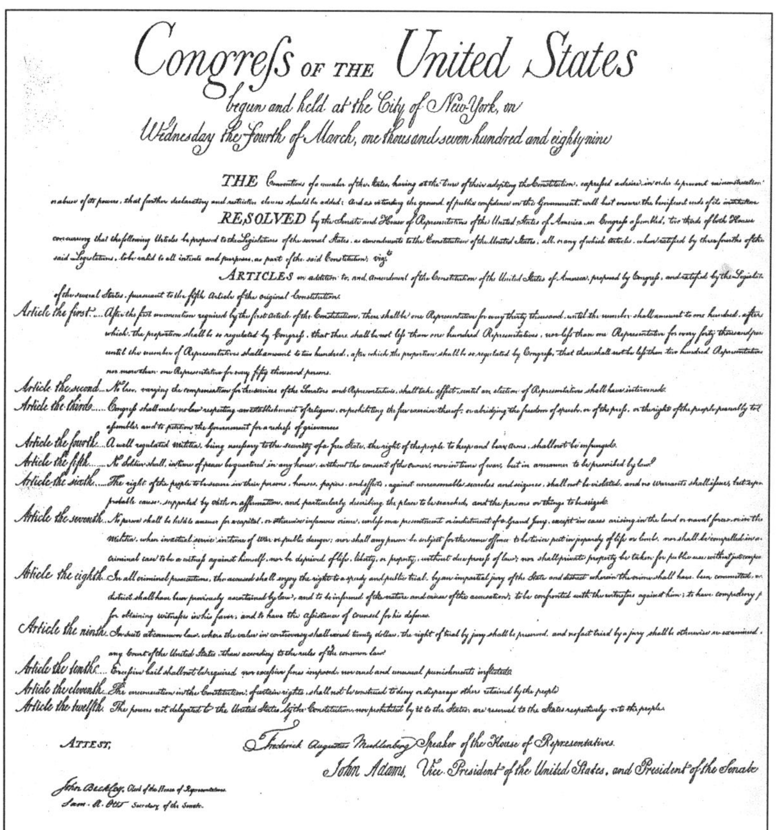

Congress of the United States
begun and held at the City of New-York, on
Wednesday the Fourth of March, one thousand seven hundred and eighty nine

THE ...
RESOLVED ...
ARTICLES ...

Article the first ...
Article the second ...
Article the third ...
Article the fourth ...
Article the fifth ...
Article the sixth ...
Article the seventh ...
Article the eighth ...
Article the ninth ...
Article the tenth ...
Article the eleventh ...
Article the twelfth ...

ATTEST,
Frederick Augustus Muhlenberg Speaker of the House of Representatives.
John Adams, Vice-President of the United States, and President of the Senate

The original Bill of Rights, passed by Congress in 1789, contained twelve amendments. By 1791, the states had ratified ten of the amendments, rejecting two dealing with formulas for the election of members of Congress and their reimbursement.

The Price of Religious Freedom

Freedom of religion means more than having a Catholic cathedral on one corner, a Lutheran chapel on the next, and a Jewish synagogue across the street. Where there is true religious freedom there will also be a vibrant, colorful, and sometimes rowdy religious fringe. Americans know that well enough. They see Hare Krishna monks on the sidewalks, read ads on kiosks for Eckankar, and peruse articles in the *National Enquirer* about swamis, gurus, and fortune-tellers. It is entertaining to think that Los Angeles County has an official witch on the payroll, that San Francisco's Glide Church feature rock music and light shows, or that a group called "The Celestials" believes Mt. Shasta is an intergalactic UFO base.

Yet religion of the fringe poses painful dilemmas for democratic societies. There is, for instance, simple nuisance to think about. The Hare Krishnas can raise a frightful racket with their chants and drums. Other groups panhandle shoppers on the sidewalks or confront them with hell-fire preaching. Among the beliefs of a Philadelphia cult called MOVE was the importance of depositing refuse on the street. But nothing compares with the nuisance created by a self-styled maharishi called Rajneesh Bhagwan Shree, whose followers moved in on the Oregon farming community of Antelope, registered as voters, and proceeded to reshape the town in their own image.

Or there is the issue of fraud. Much of the astrological, fortune-telling, and prophesying activities of the fringe religions—for which millions of dollars changes hands yearly—would be fair game for bunko squads in another setting. Many evangelical churches engage in high-pressure solicitation through their radio and television stations. Still others trade in miracles. A Portland woman recently sued the Church of Scientology for $39 million for falsely promising to alter her personality and raise her IQ.

A number of religious groups run afoul of fundamental mores. Polygamy, complex marriage, and outright free love have eroded family values and created chaos in domestic relations. (When everyone is married to everyone else, who inherits what?) The use of drugs is claimed as necessary to some religious observances, and in the case of the Navaho Indians, drug laws have had to be altered accordingly. Some cultists have denied their children education, deprived them of medical attention, or forbidden them to participate in citizenship.

The question of mind control raises another raft of issues. L. Ron Hubbard's practice of "dianetics"—a sort of do-it-yourself psychoanalysis—was in constant legal trouble and pronounced "a serious threat to health" by the American Medical Association, but after Scientology was converted to *The Church* of Scientology, its problems diminished. Followers of the Rev. Sun Myung Moon are widely accused of brainwashing recruits, but this is mild compared to the starvation, beatings, and torture inflicted by other groups. Such was the degree of mind control achieved by Jim Jones over his followers in Guyana, that when he ordered them to commit suicide they willingly obeyed.

Churches, of course, get into politics, too. This seems tolerable in the context of Republican versus Democrat or liberal versus conservative, but what about in the context of white versus black? Certain fundamentalist groups have encouraged racism among whites, while extremist black sects, such as Louis Farrakhan's in Chicago, have fostered the same sort of hatred in reverse. Then, too, Irish-American Catholics have gathered money for the IRA and Indian-American Sikhs have solicited funds for the Punjab—knowing full well that the result would be death and destruction.

And occasionally religions of the fringe have gone all the way. Anton LaVey's First Church of Satan may be harmless enough with its theatrical bohemianism, but some forms of devil worship indeed call for bloodletting. A teenager murdered his foster parents after claiming to have received satanic orders to do so, and in a spectacular Utah case, two self-proclaimed prophets embarked on a murder spree after God supposedly provided them with a hit list. In Idaho, a group calling itself "The Covenant, the Sword, and the Arm of the Lord" is wanted for multiple homicides.

Such difficulties are not experienced when there is no religious freedom. In Saudi Arabia, to name just one example, there exists near perfect religious harmony. Non-Muslims are welcome in the country, under carefully controlled circumstances, but their religion is not: all baggage is subject to inspection at the Rijadh Airport, and foreign religious items (such as family Bibles) are routinely confiscated. The only problem is that where there is no religious freedom—excesses and all—there is no other kind of freedom either.

Found in America: A Study in Heritage, *Kendall/Hunt, 7th Edition, p. 168–69.*

beyond utilitarian value and extending equitable treatment to groups who would otherwise be denied it. This is not perfect justice either—as witness the examples given above—but it could and did move American society in the right direction. One might say that the Bill of Rights helps *create* justice as an ideal.

A final part of the answer has to do with dignity. By affirming the legitimacy of the unpopular cause, the other point of view, the alternative lifestyle, we enhance respect for diversity and warrant that the choices we make have meaning and value. We affirm human dignity.

The Rights of Mankind

So understood, the Bill of Rights became an integral part of the Founding. In the Declaration of Independence, Americans submitted that government must protect the rights of its citizens. They went on to draft and approve a Constitution they thought capable of doing just that. The Bill of Rights reached back to the original founding document and reasserted its truths, then cast the Constitution's moral text into strong normative language. It tied the Founding together and made it a coherent whole.

The Bill of Rights reflects the "deeper wisdom" we have seen elsewhere. Like federalism, it was forced on the Framers by circumstances and then, slowly, embraced by them as its implications emerged. If they were slow to grasp its moral significance, it was because the moral significance of the Constitution was obvious to them and they saw no need for restatement. What they did not see was the view from two centuries hence, our own time, when the meanings of words had changed and public morality would need all the help it could get.

In 1948, the UN General Assembly proclaimed a Universal Declaration of Human Rights. It is to this document that we refer when we allege the violation of "human rights" today. The debt of the Universal Declaration to the US Bill of Rights is pretty clear. For example:

> **Article 18**
> Everyone has the right to freedom of thought, conscience and religion.
>
> **Article 19**
> Everyone has the right to freedom of opinion and expression.
>
> **Article 20**
> Everyone has the right to freedom of peaceful assembly and association.

We can think of such expressions in terms of Thomas Reid's "moral sense." What is it, asked Reid, that sets human beings apart and make them fully human? Their common intuition of right and wrong. Even where the content of that intuition may vary from culture to culture, the concept of *rights* can fuse it all together. True, we can't all agree that human beings have a right to financial success or happiness in marriage, but we can all agree that they deserve to live their own lives harmlessly or enjoy the fruits of their own labor. The American Founders' view of rights has largely become that of mankind.

CHAPTER 11 STUDY AID

1. Need for a Bill of Rights:

- Federalists: No, people already sovereign; bill of rights inherent in constitution; it would open a Pandora's Box of implied powers; its enforcement problematic
- Anti-Federalists: Yes, people needed to be protected from each other; enumerated powers needed to be kept in check; all rights needed to be specified, not just implied; symbolic value

2. Uses of a "bill of rights":

- Assertion of peoples' rights; reminder of those rights
- Explicit expression of universal natural rights; the Good; freedoms specified, not assumed
- Addressing the "Great Oughts"
- Downside: parchment barrier; juxtaposes individual v. society; enumeration—what to include or leave out; unforeseen implications or interpretations

3. James Madison and the Bill of Rights:

- Bill of rights to insure minority rights; not contradict Constitution; express civil rather than natural rights; be enforceable in court of law
- Balance broad v. narrow language; limit legally open-ended language
- Express "Great Oughts" such as freedom of conscience, expression, and right to privacy
- Imply accountability

4. The Supreme Court

- Judicial review
- Elasticity
- The meaning of words such as speech, religion, etc.
- Intent v. application

5. American Freedom

- Expansion of rights: expanding democracy in terms of justice and equality
- From civil rights to absolute rights: duties and privileges
- Conflicting rights
- Abrogating responsibility
- Loss of protection; parchment barrier; unenforceable
- Moral character of rights; protect those out of favor; influence the world by affirming human dignity
- Rights of mankind and the "Good"

CHAPTER 11 STUDY GUIDE

NAME: ______________________ **SECTION:** ____________

Key Terms

Bill of Rights

Natural Rights

Great Oughts

Civil Rights

Draftsmanship

Conscience

Expression

Privacy

The Supreme Court

Multiple Choice Questions

1. Who authored the Bill of Rights?
   a. George Mason
   b. Thomas Jefferson
   c. James Madison
   d. Gouverneur Morris
2. Narrow draftsmanship is associated with which of the following words:
   a. Vague
   b. Specific
   c. Simple
   d. General
3. Which of the following is not a right protected by the Fifth Amendment?
   a. No double jeopardy
   b. No self-incrimination
   c. Trial by jury
   d. Grand jury indictment for serious crimes
4. The texts lists each of the following as Great Oughts except:
   a. Freedom of Conscience
   b. Freedom of Religion
   c. Right to Privacy
   d. Freedom of Expression

Review Questions

1. Why was a bill of rights not included as part of the original Constitution?

2. Why did the Founders fear that too much focus on individual rights might lead into the Human Predicament Cycle?

3. What are the inherent difficulties in drafting a bill of rights?

4. Why were certain rights drafted narrowly and others broadly? Why not draft them all one way or the other?

5. What role does the Supreme Court play in defining what the Bill of Rights means?

Review Exercises

1. Explain the relationship between natural rights and civil rights.

2. Explain the idea that natural rights are "unenforceable."

3. Briefly describe the interplay of rights and responsibilities as the Founders understood it.

Big Essay Question

Think philosophically about individual rights within a society. Does a community have rights the same way a citizen does? Are a group's rights paramount over individual rights? Create a scenario where an individual and a group each feel their rights are being violated by the other party. In a thoughtful essay resolve the dispute while addressing the issues raised above.

CHAPTER 12

From the Constitution to the Republic

On April 30, 1789, George Washington took the oath of office in New York City. His procession to the temporary capital sparkled with cheerful symbolism. He was given a public banquet in Philadelphia. He passed through a triumphal arch at Trenton. And he was rowed across the Hudson in a specially constructed barge with thirteen pilots at the oars. It was fitting that the new republicanism should make its debut amid fanfare. Beneath the celebration, however, deep currents swirled and eddied—as an end and a beginning came together.

Perils and Possibilities

In ratifying the Constitution, Americans had accepted a system with unprecedented power over their lives. Scarier still, the system was of their own creation—unknown and untried. Republican governments of the past had come to reveal all sorts of dangerous inclinations. Would the same thing hold true here?

Most statesmen could have dashed off a short list of challenges. There were disparate interests in political society—tidewater versus backcountry, North versus South, commerce versus agriculture. There was the divisiveness of local attachments. And there was nothing but the sovereignty of the people to bind Americans together—no king, no lords, no official church, no ivied ruins, no mystique of ancient nationhood.

On the other side, there was an empowering sense of adventure. The American people, *as* a people, had met their difficulties head on, adopted a plan to rectify them, and now stood willing to forge ahead. From antiquity they took a new sense of comfort. Where they had often spoken fretfully of the plight of Athens, they now began to emphasize the triumph of Rome. They were beginning to see themselves as modern-day heirs of the Roman Republic.

Ambiguity, Precedent and the General Constitution

Americans had lived under the British constitution long enough to know that their own founding document was but a rough sketch. What might be called the *general constitution* of any political society included many things that could never be reduced to parchment—traditions, precedents, established procedures, tacit understandings, behavioral codes, and most importantly, working institutions of government. If all of this represented the living, breathing body politic, the document written in Philadelphia was only a skeleton. Everything else must now be filled in.

Much of the filling in would be accomplished by *precedent*. English culture had always been sensitive to tradition and Americans were inclined to follow suit. In consequence, many facets of the General Constitution would come to exist for no better reason than that's the way it was done way back when. Here was one of the great benefits of having George Washington as President. A person of lesser stature could not have established precedents as easily as Washington did, nor could he have made them as binding.

In seemingly small matters, there was large potential significance. Consider the manner of announcing the President. John Adams' suggestion of "His Majesty, President of the United States and Protector of Their Liberties" might have steered the executive office toward monarchy. Simpler forms were deemed more republican.

In some cases, the setting of precedents actually altered the way the Constitution worked. For example, the wording of Article II, Section 2 states that the President "shall have Power, by and with the Advice and Consent of the Senate, to make Treaties, provided two thirds of the Senators present concur." The clause seems to require the President to seek the "advice" of the Senate while treaty negotiations are in progress. The first time Washington did this he received more advice than he wanted, and much of it was confused and contradictory. He vowed never to bring the Senate to the negotiating table again, and his successors have remained as true to the vow as he did.

We have seen that the Constitution, while clear in most respects, was a study in ambiguity when it came to certain issues. The Framers could walk away from that little difficulty but the government couldn't. The President and his Cabinet had to decide for themselves how far the Constitution's "stretch-points" could be stretched, and when questions of interpretation arose—as surely they must—someone had to work out how and by whom they would be settled. Language could go only so far in creating a political society; beyond that it was up to reason, wisdom and virtue.

The Interplay of Personalities and Events

Republican theory required leadership to be ***statesmanship***. The true statesman, in the thinking of the time, was one who always took the broader view and considered long-term consequences. He thought of the public good, not that of his friends or his "party," and thus stood above the fray of politics. He recog-

The Power of Precedent

How should citizens address the President of the United States? We know now, of course, but only with the benefit of hindsight. In 1789, when the first President took office, no one knew the answer and the question was vigorously debated. After all, in a seemingly small matter such as titles and forms of address, precedents could be established which might push the evolution of the President toward, say, monarchy. The question was not insignificant.

John Adams, for one, believed that the American President should possess as much dignity as any European monarch. "If the *state and pomp* essential to his great department are not, in good degree preserved," he wrote, "it will be vain for America to hope for consideration with foreign powers." Accordingly, Adams proposed that the President be addressed as: "His Majesty, President of the United States and Protector of Their Liberties."

That was a mouthful, especially to those who believed in republican simplicity. Indeed, after the Revolution it was quite fashionable for a time to dress without adornment, to avoid fancy speech, and to build houses of dignified plainness. When the republican Mr. Jefferson heard about Adams' proposal in Paris, he repeated Ben Franklin's assessment of that worthy: "Always an honest man, often a great one, but sometimes absolutely mad."

The advocates of republican simplicity won the day. The chief executive was called simply "The President of the United States" and addressed even more simply as "Mr. President." For his part, Adams was injured by the affair, for it was now whispered that he secretly desired monarchy. If so, he must have disliked the title someone hung on *him*: "His Rotundity."

Found in America: A Study in Heritage, *Kendall/Hunt, 7th Edition, p. 211.*

nized his own limitations as well as those of others, and was never so sure of his principles that he must advance them at gunpoint. Above all, he possessed those elusive qualities described by the word *areté*.

Events in the new republic were shaped by five personalities. All five had been frontline leaders in the Revolution, and three had been delegates to the Constitutional Convention while the remaining two had represented the United States abroad. While all five were staunch patriots, they embraced different strains of republicanism and in consequence had different ideas about the General Constitution. Thus, some would flesh out the bones of the Philadelphia document one way and some quite another. If any one of the five had been someone else, or if they had interacted differently, American history might well have been otherwise.

George Washington. First in the hearts of his countrymen did not say quite enough for him. By 1789 Washington was accepted as *the* Founding Father. His eager return to Mount Vernon after the war and his obvious reluctance to abandon it for public life only fortified the esteem in which he was held. While he was not a man of deep thought or wide learning, he was reasonably well read and held strong ideas. His support for independence from England and for the new federal Constitution had in all likelihood made both of these possible. He was no political fool either. He had had much to do with Congress during six years of war and knew its ways only too well. He realized that in order to maximize his effectiveness as President, he must remain above ordinary politics—while at the same time being very much a part of it.

Alexander Hamilton. His illegitimate birth in the West Indies may have been the great shaping force in his life, for it made Hamilton ever the outsider and lone wolf. He unabashedly admired English institutions, as he admired the aristocracy that supported them, and he envisioned some version of both for America. A man of personal charisma and intellectual vigor, Hamilton was not adept at politics. He was too brash and impetuous, too disdainful of lesser men, and too scornful of ideas with which he disagreed. Once in a while his judgment failed him, and he said and did incomprehensible things. For the most part, however, his vision was clear and his ideas exceptionally strong. He knew precisely what he wanted for the American Republic and precisely how to get it. "To confess my weaknesses," he wrote as a boy, "my ambition is prevalent."

Hamilton had an almost filial relationship to Washington, and he made the most of it. As aide-de-camp, he had been the general's right-hand man during the Revolution. Then, inexplicably, they had quarreled and Hamilton had severed all ties. The rift was not repaired until much later, and Washington remained leery of Hamilton's volatility. Still, he made the young man his Secretary of the Treasury, the Cabinet post he deemed most important, and after that his support never wavered. Hamilton was portrayed by enemies as a cynical manipulator, spinning his webs out of flattery, but his influence with the President was far more substantial than that. The two men saw the world alike.

Thomas Jefferson. If Alexander Hamilton was one man of mystery, Thomas Jefferson was another. Through a combination of luck and skill, Jefferson had married well, acquired a fortune, and worked his way into the dress circle of

Virginia society. Jefferson's mind ranged over broad expanses of the human experience, from art and architecture, to music and literature, to botany and zoology. History and political science were at the center of his attention, and he had studied law at William and Mary. He was by reputation, if not always by accomplishment, an eighteenth-century Renaissance man.

While still a young man, Jefferson had been catapulted into fame by his authorship of the Declaration of Independence. He failed to distinguish himself as Virginia's wartime governor, however, nor as America's second minister to France. There was something deeply conflicted in his personality—a mind divided against itself—which made it difficult for him to execute effectively.

Jefferson's contribution to the new republic was that of visionary. He had articulated a rudimentary concept of democratic republicanism in the Declaration's opening paragraphs. Later on, while serving as Washington's Secretary of State, he found himself embroiled in controversies which prompted an elaboration of that concept, and later still, as Vice President, he continued to fill in the blank spaces. "Jeffersonianism," as it would be called, was part thought and part feeling, and if it wasn't quite a full-fledged political religion, millions of Americans would embrace it as such all the same.

James Madison. Both as "Father of the Constitution" and "Father of the Bill of Rights," Madison had already won fame enough for two Founders. During the Washington and Adams Presidencies he would achieve his third degree of glory as a "Father of American Politics." The friendship of Jefferson and Madison was of as great a moment in the new republic as that of Washington and Hamilton. Much of what would become known as "Jeffersonianism" would be worked out in an ongoing dialogue between them. It is a tribute to Madison's intellectual integrity that in the unfolding of real-world events he was able to rethink some of his key ideas and draw a different set of conclusions from them. Madison the Federalist was to become Madison the Democratic-Republican.

John Adams. Adams was every bit as capable as his fellow statesmen, but he lacked Hamilton's dash, Madison's depth and Jefferson's way with words. Accordingly, while they grew into marble heroes, "Honest John" never quite took his place beside them in the pantheon. That he was occasionally dismissed as "His Rotundity" tells us much about the inner man. No one had a deeper understanding of politics and government, history and law, human nature itself, yet Adams lacked some essential quality of leadership. "He is always a wise man, often a great one," Franklin said of him, "but sometimes he is absolutely mad." He was politically disadvantaged by his Yankee background and, as some would have it, Yankee character as well, for he could be brittle and obstinate in situations requiring finesse. Yet were it not for his personal integrity—and his ability to rise above partisanship—matters in the new republic may have turned out far worse than they did.

All five of these Founders knew the rough-and-tumble of politics. Yet all five may have been blinded into supposing that republican politics would somehow be different. They spoke wistfully of statesmanship as something over and above what they had experienced in the British Empire—as if the Revolution could somehow redeem them from the human condition. Events soon proved that to be wrong.

Hamilton: Politics and Ideology

Alexander Hamilton:
http://xroads.virginia.edu/~CAP/ham/hamilton/html
http://law.umkc.edu/faculty/projects/ftrials/burr/HamiltonBio.htm

When President Washington appointed Alexander Hamilton to be his frontline economic officer, he knew that the new republic faced do-or-die challenges on precisely that front. It needed to establish foreign credit. It needed to work out favorable trade relations. It needed to put the government itself on a paying basis. It needed to establish a sense of legitimacy both at home and abroad. And it needed to untangle a skein of difficulties left over from the Revolution. For the British continued to occupy forts on US soil, pointing to the debts still owed London merchants, while Americans justified the debts by pointing to a want of trade. It was a barrow-full.

In Hamilton's mind, these problems were not isolated. They were tied together by what we saw in chapter 6 as an *ideology*. We recall from that discussion that coherent systems of ideas are often found in the political world. They bring order to otherwise chaotic situations—and they usually benefit a certain group or interest in society. Hamilton's ideology fit this description to a tee.

Political-economy: Social science dealing with the interrelationship of governmental and economic processes.

He had long been a student of British political-economy. The British had worked out their mercantile system over a period of some three hundred years and they had had ample opportunity to correct and fine-tune it. Hamilton had no doubt how the Lords of Trade in London would handle the Americans' present difficulties, and he proposed to handle them the same way. But first—or rather in the process—he must help to bring forth the sort of polity that would be capable of handling them. That was where ideology came in.

To begin with, Hamilton thought, America must become a fully sovereign nation: it must think and act like a unified commonwealth, not a gaggle of quarreling fiefs. It must develop its own identity, its own culture, its own traditions. And the integrity of the nation must be the prime consideration in determining policy. Individuals should be willing to sacrifice for the good of the whole.

Second, the various elites of society—the "rich and the wellborn"—must be weaned away from their local attachments and brought into partnership with the federal government. The quasi-aristocracy that Hamilton had in mind would require certain modes of thinking and a certain style of governance. It would need ruffles and lace to awe the many and great projects to challenge the few. Only when aristocratic élan was infused to political leadership, Hamilton believed, could republicanism truly succeed.

These ends could all be achieved by a robust central government with extensive authority and ample revenue, a government able to command respect. But such a government was impossible to create under the Constitution with its carefully enumerated powers. At the Philadelphia Convention, Hamilton had argued for a general grant of legislative power, which he deemed absolutely essential for nationhood—and had been resoundingly outvoted. Yet he remained undaunted. By broadly interpreting constitutional language, Hamilton believed he could assemble the grant he needed from three enumerated sources: the taxing and spending clause, the general welfare clause and the necessary and proper clause.

The latter was the real key, for if its loose wording were construed broadly enough, the necessary and proper clause could tap power almost anywhere in the Constitution. Hamilton's strategy soon became evident. In a landmark memorandum interpreting the clause in question, he argued that the word *necessary* need not be limited to that which was absolutely indispensable. "Necessary often means no more than needful, requisite, incidental, useful, or conducive to. . . ." Congress, he urged, should not be hobbled as to its choice of means for a given legislative end—the means ought to be up to the discretion of the legislators. He went on to pronounce a soon-to-be-famous criterion: "This criterion is the *end*, to which the measure relates as a *mean*. If the *end* be clearly comprehended within any of the specified powers, and if the measure have an obvious relation to the end, and is not forbidden by any particular provision of the Constitution, it may safely be deemed to come within the compass of the national authority."

Hamilton's doctrine of ***implied powers*** became the basis of his entire legislative program, which emerged in a series of Treasury Department reports. The federal government should, Hamilton said:

1. Fund its entire debt at par value. Only thus would it acquire the good faith and credit of a sovereign nation.
2. Assume the debts of the states and pay them off as well.
3. Enact a tariff to protect certain domestic manufactures. This would create a balance among agriculture, commerce and industry, and would provide for a growing economy.
4. Negotiate a trade agreement with Great Britain, securing the best advantages possible.
5. Establish and operate a national bank, in cooperation with private financial interests, along the lines of the Bank of England. Such a bank would provide capital for economic expansion.

Hamilton's proposals were designed to address the problems described above. Yet they were also designed to advance his ideological agenda—and they fairly crackled with controversy:

1. The federal debt existed in the form of government notes—IOUs—dating from the Revolution. Most of these were no longer in the hands of their original owners (soldiers of the army, suppliers, et cetera) and were trading among speculators at a few cents on the dollar. If the government were to pay them off at par value, the speculators—many of whom had friends in high places—stood to make a stupendous killing.
2. Some of the states had already retired their revolutionary war debt, often at considerable sacrifice. These states, through general taxes, would now be required to fund the debts of those that had not paid off their obligations.
3. During the Constitutional Convention there had been stormy debates about tariffs. Agricultural states, especially in the South, feared retaliation against their own exports if the United States now began slapping tariffs on European manufactured goods. Besides, these goods would now be more expensive to buy—along with those of domestic manufacture. Hamilton was in effect dunning agriculture to subsidize industry—and consumers to profit producers.
4. Many Americans had grave reservations about seeking a trade agreement with Great Britain. They weren't sure that the new republic should cozy up to a power that was in the habit of exploiting it. Nor did they like the idea of

being drawn into European quarrels as a lackey of the British. The United States, they said, should establish trade relations with *all* countries and remain aloof from special attachments.

5. The proposed national bank drew the most fire of all. It contemplated an alliance between government and the "wellborn"—to the possible disadvantage of everyone else. It might prove to be the opening wedge of aristocracy. Worst of all, there was absolutely no constitutional authority for it. Any argument that would enable the government to establish and operate a bank would enable it to do virtually anything.

Washington knew that Hamilton's proposals would be hard to swallow. He also knew that their audacity would make it impossible for him to give them lukewarm support: he must either quash the entire program or back it to the limit. And he knew that a great deal depended on the outcome. Hamilton's proposals would begin fleshing out the Constitution in a certain way. They were, accordingly, as momentous as the Constitution itself.

The most troubling issue for the President was Hamilton's loose-jointed, free-wheeling construction of constitutional language. While the authority for most of the Treasury Secretary's program fell more or less within the Constitution's enumerated powers, the national bank clearly did not. And justifying it by means of Hamilton's "necessary and proper" argument left no logical stopping place, as critics were quick to point out. If a national bank was "necessary" to the coining of money and regulation of its value, why not a bank monopoly? Or state-operated gold mining? Or commodity price controls? A lot of things were arguably "necessary."

Washington solicited an opinion from Thomas Jefferson, his Secretary of State, knowing in advance that Jefferson would vehemently oppose Hamilton's position. Nor was he disappointed. For in his written opinion, Jefferson argued that the word "necessary" could only be defined one way—*that which was indispensable*. Congress, in other words, was limited to those means that were absolutely required for the performance of an enumerated task. Nothing in its constitutional role of coining money or regulating its value remotely required Congress to set up a bank.

Within the principles of "**strict construction**," as Jefferson's position soon became known, was the kernel of another political ideology—one almost diametrically opposed to that of Hamilton. And in the hothouse atmosphere of the new republic, that competing ideology was very soon to sprout. For the President, after much soul searching, decided to back Hamilton's program all the way.

Jefferson: Politics and Ideology

Like Hamilton, Jefferson had had little direct input into the Constitutional Convention. He was serving as US minister to France at the time. Nor was Jefferson's initial reaction to the document very favorable. He feared that there was too much power in the federal engine and too few brakes. It was only by means of his friendship with James Madison that he was eventually brought around to the Constitution's support.

If old-fashioned Toryism was the starting place for Hamilton's ideology, old-fashioned Whiggism was the starting place for Jefferson's. He truly, if a bit naively, believed in an lost golden age, a time of peace and plenty when sturdy English yeomen tilled their own soil and governed their own lives. Even the king in this mythical Saxon democracy was but an elected representative of the people.

Unsurprisingly, Jefferson subscribed to the "commonwealth ideology" discussed in chapter 3. Whigs believed that the ancient freedoms had been undermined, little at a time, by the "court party," that cabal of would-be aristocrats always drawn to the center of power. And these must ever be resisted by the "country party," whose identification with, and attachment to, the land is a source of sublime independence and great moral strength. If one would look for republican virtue, said the Whigs, look for it here.

Jefferson's own observance had deepened these ideas. His attachment to the soil was romantic, almost mystical, and in his *Notes on Virginia*, he had written enthusiastically about the natural world. Naturalism was one of the great themes of the time. Its underlying premise was that civilization corroded human character, and as a result the "natural man" was stronger, quicker, wiser, more alert, and more wholesome than his city-bred counterpart. At the foot of the Blue Ridge Mountains, where Jefferson's Monticello looked out over a heart-stopping vista, he was convinced that he saw the corruptions of the centuries fading away and the natural man coming back into his own. For there stretched before him a vast expanse of rich land in moderate-sized freeholds, owned by families on the rise. The work ethic was strong in most of them, and their actions bespoke moral self-governance. Jefferson compared this highly idealized picture with the misery and squalor he had seen in France. Here, he thought, were the makings of a true republican society.

If this was not the first democratic vision of America, it was perhaps the most consequential. But its fruition depended on an approach to government that was almost exactly the opposite of Alexander Hamilton's. Power must be dispersed throughout society, not concentrated at its center. Politics must be local, not national, in character. And national authority, with its centralizing tendency, had to be rigorously curtailed by constitutional means—lest the "court party" resume its age-old machinations. Only the strictest possible construction of the Constitution would keep the government small, custodial and in its place.

Jefferson wrote his memorandum to the President with an acute sense of desperation. Strict construction was the fulcrum on which his vision of America turned. If he lost on that point, all else stood in peril. And when Jefferson did lose, he recognized the dreadful significance of it. Washington's adoption of the Hamilton program, given the precedent-setting circumstances, risked nothing less than a betrayal of the Revolution.

Foreign Intrigue

The rivalry between Jefferson and Hamilton on domestic policy may have turned out very differently at any other time in history. As it happened, while the two of them were squaring off, the calm world of the European Enlightenment came to an abrupt and shattering end.

Much of the early history of the United States took place against the backdrop of the French Revolution. At times, events in Europe threatened to overwhelm the young American republic.

The French Revolution began in 1789, the year George Washington took office. It would continue to lurch and careen through the entire decade of the 1790s—then erupt into a global war. Americans watched the unfolding of events with fascination. To many, it appeared that the French people had chosen to follow their own lead, hacking off the fetters of monarchy and stepping into a glorious republican future. But other readings were possible, especially after the execution of the king and the commencement of the Terror. Edmund Burke, a member of Parliament who had warmly supported the American cause, was among the first to point up the radical nature of the French Revolution, and suggest that it might not lead to a glorious anything.

By 1793 England and France were at war. The latter sent emissaries to the United States—the first of them being "Citizen Genêt"—and called on the American people to support the cause they themselves had inspired. While President Washington counseled strict neutrality—the last thing the US needed was a foreign war—Americans found such a feat impossible.

They divided on the French Question more or less as they were dividing over the issues posed by Hamilton's economic program. Hamilton's followers strongly identified with the British. England, they said, was fighting for the established order, political moderation and plain sanity. It was also fighting for a dominance of world trade that was more than incidental to Anglo-American relations. Jefferson's followers identified equally strongly with the French. Excesses notwithstanding, they said, the Revolution truly was about liberty, equality and fraternity. The revolutionary experience was bound to be a painful one because the French people had been downtrodden for centuries.

The situation would become ever more precarious for the American Republic. Every domestic issue would come to be seen in the light of the French Question and interpreted or misinterpreted accordingly. Foreign affairs are always capable of arousing strong passions, and the present case was no exception.

The Emergence of Political Parties

Virtually none of the Founders expected political parties to develop in the American Republic. Despite the existence of Whigs and Tories in England, the modern concept of political parties simply did not exist in the 1790s. For it depended on another concept, also not yet fully in existence—that of ***loyal opposition***.

The very words were an oxymoron. "Opposition" could not be "loyal," just as "rain" could not be "dry." The reason why had to do with the assumption, still flickering in the English mind, that monarchy had been anointed by God. Accordingly, the English common law had a doctrine known as ***seditious libel***, holding that the state, in the person of the king, had a right to protect itself from sedition—that is, the desire to destroy or overthrow it—by punishing those who spoke against it with malice. The doctrine also came to protect the king's ministers, his officers, and eventually even minor functionaries who acted in his name—the government in other words. Without being able to criticize the government, of course, politics as we know it could not exist, for that is what parties do.

That was only one obstacle. There was a deeper one rooted not in monarchies but republics, the concept of a public good. If you really believe in such a thing—and the American Founders did—it is almost impossible to believe in the sort of politics we practice today. For the public *as a whole* can have but a single good, not a multiplicity of them, and the political process ought to seek out that good alone. Any other will be the good of some group or party or class, not that of the entire body politic. If George Washington were asked if he represented Alexander Hamilton and his friends as opposed to some other interest, he would have been horrified. A President must represent *everyone*.

And there was yet another reason for being leery of political parties. In the Founders' minds, a "party" was merely a large and dangerous "faction." We recall that in *Federalist* 10, Madison had argued that factions would always exist when interests were in conflict. The trick was not to obliterate them or deny their existence, but to neutralize their effect through counterpoise. The more factions contending for power, the less likely it would be that any one of them could achieve hegemony. But for Madison, indeed for all the Founders, a *majority* faction—that is a faction capable of dominating a legislative assembly—spelled the end. For, by definition, such a faction had won the political game and could enact its own will into law.

One might argue, then, that from the Founders' perspective, the Constitution had been designed to *prevent* the formation of political parties. In the operation of the electoral college, for example, we see the hope that the political game would be played by individuals and small groups, sometimes in contact with one another and sometimes not, acting out of self-interest perhaps but also acting out of civic virtue—and never acting as a "party." It seemed at least possible anyway.

Until Hamilton's program came along.

Hamiltonians

Any proposals as bold and controversial as Hamilton's would require skill to steer through Congress. Accordingly, Hamilton and a few close associates began meeting informally with Senators and Congressmen to discuss problems, plan strategy, and cultivate political support. Of course these "Hamiltonians" had engaged in such kibitzing for years, now to seek one end and now to seek another. The objective here, however, was not just to achieve a single victory but to enact an entire legislative program and see to its implementation—to govern.

Hamilton's group enjoyed a number of advantages. It was organized around a coherent purpose and specific goals. It shared a vision of American nationhood that was steadily winning converts. And it had the support of George Washington. Clumsy at politics himself, Hamilton had the good sense to keep in the shadows and place other, more engaging, personalities in the limelight. Since many of these had been supporters of the Constitution in the first place, it was only natural that they appropriate the name "Federalists."

Jay's Treaty: Negotiated with Britain in 1794, US made major concessions to avert war over seizure of US ships. The terms were extremely unfavorable to US interests.

It is important to remember that the Hamiltonians did not see themselves as a "party." They were "the government." They not only dominated Congress—a fact that became all too clear as the individual measures came up for passage—they wielded influence in the executive branch as well. For instance, recall that one of their objectives was to secure a trade agreement with Great Britain. In order to accomplish this aim, the Hamiltonians, working behind the scenes, succeeded in scuttling talks with the French and then proceeded to negotiate **Jay's Treaty** with Great Britain. Hamilton personally (and inappropriately) intruded into the negotiations. As a result, Jay's Treaty was more favorable to British interests than American.

Public reaction was sharply divided. Those of pro-British sympathy applauded the treaty and soft-pedaled its controversial features. Many others, however, denounced the treaty with unwonted vehemence. "Damn Jay's Treaty and damn anyone who *won't* damn Jay's Treaty," they cried. John Jay was ubiquitously burned in effigy.

Equally disconcerting was the way Hamilton dealt with the so-called Whiskey Rebellion. To secure government revenue, Congress had passed an excise tax on the production and sale of whiskey. The effect of the tax in western Pennsylvania, where farmers reduced their grain to distilled spirits in order to transport it over the mountains to market, was to destroy the profitability of an entire industry. In 1794 some 3,000 farmers united in opposing the tax and disrupting its collection.

Hamilton had been waiting for just such an opportunity to flex the new government's muscles. He persuaded the President to call out the militia in four states and march on the rebels forthwith. Owing to Washington's ill health, Hamilton himself commanded the forces, some 15,000 strong, and saw to the capture of eighteen rebels. Even though only two of them were convicted (and both soon pardoned), the episode marked another bitter division. Farmers up and down the frontier denounced the "American Caesar."

Jeffersonians

Jefferson and Madison were only two of those who found these events appalling. There were many others. Some were former Anti-Federalists, still nursing their doubts about strong government. In their view, the double disaster of Jay's Treaty and the Whiskey Rebellion was but act one of a drama written years before at the Constitutional Convention.

The old Anti-Federalists found allies among those who had favored the Constitution's adoption but were coming to have misgivings about Alexander Hamilton. Madison was one of these. He had been among the earliest and most energetic of the "nationalists," as we have seen, and at Philadelphia had been nothing less than Hamilton's co-conspirator. Now he saw the world through different eyes. Was this, he asked, what Hamilton's nationalism really meant?

The less self-assured might have been hard put to reexamine their assumptions. Not James Madison. His life could be described as an adventure of ideas, and he followed wherever they led. His conception of a strong national government had been equipped with mechanisms to prevent what the Hamiltonians were now bringing to pass—or so he had believed. If there was a flaw in the conception, and evidently there had been, Madison would be the first to acknowledge it and start looking for possible remedies.

It was here that Madison's intellectual partnership with Jefferson came to the fore. The two had been friends from their old days in the Virginia House of Burgesses. They had remained in contact ever since, volleying ideas at one another in their correspondence and sorting out the world's experience. They had no trouble sorting this case. To anyone brought up in the commonwealth tradition, Hamilton's program represented all that virtue must guard against: elitism, patronage, favoritism, manipulation, corruption and ultimately tyranny. It was the "court party" as it worst.

Organizing opposition was almost second nature to them—they had been doing it since the days of the Stamp Act. Besides, the opposition was already organizing itself, grass-roots style, and seeking *them* out. After Jefferson resigned as Secretary of State in 1793, visitors made their way up the Blue Ridge hillside to Monticello, where the willowy Virginian was engaged in his lifelong pursuit of architectural renovation, to sound him out about the Presidency. He had no appetite for office himself, he assured them, but he was not unwilling to hear their complaints, and somewhere in the back of his mind he did have such an appetite—if only to vindicate his idea of the *true* republic.

At no time did the "Jeffersonians," as they were beginning to call one another, see themselves as creating a political party. They were simply dealing with a problem, and when the problem was solved that would end the matter. Still, the movement picked up speed. Jefferson and Madison provided it with inspiration as well as leadership. Jefferson, who was the better writer, penned hundreds of letters to encourage the faithful. Madison worked the halls of Congress. All were aware that President Washington would step down at the end of his second term in 1796, and Jefferson was mentioned ever more frequently as a worthy successor.

Monticello, home of Thomas Jefferson near Charlottesville, Virginia. Considered the most famous private home in America, it was designed by Jefferson, who spent decades refining the structure and grounds.

This left the Hamiltonians with a problem. Which of them ought to be put forward as their own candidate? Hamilton himself was out of the question. He was too cold and brittle for the campaign trail, and his well-known view of democracy—"Your people is a beast!"—did not commend him to ordinary voters. After a series of caucuses, the Hamiltonians came to agree that John Adams might fill the bill. While not precisely one of their number, Adams was a staunch supporter of the Constitution, and he generally concurred with Hamilton's legislative program.

When the votes were counted in the electoral college, John Adams had the greatest number, by a very slender margin, and therefore was elected President. And as Vice President? Thomas Jefferson of course.

The Alien and Sedition Acts

Alien and Sedition Acts/Kentucky and Virginia Resolutions: http://www.yale.edu/lawweb/avalon/alsedact.htm

The Presidency of John Adams proved to be pivotal for the new republic. If George Washington was the first President, John Adams was the first ordinary President, the first who was truly elected (Washington had simply been acclaimed), and the first to be regarded as politically mortal. When Washington took a position, as he did on Hamilton's legislative program, opposition was obliged to be circumspect. When Adams took a position, it was just a position. With the Adams administration, political life settled into a bumpy normality.

Hamilton retained the allegiance of the Federalists and through them continued to influence the President. Hamilton also played his Washington card for a final, triumphant time in his drafting of the "**Farewell Address**." For anything that President Washington would say in benediction was bound to have scriptural effect.

The address spoke movingly against the "spirit of party." It affirmed the Founders' original assumption that political life in a republic need not and ought not be partisan. It also cautioned Americans against a too ardent attachment to foreign powers. The advice was unquestionably sound on both points—yet it was also unquestionably Hamiltonian. For by cautioning fellow Americans against the spirit of party, the retiring President was saying in effect that the Federalists were not a party at all but simply the government. And by warning them away from entangling alliances, he was not condemning trade relations with England, which were peaceful and commercial, only relations with firebrand France.

But the French Question would not be banished so easily. Indeed, it would return and return, like the proverbial bad penny, and eventually prove the undoing of Federalism in general, John Adams in particular, and all of the Founders' assumptions about politics.

It soon became clear to most Americans that right or wrong, the French Revolution was out of control. Nor did its partisans help things by mixing into US politics. Citizen Genêt, after meeting his cool reception from George Washington, tried circumventing the government and appealing directly to the people. At the same time, a number of French republican journalists, crossing the Atlantic to stir up sympathy, called upon Americans to support the French economically and militarily, as one republican people to another. Specifically, they urged, the US should honor the Treaty of Alliance dating back to its own Revolution, break all ties with Great Britain, and throw the Federalists out of office.

Americans responded with ambivalence. Many of them did feel a sense of republican kinship. At the same time, they agreed with their President that foreign entanglement could prove their undoing. They regarded the old French alliance as a dead letter and Jay's Treaty, no matter how unpopular, as the law of the land. And increasing numbers of them viewed events in France with alarm. The Federalists capitalized on these doubts in fighting their own political battles. It was easy, for example, to suggest that Jeffersonians were nothing but an American version of Jacobins—wild-eyed and seditious.

Relations between the French government and the Adams administration continued to deteriorate. In addition to the frictions sketched above, there was a new one in 1797, caused by the resumption of Anglo-French hostilities. For this meant that US ships trading in the British West Indies were now subject to seizure. By 1798 war fever was sweeping the United States, and on the high seas Americans ships were engaged in armed conflict.

It would be difficult to imagine a more divisive set of circumstances. With American ships "under attack" by a foreign power, there was a strong inclination to rally around the flag, and Hamiltonians exploited it to the limit. Jefferson's followers, on the other hand, viewed the hysteria as cold-blooded manipulation and said so. The temperature steadily mounted.

As President, Washington exemplified the qualities associated with statesmanship and areté. Yet, his last official utterance was greatly influenced by Alexander Hamilton and seemed to endorse the Federalist position as his own.

It was at this point that the partisan newspaper made its debut. It was a combination of local newsletter, serialized pamphlet, political broadside and town tattler. It specialized in purveying the inside dope—what was *really* going on—for which Americans have always had a peculiar weakness. Accordingly, the editors' free-swinging style, gross errors of fact, and dark allegations of scandal were all submitted with greatest gusto.

There were papers on both sides. Those on the Hamiltonian side enjoyed the moral advantage of defending the government, while those on the Jeffersonian were attacking it. And the latter, remember, was sedition. It was openly criticizing elected authorities, subjecting them to malice, with the sole intention of undermining public confidence. George Washington, who caught the merest brush of tar, found the situation completely outrageous, while poor Adams came in for the entire bucket.

It was clear to Adams that the situation must somehow be defused. He dispatched emissaries to Paris with instructions to negotiate an end to the hostilities if at all possible, but Paris was swirling with its own dark passions

XYZ Affair: Diplomatic incident in 1798 in which Americans were outraged over French demands of a bribe to negotiate with American diplomats.

and the time was anything but propitious. After the emissaries were solicited for a bribe (the notorious XYZ affair), Franco-American relations seemed headed over the falls.

Hamilton and the Federalists responded with decisive action. War abroad and sedition at home posed dire threats to the republic, they concluded, and justified whatever means were at hand. Said Harrison Gray Otis: "Government has a right to preserve and defend itself against injuries and outrages which endanger its existence. . . ." The result, after a bitter debate in Congress, was passage of the infamous **Alien and Sedition Acts**.

The Alien Act authorized the President to deport any aliens he regarded as "dangerous to the peace and safety of the United States." Adams never availed himself of the authority the measure granted him, but not for a lack of urging on the part of Hamilton, who had fashioned it as a weapon against the foreign journalists. As for the domestic journalists, they were faced with the Sedition Act.

And this too was a weapon. It outlawed any and all "false, scandalous and malicious" writing against the government or its officers, including the President and members of the Congress. By its terms, unfriendly journalists could be arrested, tried, and subjected to heavy fines and jail time. And at the instigation of bitterly vindictive Federalists like Timothy Pickering, they were. Some fifteen editors were arraigned and tried, and ten of them found guilty. All of course were Jeffersonians.

Jefferson and other "Republicans," as they were beginning to call themselves, took vigorous exception to the Alien Act, which violated several provisions of the Bill of Rights. But their heavy fire was concentrated on the Sedition Act. Indeed, that piece of legislation would soon become the fulcrum about which issues of critical importance to the new republic would turn:

Constitutional Interpretation. We have already noted the divergence between Hamilton's loose construction and Jefferson's strict construction of constitutional wording. The Sedition Act brought that clash to a head.

With the adoption of the First Amendment, the federal government was supposedly bound by the injunction that "Congress shall make no law . . . abridging the freedom of speech, or of the press. . . ." If Federalist loose construction could slip around so clear and concise a prohibition as this, said the Jeffersonians, then surely all was lost. The Bill of Rights, the Constitution itself, were worth no more than the paper they were written on.

The Federalists had a reply. The Sedition Act, they explained, merely enacted into positive law the ancient common law of seditious libel. Since Great Britain had its own celebrated freedom of expression, and since the common law of seditious libel was not held to violate such freedom, what was wrong with enacting the ancient law into a federal statute?

The answer to this question would assume crucial importance in the American experience. According to the common law of seditious libel, there was to be no "prior restraint" on public utterance. That is to say, the government could not use licensing or some other procedure to determine who should and should not be allowed to publish facts and opinions in a newspaper. Once the facts and opinions were published, however, the newspaper became fully liable for all scurrilous remarks it made about government officials. The truth or falsehood of such remarks was beside the point. For the issue was not whether the official in question actually did what was alleged but whether the government was going to come out of it with a black eye.

The Federalist position on this question led to the second pivotal issue.

Political Opposition. Granting the old concept of seditious libel its due, said the Republicans (thinking out loud), perhaps it was out of date in the modern world. For in a republic, issues must be frankly and openly discussed, and the discussions must necessarily include the character of government officials, the policies they have adopted and the actions they have taken. The freedom to speak out couldn't just be freedom from prior restraint, it had to be complete freedom—freedom from consequences later on. Seditious libel, in other words, was unsuited to a republic.

Jefferson's followers weren't quite ready to push on to the final step in this logical process, but the step was becoming clear. If citizens needed the freedom to speak against the government, then organized political opposition, on an ongoing basis, was as much a part of republican life as free speech was.

Truth and Falsehood in Politics. There was still the issue of truth to think about. If someone makes an allegation about a public official, it had darned well better be true—hadn't it? This seemingly simple question, when viewed closely, wasn't simple at all, for it disappeared into a labyrinth of unanswerables. Should the standards of the courtroom apply to political debate? Can questions of fact be neatly separated from questions of opinion? Are political slanders necessarily the same as personal slanders?

Out of the ensuing welter of discussion there emerged a whole new conception of political truth. Jefferson encapsulated it in his inaugural address:

> If there be any among us who would wish to dissolve this Union or to change its republican form, let them stand undisturbed as monuments of the safety with which error of opinion may be tolerated where reason is left free to combat it.

In other words, in republican politics, citizens must have the freedom to say pretty much anything—including "Let's dissolve the Union!"—and the voters must separate truth from falsehood themselves. In the end, Jefferson believed, truth would win out.

The Resolution of Constitutional Disputes. A fourth issue posed by the Alien and Sedition Acts had to do with constitutional disputes. Where two interpretations of the Constitution were at such odds as Hamilton's and Jefferson's, how could they possibly be reconciled?

The Republicans began thinking seriously about this problem. And the more they thought about it, the clearer it seemed that disputes must be resolved outside the federal government. After all, they reasoned, that government had been created by the sovereign states, not the other way around, so the *ultimate* sovereignty must lie with them. Besides, if constitutional interpretation were left up to some federal authority, state sovereignty could be kissed goodbye.

Accordingly, Jefferson and Madison approached state governments with what seemed to them a constitutionally sound idea. The states must take individual action to declare the Alien and Sedition Acts unconstitutional, they said. Only by such means could the federal government be kept in bounds.

The result was the **Kentucky and Virginia Resolutions**. Each of these states drew up and passed a resolution to the effect that the Alien and Sedition Acts were unconstitutional and therefore of no legal force within their respective jurisdictions. Each document also explained why this approach was the right way to resolve constitutional disputes. Jefferson, who drafted the Kentucky Resolution, and Madison, who drafted the Virginia, were coming down hard on the interpretation issue. It seemed fully capable of destroying the Union.

The Nature of the Union. The bedrock question underlying all of this had to do with the Union's ultimate character. Was it a union of sovereign states joined together by voluntary compact? Or was it a single sovereign entity in which the states played a local administrative role?

The Federalist, we recall, had argued for both alternatives at the same time, a position termed *divided sovereignty*. (That is, states remained sovereign in some areas while the United States assumed sovereignty in others.) Skeptics had argued that "divided sovereignty" was nonsense—an oxymoron—and predicted that sooner or later an issue would come along to prove their point. Well, here it was. And James Madison, who had come up with divided sovereignty in the first place, was turning to *the states* for a solution—rejecting his own doctrine!

The Kentucky Resolution, which was the more radical of the two, took the notion of ultimate state sovereignty a step further and for the first time invoked the "n" word. "Nullification" would cast a long shadow into the future. Yet under the press of the Alien and Sedition Acts it seemed the most logical term in the world. For if Congress did pass an unconstitutional law, shouldn't the sovereign states have the power to "nullify" that law within their own jurisdictions? The deeper question, unasked at the time, was that if a state had the power to nullify federal laws, didn't it also have the power to secede from the Union entirely?

For the time being, the difficulties posed by the Alien and Sedition Acts were resolved otherwise. Jefferson and his followers won the presidential election of 1800 and swept into power. And the notorious legislation lapsed into oblivion.

The laws had a "sunset provision" that would have let them automatically lapse if they were not renewed. Jefferson allowed them to lapse in 1801.

The "Second American Revolution"

The institutional effect of the Alien and Sedition Acts was to catalyze the formation of political parties. At some point along the way, it was no longer just a matter of writing letters, publishing newspapers, or meeting together as alarmed citizens. The citizens organized themselves into clubs and societies, often using the words "Republican" or "Democratic" in their name. A sense of camaraderie burgeoned among them, and they began to consult with one another. While there was no polite name for what was emerging, most people knew it was "party."

Some still shrank from the word—Thomas Jefferson among them. "If I could not go to heaven but with a party," he wrote scornfully, "I would not go there at all." It required mental gymnastics to make such a vow while building a party of his own, but as Joseph J. Ellis has noted, Jefferson "was psychologically up to the task." His ambivalence bespoke that of most Americans.

An important stage in the development process was political self-awareness. The Federalists came to think of themselves as supporters of the Constitution and defenders of sound government, while in their own mirror the Republicans saw the legitimate heirs of the Revolution. The latter also identified with those rough-hewn tillers of the soil Jefferson had written about, who alone knew the meaning of republican virtue. Soon they had hyphenated their name to "Democratic-Republicans."

Brawl in Congress. *Though the Founders saw political parties as politically evil, partisanship manifested itself immediately in the young republic. Though it would be a latter generation that established the American two-party system, the Founders themselves set the process in motion.*

Each party also drew an ever clearer picture of the other. As Jeffersonians studied Hamiltonians, they saw loose construction of the Constitution, a narrow conception of rights, a government of would-be aristocrats, a dim view of political opposition, and the personal ambitions of an American Caesar. And as Hamiltonians swung the telescope around, they saw strict construction of the Constitution, a broad definition of rights, a government too weak to prevent anarchy, a rabble of self-styled democrats in sympathy with French radicals, a taste for seditious libel, and the leadership of an avowed Deist. Both pictures were scary.

Yet we should note that while the two parties demonized one another, they kept within certain limits. It was not a case, say, of one party being *for* the Revolution while the other was *against* it—as would often occur in European politics. They might resort to ridicule and name-calling but they wouldn't resort to the guillotine, for beneath their disagreements they had a common ground, or *consensus*, in the Founding itself. What they were disputing was how the Founding was to be understood and built upon. Consider the specific issues:

- Contrasting political ideologies.
- Contrasting ideas of constitutional construction.
- Contrasting notions of free expression and political opposition.
- Contrasting ways of understanding the Union.
- Contrasting visions of an American nation.

These issues have marked many of the paths crisscrossing American history, and every major political movement has come to bear some relationship to them. For each grows out of the Founding and seeks to define its true meaning.

Election of 1800: http://library.thinkquest.org/12587/contents/timeline/1800/1800.htm

The election of 1800 was a jarring contrast to the three preceding it and one of the most unruly in US history. Reckless charges flew back and forth between the candidates—and they really were *candidates* now—accompanied by

The Election of 1800

Although there had been presidential elections before in the American republic—three of them to be exact—there had never been anything like this. It was fortunate, some said, that Washington had passed away the year before, for he would have abominated the very sight of it. He had repeatedly warned his countrymen against "the spirit of party" and now here it was in full flower.

There were plenty of issues to galvanize opposition to the ruling Federalists. Jay's Treaty with England, Hamilton's controversial financial program, the war scare with France, the specter of a standing army, and, touchiest of all, the notorious Alien and Sedition Acts. These the Federalists had rammed through Congress expressly to silence political opposition—First Amendment or no First Amendment. As a result, for speaking "falsely and maliciously" against the government, twenty-five persons had been arrested, fourteen indicted, and ten (including a US Congressman) tried and convicted, all of them admirers of Thomas Jefferson.

Jefferson, as Vice President, was oddly enough a part of the government himself. Before the Twelfth Amendment was passed, the Vice President was the presidential candidate receiving the second highest number of electoral votes; thus, he naturally represented the opposition. And Jefferson had certainly done that. Every time the Federalists had touched off a new controversy, the Virginian's facile pen had been set to framing objections. Soon everyone who was discontented, aggrieved, or just politically ambitious came to see Jefferson as a rallying point.

The Federalists were amazed at how well Jefferson's Democratic-Republicans could organize. Almost overnight, it seemed, there were committees within committees, networks of correspondence, carefully coordinated propaganda campaigns. And the new party's agents were everywhere. "Every threshing floor, every husking, every work party on a house-frame or raising a building, the very funerals are infected with bawlers or whisperers against [the] government," complained Fisher Ames. In Congress, too, every debate turned into an election skirmish. The House, wrote one observer, was like "a conclave of cardinals, intriguing in the election of a Pope."

As the idea of political opposition was new to them, the Federalists were dumbfounded by it all. They fought back stridently, viciously, as though on the brink of Armageddon. If Jefferson were elected, shrilled Hamilton to Jay, it would "the OVERTHROW of the GOVERNMENT . . . a REVOLUTION, after the manner of BONAPARTE." Worse even than that, an anonymous pamphleteer moaned. "Murder, robbery, rape, adultery, and incest will be openly taught and practiced, the air will be rent with cries of distress, the soil will be soaked with blood, and the nation black with crimes." But the Democratic-Republicans could sensationalize too. One editor referred to the President as, "old, querulous, bald, blind, crippled, toothless Adams."

There were other quirks in the campaign that might seem familiar today. When Jefferson dispatched his Philadelphia friend, Dr. Logan, to Paris to see if the French quarrel could be patched up, a poem in the Federalist press accused him of wanting to make time with Logan's wife. When Republican congressmen rose to speak against the Alien and Sedition Acts, Federalists in the House chamber were smitten with such a coughing fit that the speakers could not be heard. Elbridge Gerry, a prominent Massachusetts Federalist, dramatically switched parties and ran for Governor. And a tunesmith knocked out the country's first political song:

Rejoice, Columbia's sons, rejoice
To tyrants never bend the knee
But join with heart and soul
and voice
To Jefferson and Liberty.

Found in America: A Study in Heritage, *Kendall/Hunt, 7th Edition, p. 208.*

merciless lampooning in the press. A backdrop was provided by the naval war with France, which continued to flash and flare during the campaign. Several hundred US merchant vessels were seized by French privateers, but the American Navy, which had been expanded from three ships to fifty-five, acquitted itself well and bagged some eighty-five French prizes.

The election of 1800, referred to by some as the Second American Revolution, witnessed the peaceful transfer of power from the Federalists to Jefferson's Democratic-Republicans. One ideology yielded up power to another without revolt or bloodshed.

Federalists cashed in on the spiraling hysteria. Their campaign often took the form of marching, saluting and striking up the band. Congressman John Allen caught the spirit of it all and characterized his Republican foes as "an *insurrection* against the Government," while Senator James Lloyd expressed a wish "to lay hands on traitors." For the Republicans this was politics at gunpoint—using the emergency to "perpetuate [Federalists'] authority and preserve their present places."

When the votes were counted on election day, Jefferson had won, and his Democratic-Republicans had swept both houses of Congress. The result, said the new President-elect, was nothing less than the "Second American Revolution."

Constitutional Government

Thomas Jefferson and Aaron Burr each had the same number of electoral votes. The decision moved to the House of Representatives, with the potential for a constitutional crisis. After many attempts to the break the deadlock, a representative changed his vote and Jefferson won the presidency. The Twelfth Amendment was passed to avoid a repetition of this situation.

For a few of the surviving Founders, Jefferson's triumph marked the end of the American experiment. The "spirit of party" had prevailed after all. A hostile faction had not only swelled into a majority, it had captured control of the entire government. All of the Constitution's finely tuned mechanisms of counterpoise had apparently fizzled.

Yet the inauguration of Jefferson as President was not marked by the collapse of order or rioting in the streets—and still less of "majority tyranny." Life went on pretty much as it had. This remarkable anticlimax might well be one of the most significant events in American history, for it marked the first true experience of constitutional government. The party out of power became the party in power and vice versa. *Power changed hands peacefully.*

Only gradually did Americans sort out the full meaning of what they had witnessed. Among the dawning realizations were the following:

Constitutional Mechanisms. The mechanisms of counterpoise did do their job, but not precisely as the Framers had imagined. The theory had been to prevent any single group from gaining control, and that of course hadn't worked. Yet the mechanisms operated just as well in spite of the fact. Even when the

Democratic-Republicans virtually monopolized the central government, the separations of power, the checks and balances, the federal-state dichotomy made it almost impossible to execute a tyrannical purpose.

For instance, there was enough disagreement among the Democratic-Republicans themselves—some of whom sought support from the Federalist minority—to make despotism unlikely. In consequence, Jefferson faced many of the same old obstacles that Adams had. Controversial ideas had a way of transforming into political dissent, and dissent made their progress very difficult.

Political Ideology. A second revelation was that being in power was rather different from being out of power, especially when it came to ideology. Jefferson's strict constructionism, which had made a great deal of sense when the Federalists were in charge, made less sense with the shoe on the other foot. For example, Jefferson soon found himself on the horns of an odd dilemma. Bonaparte, who had now declared himself Emperor, suddenly found it expedient to unload the Louisiana Territory. Jefferson saw the opportunity to buy it—and thus nearly double the size of the United States—as a veritable godsend, but he lacked the constitutional authority to make such a purchase. Unless, of course, he could construe the Constitution more loosely than his ideology would allow. Ideology, Jefferson learned, must occasionally bow to practical matters. He bought Louisiana.

Loyal Opposition. A third revelation concerned political opposition. In his inaugural address, Jefferson uttered what was to become a famous line. "We are all Republicans," he said, "we are all Federalists." This statement is generally understood to mean not that Jefferson had accepted political opposition as legitimate but that he too rejected it. Like the Federalists before him, he believed that a government of the people ought not have to face organized dissent.

How did Jefferson square this belief with his *own* recent dissent? Simply by highlighting the special nature of the case. Hamilton and the Federalists were betraying the Revolution and running the country to ruin, Jefferson would have said, so desperate measures were called for. Now that the emergency was over, Americans could move forward once again. "We are all Republicans, we are all Federalists."

This sentiment did not diminish the opposition that Jefferson himself now faced from the defeated Federalists. Being turned out of office did not lessen their appetite for it, but only made them more determined to come back. Eventually the Federalists did fade away, but not without putting up a fight. They continued to battle the Democratic-Republicans on a local level through election after election, and they remained regionally strong in New England. Political opposition was here to stay.

Before the election, Republican Albert Gallatin, with sharp insight, had noted that "whoever dislikes the measures of Administration and of a temporary majority in congress" is believed by the government to be "an enemy, not of Administration, but of the Constitution." Gallatin was beginning to draw a distinction that few Americans, even in his own party, would have supported at the time: the "Administration" in power was *not* the same thing as the "Constitution." Clearly that was true. And coming to understand it thoroughly would

Swapping Issues

In the ideological world of European politics, it would be unthinkable for a party to embrace some position once held by its opponent. After all, parties are supposed to stand for something, yesterday, today, and tomorrow.

Not so in American politics. Where the two parties are hard-headed, practical minded, and basically middle-of-the-road, it is comparatively easy for them to rummage through each other's garbage cans for usable issues. Indeed, it is almost as though they sometimes sat down and just swapped. A few memorable examples:

It was the Hamiltonians (forerunners of the modern-day Republicans) who started the tradition of loosely construing the Constitution in order to augment national power. The Jeffersonians (forerunners of the modern-day Democrats) believed in strict construction in order to curb national power. Today, those positions are reversed.

In the days of Andrew Jackson, Democrats were anything but big spenders: they were known as the party of small, lean, frugal government. Jackson himself went so far as to veto construction of a national highway through the Appalachian Mountains, believing it to be an unwarranted extension of federal power. Today, of course, the Democrats are blasted as the party of "tax and tax, spend and spend."

Throughout much of the nineteenth century, Democrats, not Republicans, were the hawks in foreign affairs. They were the ones, generally speaking, who wanted to fight Mexico and expand American territory into the West. In today's world, they are usually regarded as the doves.

Republicans got their start as the party of civil rights. The left-wingers of the 1860s were to a man Radical Republicans, some of whom embraced the idea of full and complete racial equality. Democrats, by contrast, had a reputation for racial prejudice.

Finally, many of today's Americans can recall a time when it was the Democrats, not the Republicans, who were famous for unbalancing the federal budget. Considering such historical flip-flops, it is little wonder that say, Ronald Reagan could start out as a Democrat and convert to a Republican. "I didn't change," Reagan protested, "The Democratic Party did."

Found in America: A Study in Heritage, *Kendall/Hunt, 7th Edition, p. 226.*

make loyal opposition possible. If the Jeffersonians had been loyal opponents of the government before the election, then the ousted Federalists could be loyal opponents afterward.

Political Parties. A fourth dawning realization was that political parties might well have their own role to play in constitutional government. We will discuss that role in a subsequent chapter. For now, suffice it to note that parties did not behave at all in the way factions did.

Factions, as the Founders understood from antiquity, were groups in pursuit of a single narrow interest and oblivious to the public good. Neither criterion had fit the Jeffersonians. They reflected self-interest, true, but not a single one and certainly not a narrow one. And as for the public good, they believed that they and they alone represented it.

Political parties, then, were something new on the horizon. They were not "factions." They were larger, fuzzier and far more complex. Indeed, in a manner not yet fully understood, they were manifestations of the public itself.

The U.S.S. Constitution emerging victorious from battle. Given the history of the document "Old Ironsides" was named after, there seems to be something more than mere symbolism in their relationship. In the final analysis, both have been only as good as the crews that man them.

Popular Sovereignty. The transfer of power from Federalists to Republicans represented something which up to that point had existed only in theory—the sovereignty of the people. The fact was that for thousands of years the voice of the few was the only one that mattered, regardless of what the philosophers said. "The people" were seldom if ever heard.

Hamilton and the Federalists assumed that they spoke for the American people when they passed the Sedition Act and rigorously enforced it. But on sober reflection, the American people disagreed. Their election-day behavior sent a clear message. By changing the administration, peacefully and constitutionally, Americans were grasping the levers of true sovereignty. Virtually for the first time in history, the people were supervising the government.

American Patria

The Naval War with France was America's first foreign conflict. Before its outbreak, Americans had regarded the United States as a kind of abstraction—like NATO or the UN today. The sight of US frigates returning to port with French prizes stirred deeper feelings—but so did the success of the Constitution. Unlike the old Confederation, the government of the Constitution had collected taxes, paid its bills and mounted a police action against would-be violators. Its credit had been restored by Hamilton's financial program, and European merchants were willing to do business with it. It had negotiated a trade agreement with England which, however lopsided, gave it stature and credibility abroad. It had fought a foreign war and held its own.

We should credit more than the Constitution for these successes. Indeed, one might suggest that the ship of state had displayed questionable seaworthiness on its maiden voyage, and that seamanship alone had prevented the open seams and misaligned spars from sinking it. Something rather like Greek *areté*—a love of country that transcended ordinary limits—had shown itself to be as important to the Founding as structure or rights or the rule of law.

With the election of 1800, there was a sudden burst of confidence and stirrings of real patriotism. The system worked. The American Republic had arrived.

CHAPTER 12 STUDY AID

1. America—a work in progress:

- North v. South
- Town v. Country
- Commerce v. Agriculture
- A new Rome?
- Constitution: a rough sketch
- General constitution: creating precedents; titles; treaties; resolving ambiguities

2. Personality and the defining of the American nation:

- Statesmanship and areté
- George Washington—model "model"
- Alexander Hamilton—aristocrat
- Thomas Jefferson—visionary
- James Madison—architect of Constitution and Bill of Rights
- John Adams—forgotten founder?

3. Political Parties:

- Loyal opposition
- Seditious libel
- Factionalism
- Party and majority factions (*Federalist* 10)
- Hamiltonians
- Jeffersonians

4. Alien and Sedition Acts:

- President John Adams
- Citizen Genêt
- The "Press"
- The Acts
- Constitutional interpretation
- Political opposition
- Error of opinion—protected speech
- Constitutional disputes
- The nature of the Union: Kentucky and Virginia Resolutions

5. Hamilton:

- Secretary of Treasury
- Credit; trade; solvency; debt
- Sovereign nation: legitimacy and integrity
- Economic aristocracy
- Centralization: taxing and spending; general welfare; necessary and proper
- Implied powers
- Treasury Department: redeem debt at par value; assume debts of the states; tariff; trade with England; National Bank

6. Jefferson:

- Commonwealth ideology
- "Natural man"
- *Notes on Virginia*
- Democratic republicanism and strict constructionism

7. Foreign Intrigue:

- Hamilton and Britain
- Jefferson and France

8. Constitutional Government and the Second American Revolution:

- Peaceful transfer of power
- Constitutional mechanisms
- Political ideology—flexible when in power
- Loyal opposition—administration in power not the same as the Constitution
- Parties
- Popular sovereignty

9. American Patria:

- America invented
- Constitution implemented

CHAPTER 12 STUDY GUIDE

NAME: ______________________ **SECTION:** ____________

Key Terms

Statesmanship

Implied Powers

Loose Interpretation

Narrow Interpretation

Loyal Opposition

Hamiltonians

Jeffersonians

Alien and Sedition Acts

Multiple Choice Questions

1. The text describes the quality of statesmanship as possessing all of the following characteristics except:
   a. Areté
   b. Political savvy
   c. Desire to fulfill the public good
   d. Putting principle above party
2. Which of the following was not included in Hamilton's agenda as Treasury Secretary:
   a. Establish a national bank
   b. Enact tariffs
   c. Pay off all existing consumer debts
   d. Establish a trade agreement with England
3. The Sedition Act was aimed at silencing:
   a. The French
   b. Opposition newspapers
   c. Traitors
   d. The Anti-Federalists
4. Who wrote the Kentucky Resolutions?
   a. Thomas Jefferson
   b. James Madison
   c. Alexander Hamilton
   d. Elbridge Gerry

Review Questions

1. What financial problems faced the nation as George Washington took office as President? How did Alexander Hamilton propose to solve them?

2. Why did Hamilton and Jefferson view the Constitution differently?

3. How did the French Revolution impact American politics?

4. Why weren't political parties a part of the British governmental system?

5. Why was Jefferson's election in 1800 dubbed "the Second American Revolution"?

Review Exercises

1. Briefly explain the causes and consequences of the Whiskey Rebellion of 1794.

2. Explain the main reasons for the passage of the Alien and Sedition Acts.

3. Explain why the English doctrine of seditious libel did not find a place in Constitutional America.

Big Essay Question

Many European democracies have dozens of political parties competing for office. Yet America has remained basically a two-party country for over 200 years. Why? What are the advantages and disadvantages of our two-party system compared with multi-party governments in Europe? Be thorough.

CONCLUSION

The Unfinished Founding

In the afteryears, as Thomas Jefferson and John Adams grew into elder statesmen, they patched up their political quarrel and became friends once more. Indeed, their correspondence, which continued until the day they died—the same day, oddly enough, and the Fourth of July, and the fiftieth anniversary of American Independence—became a sort of founding document in itself. For in letter after letter, the two aging revolutionists tried to work out the meaning of their lives in the context of the American Founding.

There was a good deal to sort through. We might think of them—Jefferson at Monticello and Adams at his Braintree farm—as leafing through the materials we encountered in chapter 1, and asking one another how this or that element had worked its way into the United States of America. What was the significance of the ancient forms and ideas, of Enlightenment thought, of structure, counterpoise, laws and institutions, the English heritage, Puritan habits of mind? Where did self-interest come into the picture? Where did virtue? What role was played by Nature and Nature's God? Had classical republicanism been behind it all, or was it a work of modern liberalism, or even a forward-leaning democracy?

There was only one clear conclusion. America was first and foremost an idea. It had grown out of human experience, to be sure, and many kinds of reflection on that experience, but when all was said and done, it was that single powerful idea that counted the very most. Human beings were meant to live better than they did. It was possible for them to create and maintain a Good Society.

America-as-idea can never be fully wrapped up. Its very nature resists final packaging—for it is an idea about perfectibility. As long as people on the planet believe there may be better ways of doing things, the American idea will continue to prove elusive.

The Founding as we have considered it can be broken into three parts. The building of colonies in the 1600s and the search for some general meaning of the colonial experience amounted to a kind of First Founding. Were we to step out of a time machine in the year 1730—which would be the centennial of the great Puritan migration—and ask Bostonians on the street or Virginians on their tidewater plantations about their "American" identity, they might say something about the New World and a fresh start, or about a simpler life and more responsive institutions, and their comments might have a discernably American cast. They would be good and loyal Britons, for the most part, but Britons with a difference. Unlike their cousins back in Plymouth or Bristol, they would have acquired an American past, possibly two or three generations old, and more importantly an American future. They would no longer see themselves as living in an outpost of England.

The events surrounding the Revolution—the controversy with Parliament, the decision for independence, the prosecution of the war, the working out of statehood and confederation—amounted to a Second Founding. A visit from the time machine in 1781, the year the fighting ended and the Articles of Confederation were adopted, would mark a sea change from the America of 1730. Americans were flush with their long-shot military victory. Republicanism was in the air. And while there might not be an American nation as such there clearly was an American sense of patria. Americans were sure now that they resided on the sunny side of the world—and that the future opened before them.

A Third Founding took place between the spring of 1787, when delegates to the Grand Convention began to gather in Philadelphia, and the spring of 1797, when George Washington scanned over his Farewell Address. Not only was the Constitution written and adopted in that ten-year period, the Bill of Rights was ratified, and the government described in these documents became a working reality. Americans still identified with their respective states, of course, but there was a dawning awareness among them that the new federal Constitution actually functioned as it was supposed to. "American" was no longer merely geographical.

A century later, it was commonly supposed that Americans of Thomas Jefferson's day felt the same way about their republic as, say, Theodore Roosevelt did. But there was an incident during the Jefferson presidency that belied such a thought. While the facts are a bit complicated for our purpose here, their general import was the arrest and subsequent trial of a group out in the Mississippi Valley for treason against the United States. It was alleged that the group had been plotting to capture New Orleans and effect the secession of the western states. After a spectacular courtroom appearance, the accused plotters were acquitted. It wasn't the United States they wanted to dismember, they explained, but Mexico. Yet historians are by no means convinced of their innocence. So new and wobbly was the American Republic and so uncertain its destiny, there were all sorts of schemes in the air for modifying it. And high-flying schemes at that. One of the suspected insurrectos (James Wilkinson) had commanded US military forces in the West; a second (Aaron Burr) had been Jefferson's own Vice President; while a third would go on to become President himself, Andrew Jackson.

The trial of Aaron Burr before Chief Justice John Marshall laid down some important precedents about what did and didn't constitute an "act of treason." The Framers in Philadelphia had fretted about that very question, and as in so many other instances, rested unsatisfied with their conclusion. So in a way, Marshall's decision in the case might be seen as adding to the Founding. There would be many such small additions. For the American attitude toward the Founding has

While historians are still disputing Andrew Jackson's role in the Burr conspiracy, the consensus is that Jackson was probably innocent of any conspiratorial intent.

Aaron Burr:
http://gi.groliers.com/presidents/ea/vp/vpburr.html

been at once to revere it as a cosmic given and at the same time to massage it toward perfection. Let's consider some examples from our own century.

In Theodore Roosevelt's America, the Progressive movement sought to rethink the Founding and bring it up to date. The ideas of the Founders may have been good in their time, Progressives reasoned, but that time was now passed. So in a burst of creativity, Progressives altered the way Senators were chosen, substituting popular election for the process described in the Constitution. They also amended the Constitution to provide for an income tax, reorganized American banking around the Federal Reserve, increased government supervision of the economy, and launched the "noble experiment" of Prohibition. With the Nineteenth Amendment, Progressives gave women the vote.

Franklin Delano Roosevelt and the New Deal: http://geocities.com/Athens/4545/

Another member of the Roosevelt family, Franklin Delano, sought his own perfection of the Founding later on. His response to the Great Depression of the 1930s aimed at rethinking American institutions still a different way. Roosevelt spoke of the "forgotten man"—the Depression's most hapless victims—and suggested that a "new deal" of the political cards was necessary to address his plight. FDR looked to the Bill of Rights, for example, and argued that the rights listed there ought to include minimal economic security. Such innovations as government welfare, the Social Security system, and federal insurance against bank failure resulted. So did systematic intervention into the market system.

Brown v. Board of Education of Topeka: The 1954 Supreme Court decision that declared "separate but equal" schools violated the Constitution.

What became known as the civil rights revolution provided yet another revision of the Founders' work. The Supreme Court led the way this time. Its decisions in such breakthrough cases as *Brown v. the Board of Education* reinterpreted the Constitution to mean that states must protect the freedoms of all citizens, not just those of a privileged caste. For the first time in history, African Americans could dine in white eating establishments, sleep in white hotels, and move to the front of the proverbial bus. Black voters were registered by the millions and their power transformed US politics. When Martin Luther King Jr. said, "Thank God, we are free at last!" he might as well have said, "Thank God, the Founding now includes everyone!"

The list goes on. Ronald Reagan was elected President in 1980 on the promise that he would return America to its founding principles. George W. Bush pledged to cleanse the atmosphere in Washington and restore the Founders' sense of hopefulness. Radical feminists, gay liberationists, spokespersons for the disabled all raised their voices in the often clamorous national debate to assert that the Founding needed to be tweaked their way—or even redone from scratch.

Each of these developments illustrates the dynamic, ongoing and unfinished nature of the American Founding. America is an idea. It can never be pinned down for good.

President Franklin Delano Roosevelt adopted remembering the "forgotten man" as a theme as he led the nation during the dark days of the Great Depression during the 1930s.

Filling Out the Founding

In the human life span between Jefferson's election and the end of the Civil War, the American Founding underwent a period of systemic development. *Systemic*, in this sense, refers to filling in the institutional blank spaces to create an entire whole, or system. The process had begun in the Washington presidency, as advisors had clashed over the meaning of this word or that phrase in the Constitution. The elements that would now be added were of a different sort.

We might say they were foreshadowed, though not explicitly described, in the Founding documents. They were extensions of the republican idea or were necessary to its full realization. The party system that began evolving in the 1790s was a case in point. If parties hadn't come into being when they did and how they did, they would have evolved later on and been catalyzed some other way, for they were a systemic outgrowth of the Founding.

And there were several others.

Judicial Review

What we have come to call *judicial review* is older than the American Founding. Back in colonial days, for example, the acts passed by local assemblies were always reviewed by the Privy Council in London to make sure they didn't contravene fundamental law, and any measure failing the test was pronounced null and void. After the Revolution, several American states included judicial review in their constitutions, requiring that all bills passed by the legislature be submitted to the state supreme court before final acceptance. The need seemed only logical.

The Framers in Philadelphia had discussed judicial review themselves and thought about including it in the Constitution. Yet they hadn't done so. Several of them supposed that federal judges would review the laws anyway as a matter of course. In supposing that, however, they might only have considered the review of *state* laws. The more ticklish case of federal courts reviewing *federal* laws might not have occurred to them. But that is precisely what the theory of judicial review would require—someone to hold *Congress* to the Constitution.

Judicial review eventually came up anyhow and in a rather dramatic way. One day, more or less out of the blue, the United States Supreme Court declared a federal law "unconstitutional." The case at hand was made to order. It had to do with the question of whether Jefferson's Secretary of State, James Madison, could refuse to process an appointment made by John Adams on the last day of his Presidency, as he was mightily tempted to do. The precise issue was a technical one and need not detain us here. The point was that Congress had made a slipup in drafting the legislation that set up the federal court system, and one provision of the law exceeded constitutional authority, or at least so it could be argued.

The person doing the arguing was John Marshall, Chief Justice of the US Supreme Court—himself an Adams appointee. He and the Federalist party stood to gain politically, of course, if the case were decided against Madison, for the outgoing President had fairly stuffed the federal bureaucracy with appointments. However, with razor-sharp cunning, Marshall decided the case of *Marbury v. Madison* the other way, losing the battle of the moment but winning a very big war. For he realized that the Jeffersonians were not likely to contest his legal reasoning when *they* benefited from the outcome. And in the process, judicial review became a reality.

It was not a power the Supreme Court would invoke often, at least not in the early years, but the precedent had been set all the same, and Americans came to accept judicial review as part of the system. They liked the logic of it. It obviously helped maintain the separation of powers, for if Congress began encroaching upon the executive or judicial branches, the Supreme Court could slap it down. It also operated as yet another check and balance. Since the Supreme Court couldn't carry out its own edicts, it would have to depend on the cooperation of the executive, each offsetting the other. Finally, Americans seemed to agree with Madison (*Federalist* 78) that the Court was in no position to wield the review power tyrannically: it could declare laws unconstitutional but it couldn't pass other laws in their place.

In time, some of that optimism faded. For over the years the Supreme Court grew ever more expansive in its use of the negative. Furthermore, some Justices extended the logic of judicial review beyond determining whether laws were constitutional or not, and began to interpret the Constitution itself. We have already seen, in the case of Hamilton's economic program, how great the power of interpretation could be and to what ends it could lead. If federal Justices were free to construe constitutional wording, and do so without practical limitations, they had the power to twist the original document far out of shape. After the Civil War, for instance, they began interpreting the phrase "due process of law" to mean something entirely different from its traditionally accepted usage—and then using the new interpretation to hobble the power of state governments.

In a still later time, the Justices proved they could turn the power to expound into the power to legislate. The Court's assertion, for example, that a "right to privacy" was implied in the wording of several amendments, that such a right could be used to shield, say, the practice of abortion from state control, and that abortions were legal or illegal according to standards laid down by the Justices themselves struck critics as full-blown "judicial legislation." In other words, by interpreting the Constitution's vague words and phrases according to their own ideas, the Justices were doing what legislatures were supposed to do—make public policy.

For some constitutional scholars, then, judicial review opened Pandora's box. The fact remains, as Christopher and James Lincoln Collier have pointed out, "somebody had to do it." A republican government based on a written constitution would seem to require that some authority (other than the legislative of course) watch over the constitutional text. Charles R. Kesler has argued that the Constitution did not operate as the Framers themselves intended until the separation of powers was backed up by judicial review. Then and only then did republican governance become republican liberty.

Jacksonian Democracy

Just as judicial review may have inhered in the republican idea, so too, in all likelihood, did the political system we have come to call "democracy." It bears little resemblance to the democracy of ancient Athens, where citizens exercised day-to-day control over their government. Ours is a *representative* democracy, much like that of colonial Virginia, where elected representatives make the important decisions. Except that in our system (*un*like colonial Virginia's) ordinary citizens are free to seek office and participate fully in the political process, not just show up on election day.

Was the advent of modern democracy revolutionary or only evolutionary? Historians are still not sure. It is probably fair to say that some of the changes were evolutionary and others quite revolutionary.

We have already examined the democratic implications of the American Revolution. They were manifest in small ways before the shots were fired on Lexington Green, and in much larger ways after that. Many of the new state governments came to suffer from what James Madison saw as an excess of democracy, which is one reason why they found themselves in such confusion after the war. Democratic politics, with its coarse appeals and strident demands, was more than a little unsettling to those who were used to aristocratic decorum in public life. Yet Madison was no enemy of popular government. He merely wished to insure that it didn't succumb to its own excesses.

By the time of Adams' presidency, Madison was encouraging those excesses himself, for he and Jefferson were breaking all sorts of rules in their war against the Federalists. And Jefferson was thinking about the common people in imaginative new ways. It wasn't yet clear in his own mind what he meant by his oft-repeated phrase "pure republicanism," but he knew it had something to do with the uncorrupted nature of ordinary Americans. The more he thought about it the more convinced he became that the common people—those tillers of the soil—could bring a kind of virtue to the political process that the wisest sages might lack.

As President, Jefferson neither appointed common men to high office nor sought their political approval. He hardly needed to. His own party, heeding democratic logic, was already appealing to an ever wider electorate and for the most down-to-earth of reasons. Fending off the Federalists meant knowing the desires of grass-roots voters, mobilizing popular energy, dabbling in crowd psychology, and above all, getting out the vote on election day. Bastion after bastion of drawing room politics tumbled down as politicians embraced the new logic. For example, why should candidates be nominated by a caucus of fellow officeholders? it was asked. Nomination by the whole party, meeting in a convention or holding a primary election, whipped up far more enthusiasm and gave the voters a sense of empowerment.

Andrew Jackson: http://gi.groliers.com/presidents/ea/bios/07pjack.html

While Jefferson symbolized these developments, Andrew Jackson was their true beneficiary. His presidency followed Jefferson's by twenty years—a time of seismic change. After two terms in office, Jefferson more or less handed over the White House to James Madison, who passed it along to yet another Virginian James Monroe. By then the Federalists had ceased to exist as a political force, but party politics was far from over. Indeed, after the traumatic election of 1824, in which Jackson was narrowly defeated by what his followers called a "corrupt bargain," party took on a new life. The "Jacksonians" vowed they would return in '28. And to make good on the vow, they employed all the new techniques of democratic organization and follow-through plus a few more of their own. John Quincy Adams must have known his presidency would be short.

Beginning with the election of 1840 it became necessary for candidates to appear as "men of the people" rather than statesmen or exemplars of areté.

In office, the Jacksonians pursued the democratic idea to its logical conclusions. They not only organized their party from the upper echelons to the rank and file, they organized it *as* a party, a permanent association of people and interests bound together by ideology. (The Jeffersonians, by contrast, had simply rallied together under the press of circumstances to oust Hamilton.) And as such, they planned for their own future. Party functionaries divided and subdivided responsibilities and built a system of rewards into party service, with the result that, say, a faithful "ward heeler" of some local district, having mustered out the vote in election after election, could expect a government post requiring only minimal attention and returning a reasonable reward. This, of course, meant that the winning of an election by the party out of power would compel the replacement of civil servants up and down the line. "To the victor go the spoils," proclaimed Jackson.

Two things now became clear. The first was that two-party politics was here to stay. There simply couldn't be a party in power as vigorous as the Jacksonians without exciting an organized opposition. Indeed, various foes of "King Andrew I" were brought together almost solely by their aversion to his "tyranny"—they even swiped a term from the history books and called themselves "Whigs"—and pledged to work for his overthrow. There might have been three or four of these countervailing parties, as would be the case in other emerging democracies, but subtle features of the American Constitution tended to narrow the field down. For example, the way votes were apportioned in single-member legislative districts meant that small splinter parties were left with virtually no influence in the government, so there was constant pressure for smaller parties to combine together until they had a real chance of winning.

The second point of clarity was that both parties would have to employ similar tactics. If popular appeals were the way to win elections, then *both* parties must make use of them. In the election of 1840, the Whigs, stung by the charge that their candidate was not truly a man of the people, focused their campaign on his expressed desire to retire to a log cabin someday and drink hard cider. They trundled makeshift log cabins through the streets and of course passed out hard cider on election day. What was good for the goose was good for the gander.

Democracy plainly had its drawbacks, and critics were not slow to point them out. Yet it had its strong points too. Looking back from the twenty-first century, we can see that in both strengths and weaknesses democracy grew out of the Founding as surely as oak trees grow from acorns. When the Founders made their commitment to popular government, and more importantly when they found ways to make popular government safe and sane, democracy—log cabins, hard cider and all—lay dead ahead.

The Civil War

Among the Founding's imperfections, the most serious was that it established institutions of freedom in a political state that recognized slavery. The Founders weren't directly responsible for what followed, to be sure, but by making the compromises necessary for nationhood, they made future conflict a likely possibility.

Later on, as war clouds gathered, it would be alleged that the Constitution was meant to protect slavery in perpetuity (a charge still heard today) or, alternatively, that it was meant to bring about slavery's eventual demise. John C. Calhoun, a perceptive constitutional scholar, came to realize that neither allegation was true. The fact was, the system set up by the Founding was neither wholly friendly nor wholly unfriendly to slavery, and as a consequence slaveholding could grow and wax strong in the republic's early decades—yet could never be completely secure. If slave states were eventually to lose parity in the House of Representatives, then in the Senate, and finally in the Supreme Court, the day might come when, given the logic of democracy and the climate of world opinion, slaveholders of the South might be forced to give way. Calhoun, who was a southerner himself, came up with one answer to the problem: the Founding would have to be reworked with the preservation of slavery in mind. But other answers were also possible.

One of them was civil war. It was a dreadful thought to contemplate but so was World War II. Some will shudder at the comparison but one could certainly be made. In both cases, there was a massive wrong to be dealt with, not a simple disagreement or conflict of interest, and no amount of compromise could make it go away. The American South grew into a closed society, not wholly unlike Nazi Germany, defending a practice that seemed ever more deplorable. As a result, southerners, through their democratic institutions, expressed a well-founded fear that stemming the expansion of slavery would undermine their cherished way of life; while northerners, through *their* democratic institutions, expressed an equally well-founded fear that the expansion of slavery would undermine all freedom. Democracy, in other words, had tied itself into a knot.

Abraham Lincoln was one of the few who could see the knot clearly. As he looked back upon the Founding, he understood that democracy absolutely and necessarily depended upon a belief in human equality and natural rights. There was no other logic by which the American colonies could have opposed British tyranny, fought and won a revolutionary war, and stood by a Constitution they themselves had written. Slavery, by contrast, absolutely and necessarily depended on *in*equality and natural privilege. How else to justify human bondage? Alexander Stephens, Vice President of the Confederacy, saw the knot as clearly as Lincoln did. In his inaugural address, he said:

> Our new government is founded on the opposite idea of the equality of the races. Its cornerstone rests upon this great truth: the Negro is no equal to the white man. This government is the first in the history of the world based upon this great moral truth.

Given the knot, here was the problem. If America continued to countenance slavery, it must eventually undermine its own Founding. Lincoln saw signs of that already. For example, when Illinois Senator Stephen A. Douglas proposed to settle the slavery-expansion controversy by allowing the settlers in each new territory to decide the issue for themselves, millions of Americans embraced it as an inspired solution. But Lincoln was aghast. How could they possibly believe in a natural right to liberty, he asked, while deciding for or against slavery *at the ballot box*? Would they think of deciding for or against murder or cannibalism that way? And if they didn't believe in a natural right to liberty, how could they possibly believe in the United States?

In the dolorous march of events toward Fort Sumter, two unpleasant truths about the Constitution came to light. The first was that its elaborate system of springs and balances, however good at countering tyranny or anarchy, was not good at resolving sectional conflict. Various constitutional mechanisms were called into play in vain attempts to resolve the slavery-expansion problem. Some helped a little, most didn't help at all, and a few actually made things worse. Sectional jealousy had been the bane of the Philadelphia Convention, we recall, and now it was the bane of the republic.

The second unpleasant truth was that the Constitution was subject to widely dissimilar interpretations. This was not really new. Federalists and Anti-Federalists had had such disagreements. So had Hamiltonians and Jeffersonians. Now the dispute was between northerners and southerners—who unlike the others could call out armies. The question centered upon the ultimate nature of sovereignty in a federal system, and it was put this way: should there

The "Founding" left unresolved several critical issues such as slavery and federal-state relations. It took the trauma and carnage of a civil war to seemingly lay these matters to rest.

arise an all-bets-off showdown between state and national authority—which of the two must finally give way? The answer depended on whether one saw the Union as an association of sovereign states or itself a final sovereignty. Unsurprisingly, the South answered the question one way and the North another. If push came to shove, southerners said, the sovereign states could always secede from the Union and reclaim their former independence. No, northerners replied, if push came to shove, the states would have to acknowledge the ultimate sovereignty of the Union. There was no "right" answer, of course. There was only an appeal to arms.

For these and other reasons, Lincoln came to realize that whatever else happened, slavery must be ended. "A house divided against itself cannot stand," he said, meaning that a self-conflicted nation could not survive for long. Accordingly, as the horrors of battle mounted during the first years of the war, the President finally determined to take history into his own hands and set the slaves free. He didn't do it for political advantage or diplomatic expediency (although both came into play), but to straighten out that catastrophic kink in the American Founding. In the Gettysburg Address, Lincoln termed the war's result "a new birth of freedom," and it certainly was that for four and a half million slaves. But he intended the phrase to have another meaning as well. It was a new birth of *American* freedom—a refounding of the republic.

As such, the final victory of the northern armies in 1865 led to a whole new set of constitutional arrangements. What had caused the war originally must never cause it again, vowed the Republican leadership. The freedmen must never be reenslaved (by political chicanery, perhaps, or legal subterfuge), and fuzziness about the Union must be cleared up once and for all. They pushed a series of Reconstruction Acts through Congress to insure both results. And then, when they realized that the legislation might be overturned by the Supreme Court as unconstitutional, they decided to recast it in three constitutional amendments.

Unlike some refinements of the Constitution, the Thirteenth, Fourteenth and Fifteenth Amendments did not contemplate minor tinkering but a major overhaul. We have observed that the Philadelphia delegates worked out a government for an existing society, not a better society. The framers of the new amendments had no such qualms. They believed that the South would subvert the war's outcome if it could, and thus they resolved to take the South apart and put it back together. There was no way to accomplish this "reconstruction" without reconstructing the rest of the country as well.

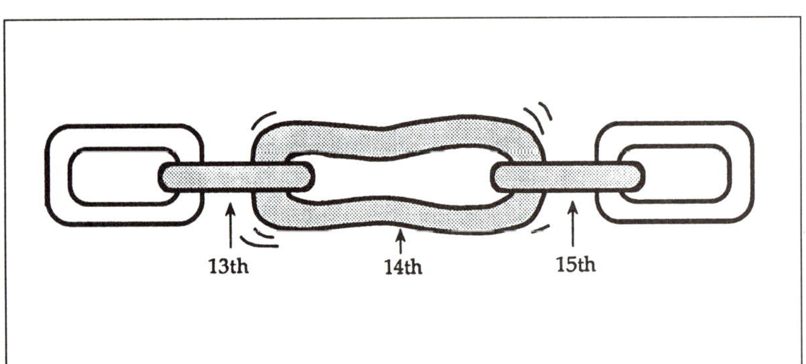

Three new links were added to the Constitutional chain following the Civil War. The Fourteenth Amendment constituted a significant new stretch-point for constitutional interpretation.

Such was the intention behind the Fourteenth Amendment. Its language was deceptively simple.

Gettysburg Address

Fourscore and seven years ago our fathers brought forth upon this continent a new nation, conceived in liberty, and dedicated to the proposition that all men are created equal.

Now we are engaged in a great civil war, testing whether that nation, or any nation so conceived and so dedicated, can long endure. We are met on a great battlefield of that war. We have come to dedicate a portion of that field, as a final resting-place for those who here gave their lives that that nation might live. It is altogether fitting and proper that we should do this.

But, in a larger sense, we cannot dedicate—we cannot consecrate—we cannot hallow—this ground. The brave men, living and dead, who struggled here have consecrated it, far above our poor power to add or detract. The world will little note, nor long remember what we say here, but it can never forget what they did here. It is for us the living, rather, to be dedicated here to the unfinished work they who fought here have thus far so nobly advanced. It is rather for us to be here dedicated to the great task remaining before us-that from these honored dead we take increased devotion to that cause for which they here gave the last full measure of devotion—that we here highly resolve that these dead shall not have died in vain—that this nation, under God, shall have a new birth of freedom—and that government of the people, by the people, and for the people, shall not perish from the earth.

Version used is from the National Park Service website for the Lincoln Home National Historic Site: http://www.nps.gov/liho/writer/address.htm

But a quick review of the broadly-framed amendments in the Bill of Rights—and a reflection on how *they* turned out—should help to make the framers' approach clear. Henceforth, the amendment declared, citizenship would apply to the United States as a whole, not just the individual states, and the federal government was obliged to make sure that US citizens enjoyed equal privileges and immunities, equal protection of the laws, and equal enjoyment of their fundamental rights. The nation, in other words, was the final sovereignty after all.

Here is a way of seeing what the framers had in mind. In the Constitution of 1787, federal power had applied to individuals only in narrowly prescribed areas—such as the punishment of counterfeiters—and most people never felt governed at all above the state level. In the Constitution of 1868, by contrast, federal power impinged on individuals in a hundred new ways, for the state-versus-national balance had been sharply tilted. The war was responsible for some of the change, true enough, but most of it was due to the Fourteenth Amendment, whose loose wording stretched federal power like another "necessary and proper" clause.

Civil Rights Movement: http://seattletimes.nwsource.com/mlk/movement/Seatimeline.html

In the new constitutional reality, the federal government could protect citizens from their own states if need be. That wouldn't happen in the 1860s, for there was much dithering and foot-dragging among postwar leaders, but it *would* happen in the 1960s. Only when the civil rights movement got under way and Martin Luther King began to dream his dream would the meaning of the Fourteenth Amendment become fully manifest. And, yes, it would bring forth a whole new America.

An American People

In subtle ways, the success of the Founding helped bring an American people into being. They were hardly *a* people when they ratified the Constitution. And for decades afterward, they thought of themselves as New Yorkers or Rhode Islanders or South Carolinians first, and as Americans only secondarily.

In the 1830s, however, when a clear-sighted young Frenchman by the name of Alexis de Tocqueville toured the United States, he found all sorts of indications that an American people was emerging. Their habits of mind, patterns of thought, social structure, value system, and much else were becoming observable. Tocqueville attributed this fact to democracy, which certainly played a role in it, but he might just as logically have attributed it to the Founding.

For nothing forges the identity of a people quite like a sense of patria. Canadians, Americans and Mexicans share a continent together but the element of *patria* is lacking among them. It was lacking among Americans too when they lived in separate colonies and even when they lived in separate states. Shared experience caused it to grow, of course, and so did the presence of a common enemy, but it was the Founding that brought real patria into being. The give-and-take of politics, the excitement of shared dialogue, the response to common leadership were what fashioned Americans into a nation.

Only when the house had been framed up, so to speak, did its owners catch a sense of themselves as family.

Much the same could be said of economic life. As colonists, Americans had worked within the British imperial system. That hadn't promoted economic integration, exactly, but it had promoted peace. The moment British authority was removed, the peace went along with it and the colonies, now states, began working at cross-purposes. The Founding altered that situation dramatically. It consolidated Americans into *an* economy just as it consolidated them into *a* nation, by creating a single vast market. Hamilton sought to integrate this continent-sized trade zone, promoting manufactures here, banking there, agriculture somewhere else, the parts working together as a whole. He succeeded too. Cotton from the South was soon making its way to northern mills in cars rattling along western railroads, the whole operation funded by eastern bankers. All sections gained mutual benefit, and all came to identify with one another more closely.

At the same time the new American republic gradually turned more and more democratic, it also became increasingly commercial. Jefferson's nation of virtuous yeoman farmers was eventually eclipsed by a Hamiltonian market economy.

There was another benefit. Without tariff walls dividing one state from another, other walls began to crumble. Local prejudice, for instance. It was harder to believe that all Yankees were skinflints or all Georgians bumpkins when you had profitable business ties to both. Sectarian antagonisms moderated too, as did interstate political disputes. Without trade wars to intensify such frictions, the frictions were easier to smooth out in the courts, and the fighting and bickering between states eventually subsided. Economic peace brought all sorts of advantages.

So did economic growth, which itself became visible in the early republic. It was based on a different kind of founding. By the time of Jackson's Presidency, the United States was becoming ever more closely identified with the market system of economics, and there were a number of parallels to be drawn between free political and free economic institutions. Mercantilism, which was the economic model favored by the British Empire, had been a serious possibility for the American nation as well, and Alexander Hamilton wasn't its only warm advocate. But the Founding, with its republican principles and democratic outlook, rubbed against the grain of command systems like mercantilism. Americans simply didn't like privilege, political or economic, and it took privilege to make a command economy run. As the implications of the Founding continued to work themselves out, the various franchises and monopolies of English mercantilism began to seem more and more like Old World tyranny. And the market system gained an ever stronger claim on the future.

By similar means, the United States came to acquire a kind of spiritual core as well, based on those old Puritan myths. Americans began making allusions to that Eden-like wilderness in the West, to the Promised Land of God's favored people, and to the New Jerusalem of future possibility. One needn't belong to any particular church (or any church at all) to identify with these images—they transcended all affiliations. Americans waxed confident in the world because their sense of chosenness seemed to have been validated by history. Credit the Founding once again.

That was what the Founding really meant: that things started coming together. Visitors could make jottings about the "American style" or the "American spirit" because the Founding had truly begun to take. Traits such as optimism, individualism, boldness in risk taking, even "Yankee ingenuity" took on meaning only partly as evidence of an emerging national character—they were by-products of the Founding too. For the Founding had given people new ways of looking at the world.

In a word, America as idea had begun to correspond to America as reality. And while the realities would always be a little disappointing—and the ideals always a little elusive—a nation-state had come into being that, whatever else, took the notion of the Good Society seriously. There was a USA now and it promised to be a going concern.

This tattered, battle-worn flag flew over Fort McHenry in Baltimore Harbor during the War of 1812, where it gained immortality as the Star-Spangled Banner *(note the fifteen stars and stripes, one for each state at that time). Francis Scott Key felt prompted to pen the lines that were later incorporated into the United States national anthem. He did so while detained aboard a British ship during the unsuccessful British bombardment of Fort McHenry in 1814. The seldom sung third stanza of the anthem contains the following verses:*

> Blest with vic't'ry and peace, may the heav'n
> rescued land praise the Pow'r that hath made and
> preserved us a nation!
>
> Then conquer we must, when our cause it is just,
> and this be our motto: In God is our trust!
>
> And the star-spangled banner in triumph shall wave
> o'er the land of the free and the home of the brave!

CONCLUSION STUDY AID

1. The Founding:

- America: an idea and a work in progress
- First Founding: the colonial experience
- Second Founding: the Revolution era
- Third Founding: the Constitution, Bill of Rights, and Washington's administrations

2. Adding to the Founding:

- Aaron Burr trial
- Progressives
- Forgotten man and the New Deal
- Civil Rights movement
- An ongoing and dynamic process

3. Filling out the Founding:

- John Marshall and judicial review
- Jacksonian democracy and the two-party system
- The Civil War: an end to slavery and a new birth of freedom
- The Thirteenth, Fourteenth, and Fifteenth amendments

4. An American People:

- Alexis de Tocqueville
- Emerging political identity
- Economic life
- American mythology, style, and spirit
- A Good Society, if we can keep it

CONCLUSION STUDY GUIDE

NAME: ______________________________ **SECTION:** ______________

Key Terms

Progressive Movement

Judicial Review

Judicial Legislation

Jacksonian Democracy

Patria

Review Questions

1. If America is an idea, what is that idea?

____
____
____
____

2. What were the First, Second, and Third Foundings listed in the conclusion?

____
____
____
____

3. How did judicial review become part of our governmental system?

____
____
____
____

4. How did the era of Jacksonian Democracy alter American politics?

____
____
____
____

5. In what ways did the Civil War become a part of the American Founding?

6. Explain the provisions of the Thirteenth, Fourteenth, and Fifteenth Amendments.

7. In what ways did America's economic system come to complement America's political system?

APPENDIX 1

The Declaration of Independence

When in the course of human events it becomes necessary for one people to dissolve the political bands which have connected them with another and to assume, among the powers of the earth, the separate and equal station to which the laws of nature and of nature's God entitle them, a decent respect to the opinions of mankind requires that they should declare the causes which impel them to the separation.

We hold these truths to be self-evident, that all men are created equal; that they are endowed by their Creator with certain unalienable rights; that among these are life, liberty, and the pursuit of happiness. That, to secure these rights, governments are instituted among men, deriving their just powers from the consent of the governed; that, whenever any form of government becomes destructive of these ends, it is the right of the people to alter or to abolish it, and to institute a new government, laying its foundation on such principles, and organizing its powers in such form, as to them shall seem most likely to effect their safety and happiness. Prudence, indeed, will dictate that governments long established should not be changed for light and transient causes; and, accordingly, all experience hath shown that mankind are more disposed to suffer, while evils are sufferable, than to right themselves by abolishing the forms to which they are accustomed. But when a long train of abuses and usurpations, pursuing invariably the same object, evinces a design to reduce them under absolute despotism, it is their right, it is their duty, to throw off such government and to provide new guards for their future security. Such has been the patient sufferance of these colonies, and such is now the necessity which constrains them to alter their former systems of government. The history of the present King of Great Britain is a history of repeated injuries and usurpations, all having, in direct object, the establishment of an absolute tyranny over these States. To prove this, let facts be submitted to a candid world:

> He has refused his assent to laws the most wholesome and necessary for the public good.
>
> He has forbidden his governors to pass laws of immediate and pressing importance, unless suspended in their operation till his assent should be obtained; and, when so suspended, he has utterly neglected to attend to them.
>
> He has refused to pass other laws for the accommodation of large districts of people, unless those people would relinquish the right of representation in the legislature, a right inestimable to them and formidable to tyrants only.

He has called together legislative bodies at places unusual, uncomfortable, and distant from the depository of their public records, for the sole purpose of fatiguing them into compliance with his measures.

He has dissolved representative houses, repeatedly for opposing, with manly firmness, his invasions on the rights of the people.

He has refused, for a long time after such dissolutions, to cause others to be elected; whereby the legislative powers, incapable of annihilation, have returned to the people at large for their exercise; the state remaining, in the meantime, exposed to all the danger of invasion from without and convulsions within.

He has endeavored to prevent the population of these States; for that purpose, obstructing the laws for naturalization of foreigners, refusing to pass others to encourage their migration hither, and raising the conditions of new appropriations of lands.

He has obstructed the administration of justice by refusing his assent to laws for establishing judiciary powers.

He has made judges dependent on is will alone for the tenure of their offices and the amount and payment of their salaries.

He has erected a multitude of new offices and sent hither swarms of officers to harass our people and eat out their substance.

He has kept among us, in time of peace, standing armies, without the consent of our legislatures.

He has affected to render the military independent of, and superior to, the civil power.

He has combined with others to subject us to a jurisdiction foreign to our Constitution and unacknowledged by our laws, giving his assent to their acts of pretended legislation—

For quartering large bodies of armed troops among us;

For protecting them by a mock trail from punishment for any murders which they should commit on the inhabitants of these States;

For cutting off our trade with all parts of the world;

For imposing taxes on us without our consent;

For depriving us, in many cases, of the benefit of trial by jury;

For transporting us beyond seas to be tried for pretended offenses;

For abolishing the free system of English laws in a neighboring province, establishing therein an arbitrary government, and enlarging its boundaries, so as to render it at once an example and fit instrument for introducing the same absolute rule into these colonies;

For taking away our charters, abolishing our most valuable laws, and altering, fundamentally, the powers of our governments.

For suspending our own legislatures and declaring themselves invested with power to legislate for us in all cases whatsoever.

He has abdicated government here by declaring us out of his protection and waging war against us.

He has plundered our seas, ravaged our coasts, burnt our towns, and destroyed the lives of our people.

He is, at this time, transporting large armies of foreign mercenaries to complete the works of death, desolation, and tyranny already begun with circumstances of cruelty and perfidy scarcely paralleled in the most barbarous ages, and totally unworthy the head of a civilized nation.

He has constrained our fellow citizens, taken captive on the high seas, to bear arms against their country, to become the executioners of their friends and brethren, or to fall themselves by their hands.

He has excited domestic insurrections amongst us and has endeavored to bring on the inhabitants of our frontiers, the merciless Indian savages, whose known rule of warfare is an undistinguished destruction of all ages, sexes, and conditions.

In every stage of these oppressions, we have petitioned for redress in the most humble terms; our repeated petitions have been answered only by repeated injury. A prince whose character is thus marked by every act which may define a tyrant is unfit to be the ruler of a free people.

Nor have we been wanting in attention to our British brethren. We have warned them, from time to time, of attempts made by their legislature to extend an unwarrantable jurisdiction over us. We have reminded them of the circumstances of our emigration and settlement here. We have appealed to their native justice and magnanimity, and we have conjured them, by the ties of our common kindred, to disavow these usurpations, which would inevitably interrupt our connections and correspondence. They, too, have been deaf to the voice of justice and consanguinity. We must, therefore, acquiesce in the necessity which denounces our separation, and hold them, as we hold the rest of mankind, enemies in war, in peace, friends.

We, therefore, the representatives of the United States of America, in general Congress assembled, appealing to the Supreme Judge of the world for the rectitude of our intentions, do, in the name and by the authority of the good people of these colonies, solemnly publish and declare, that these united colonies are, and of right ought to be, free and independent states: that they are absolved from all allegiance to the British Crown, and that all political connection between them and the state of Great Britain is, and ought to be, totally dissolved; and that, as free and independent states, they have full power to levy war, conclude peace, contract alliances, establish commerce, and to do all other acts and things which independent states may of right do. And, for the support of this declaration, with a firm reliance on the protection of Divine Providence, we mutually pledge to each other our lives, our fortunes, and our sacred honor.

APPENDIX 2

The Constitution of the United States of America

We the people of the United States, in order to form a more perfect union, establish justice, insure domestic tranquillity, provide for the common defense, promote the general welfare, and secure the blessings of liberty to ourselves and our posterity, do ordain and establish this Constitution for the United States of America.

Article I

Section 1.

All legislative powers herein granted shall be vested in a Congress of the United States, which shall consist of a Senate and House of Representatives.

Section 2.

1. The House of Representatives shall be composed of members chosen every second year by the people of the several States, and the electors in each State shall have the qualifications requisite for electors of the most numerous branch of the State legislature.
2. No person shall be a representative who shall not have attained to the age of twenty-five years, and been seven years a citizen of the United States, and who shall not, when elected, be an inhabitant of that State in which he shall be chosen.
3. Representatives and direct taxes[1] shall be apportioned among the several States which may be included within this Union, according to their respective numbers, which shall be determined by adding to the whole number of free persons, including those bound to service for a term of years, and excluding Indians not taxed, three fifths of all other persons.[2] The actual enumeration shall be made within three years after the first meeting of the Congress of the United States, and within every subsequent term of ten years, in such manner as they shall by law direct. The number of representatives shall not exceed one for every thirty thousand, but each State shall have at least one representative; and until such enumeration shall be made, the State of New Hampshire shall be entitled to choose three, Massachusetts eight, Rhode Island and Providence Plantations one, Connecticut five,

New York six, New Jersey four, Pennsylvania eight, Delaware one, Maryland six, Virginia ten, North Carolina five, South Carolina five, and Georgia three.

4. When vacancies happen in the representation from any State, the executive authority thereof shall issue writs of election to fill such vacancies.
5. The House of Representatives shall choose their speaker and other officers; and shall have the sole power of impeachment.

Section 3.

1. The Senate of the United States shall be composed of two senators from each State, chosen by the legislature thereof,[3] for six years; and each senator shall have one vote.
2. Immediately after they shall be assembled in consequence of the first election, they shall be divided as equally as may be into three classes. The seals of the senators of the first class shall be vacated at the expiration of the second year, of the second class at the expiration of the fourth year, and of the third class at the expiration of the sixth year, so that one third may be chosen every second year; and if vacancies happen by resignation, or otherwise, during the recess of the legislature of any State, the executive thereof may make temporary appointments until the next meeting of the legislature, which shall then fill such vacancies[4].
3. No person shall be a senator who shall not have attained to the age of thirty years, and been nine years a citizen of the United States, and who shall not, when elected, be an inhabitant of that State for which he shall be chosen.
4. The Vice President of the United States shall be President of the Senate, but shall have no vote, unless they be equally divided.
5. The Senate shall choose their other officers, and also a president pro tempore, in the absence of the Vice President, or when he shall exercise the office of the President of the United States.
6. The Senate shall have the sole power to try all impeachments. When sitting for that purpose, they shall be on oath or affirmation. When the President of the United States is tried, the chief justice shall preside: and no person shall be convicted without the concurrence of two thirds of the members present.
7. Judgment in cases of impeachment shall not extend further than to removal from office, and disqualification to hold and enjoy any office of honor, trust or profit under the United States: but the party convicted shall nevertheless be liable and subject to indictment, trial, judgment and punishment, according to law.

Section 4.

1. The times, places, and manner of holding elections for senators and representatives, shall be prescribed in each State by the legislature thereof; but the Congress may at any time by law make or alter such regulations, except as to the places of choosing senators.

2. The Congress shall assemble at least once in every year, and such meeting shall be on the first Monday in December, unless they shall by law appoint a different day.

Section 5.

1. Each House shall be the judge of the elections, returns and qualifications of its own members, and a majority of each shall constitute a quorum to do business; from day to day, and may be authorized to compel the attendance of absent members, in such manner, and under such penalties as each House may provide.
2. Each House may determine the rules of its proceedings, punish its members for disorderly behavior, and, with the concurrence of two thirds, expel a member.
3. Each House shall keep a journal of its proceedings, and from time to time publish the same, excepting such parts as may in their judgment require secrecy; and the yeas and nays of the members of either House on any question shall, at the desire of one fifth of those present, be entered on the journal.
4. Neither House, during the session of Congress, shall, without the consent of the other, adjourn for more than three days, nor to any other place than that in which the two Houses shall be sitting.

Section 6.

1. The senators and representatives shall receive a compensation for their services, to be ascertained by law, and paid out of the Treasury of the United States. They shall in all cases, except treason, felony, and breach of the peace, be privileged from arrest during their attendance at the session of their respective Houses, and in going to and returning from the same; and for any speech or debate in either House, they shall not be questioned in any other place.
2. No senator or representative shall, during the time for which he was elected, be appointed to any civil office under the authority of the United States, which shall have been created, or the emoluments whereof shall have been increased, during such time; and no person holding any office under the United States shall be a member of either House during his continuance in office.

Section 7.

1. All bills for raising revenue shall originate in the House of Representatives; but the Senate may propose or concur with amendments as on other bills.
2. Every bill which shall have passed the House of Representatives and the Senate, shall, before it become a law, be presented to the President of the United States; If he approves he shall sign it, but if not he shall return it, with his objections, to that House in which it shall have originated, who shall enter the objections at large on their journal, and proceed to reconsider it. If after such reconsideration two thirds of that House shall agree to

pass the bill, it shall be sent, together with the objections, to the other House, by which it shall likewise be reconsidered, and if approved by two thirds of that House shall agree to pass the bill, it shall be sent, together with the objections, to the other House, by which it shall likewise be reconsidered, and if approved by two thirds of that House, it shall become a law. But in all such cases the votes of both Houses shall be determined by yeas and nays, and the names of the persons voting for and against the bill shall be entered on the journal of each House respectively. If any bill shall not be returned by the President within ten days (Sundays excepted) after it shall have been presented to him, the same shall be a law, in like manner as if he had signed it, unless the Congress by their adjournment prevent its return, in which case it shall not be a law.

3. Every order, resolution, or vote to which the concurrence of the Senate and the House of Representatives may be necessary (except on a question of adjournment) shall be presented to the President of the United States; and before the same shall take effect, shall be approved by him, or being disapproved by him, shall be repassed by two thirds of the Senate and House of Representatives, according to the rules and limitations prescribed in the case of a bill.

Section 8.

The Congress shall have the power

1. To lay and collect taxes, duties, imposts, and excises, to pay the debts and provide for the common defense and general welfare of the United States; but all duties, imposts, and excises shall be uniform throughout the United States.
2. To borrow money on the credit of the United States;
3. To regulate commerce with foreign nations, and among the several States, and with the Indian tribes;
4. To establish a uniform rule of naturalization, and uniform laws on the subject of bankruptcies throughout the United States;
5. To coin money, regulate the value thereof, and of foreign coin, and fix the standard of weights and measures;
6. To provide for the punishment of counterfeiting the securities and current coin of the United States;
7. To establish post offices and post roads;
8. To promote the progress of science and useful arts, by securing for limited times to authors and inventors the exclusive right to their respective writings and discoveries;
9. To constitute tribunals inferior to the Supreme Court;
10. To define and punish piracies and felonies committed on the high seas, and offenses against the law of nations;
11. To declare war, grant letters of marque and reprisal, and make rules concerning captures on land and water;

12. To raise and support armies, but no appropriation of money to that use shall be for a longer term than two years;
13. To provide and maintain a navy;
14. To make rules for the government and regulation of the land and naval forces;
15. To provide for calling forth the militia to execute the laws of the Union, suppress insurrections and repel invasions;
16. To provide for organizing, arming, and disciplining the militia, and for governing such part of them as may be employed in the service of the United States, reserving to the States respectively, the appointment of the officers, and the authority of training the militia according to the discipline prescribed by Congress;
17. To exercise exclusive legislation in all cases whatsoever, over such district (not exceeding ten miles square) as may, by cession of particular States, and the acceptance of Congress, become the seat of the government of the United States, and to exercise like authority over all places purchased by the consent of the legislature of the State in which the same shall be, for the erection of forts, magazines, arsenals, dockyards, and other needful buildings; and
18. To make all laws which shall be necessary and proper for carrying into execution the foregoing powers, and all other powers vested by this Constitution in the government of the United States, or any department or officer thereof.

Section 9.

1. The migration or importation of such persons as any of the States now existing shall think proper to admit, shall not be prohibited by the Congress prior to the year one thousand eight hundred and eight, but a tax or duty may be imposed on such importation, not exceeding ten dollars for each person.
2. The privilege of the writ of habeas corpus shall not be suspended, unless when in cases of rebellion or invasion the public safety may require it.
3. No bill of attainder or ex post facto law shall be passed.
4. No capitation, or other direct, tax shall be laid, unless in proportion to the census or enumeration hereinbefore directed to be taken.[5]
5. No tax or duty shall be laid on articles exported from any State.
6. No preference shall be given by any regulation of commerce or revenue to the ports of one State over those of another: nor shall vessels bound to, or from, one State be obliged to enter, clear, or pay duties in another.
7. No money shall be drawn from the treasury, but in consequence of appropriations made by law; and a regular statement and account of the receipts and expenditures of all public money shall be published from time to time.
8. No title of nobility shall be granted by the United States: and no person holding any office of profit or trust under them, shall, without the consent of the Congress, accept of any present, emolument, office, or title, of any kind whatever, from any king, prince, or foreign State.

Section 10.

1. No State shall enter into any treaty, alliance, or confederation; grant letters of marque and reprisal; coin money; emit bills of credit; make any thing but gold and silver coin a tender in payment of debts; pass any bill of attainder, ex post facto law, or law impairing the obligation of contracts, or grant, any title of nobility.
2. No State shall, without the consent of the Congress, lay any imposts or duties on imports or exports, except what may be absolutely necessary for executing its inspection laws: and the net produce of all duties and imposts laid by any State on imports or exports, shall be for the use of the treasury of the United States; and all such laws shall be subject to the revision and control of the Congress.
3. No state shall, without the consent of the Congress, lay any duty of tonnage, keep troops, or ships of war in time of peace, enter into any agreement or compact with another State, or with a foreign power, or engage in war, unless actually invaded, or in such imminent danger as will not admit or delay.

Article II

Section 1.

1. The executive power shall be vested in a President of the United States of America. He shall hold his office during the term of four years, and, together with the Vice President, chosen for the same term, be elected, as follows:
2. Each State shall appoint, in such manner as the legislature thereof may direct, a number of electors, equal to the whole number of senators and representatives to which the State may be entitled in the Congress: but no senator or representative, or person holding any office of trust or profit under the United States, shall be appointed an elector.

   The electors shall meet in their respective States, and vote by ballot for two persons, of whom one at least shall not be an inhabitant of the same State with themselves. And they shall make a list of all the persons voted for, and of the number of votes for each; which list they shall sign and certify, and transmit sealed to the seat of the government of the United States, directed to the president of the Senate. The president of the Senate shall, in the presence of the Senate and House of Representatives, open all the certificates, and the votes shall then be counted. The person having the greatest number of votes shall be the President, if such number be a majority of the whole number of electors appointed; and if there be more than one who have such majority, and have an equal number of votes, then the House of Representatives shall immediately choose by ballot one of them for President; and if no person have a majority, then from the five highest on the list the said House shall in choosing the President, the votes shall be taken by

States, the representation from each State having one vote; a quorum for this purpose shall consist of a member or members from two thirds of the States, and a majority of all the States shall be necessary to a choice. In every case after the choice of the President, the person having the greatest number of votes of the electors shall be the Vice President. But if there should remain two or more who have equal votes, the Senate shall choose from them by ballot the Vice President.[6]

3. The Congress may determine the time of choosing the electors, and the day on which they shall give their votes; which day shall be the same throughout the United States.
4. No person except a natural born citizen, or a citizen of the United States, at the time of the adoption of this Constitution, shall be eligible to the office of President; neither shall any person be eligible to the office who shall not have attained to the age of thirty-five years, and been fourteen years a resident within the United States.
5. In case of the removal of the President from office, or of his death, resignation, or inability to discharge the powers and duties of the said office, the same shall devolve on the Vice President, and the congress may by law provide for the case of removal, death, resignation or inability, both of the President and Vice President, declaring what officer shall then act as President, and such officer shall act accordingly until the disability be removed, or a President shall be elected.
6. The President shall, at stated times, receive for his services a compensation which shall neither be increased nor diminished during the period for which he shall have been elected, and he shall not receive within that period any other emolument from the United States, or any of them.
7. Before he enter on the execution of his office, he shall take the following oath or affirmation:—"I do solemnly swear (or affirm) that I will faithfully execute the office of President of the United States, and will to the best of my ability, preserve, protect and defend the Constitution of the United States."

Section 2.

1. The President shall be commander in chief of the army and navy of the United States, and of the militia of the several States, when called into the actual service of the United States; he may require the opinion in writing, of the principal officer in each of the executive departments, upon any subject relating to the duties of their respective offices, and he shall have power to grant reprieves and pardons for offenses against the United States, except in cases of impeachment.
2. He shall have power, by and with the advice and consent of the Senate, to make treaties, provided two thirds of the senators present concur; and he shall nominate, and by and with the advice and consent of the Senate, shall appoint ambassadors, other public ministers and consuls, judges of the Supreme Court, and all other officers of the United States, whose appointments are not herein otherwise provided for, and which shall be established

by law; but the Congress may by law vest the appointment of such inferior officers, as they think proper, in the President alone, in the courts of laws, or in the heads of departments.

3. The President shall have power to fill up all vacancies that may happen during the recess of the Senate, by granting commissions which shall expire at the end of their next session.

Section 3.

He shall from time to time give to the Congress information of the state of the Union, and recommend to their consideration such measures as he shall judge necessary and expedient; he may, on extraordinary occasions, convene both Houses, or either of them, and in case of disagreement between them with respect to the time of adjournment, he may adjourn them to such time as he shall think proper; he shall receive ambassadors and other public ministers; he shall take care that the laws be faithfully executed, and shall commission all the officers of the United States.

Section 4.

The President, Vice President, and all civil officers of the United States, shall be removed from office on impeachment for, and conviction of, treason, bribery, or other high crimes and misdemeanors.

Article III

Section 1.

The judicial power of the United States shall be vested in one Supreme Court, and in such inferior courts as the Congress may from time to time ordain and establish. The judges, both of the Supreme and inferior courts, shall hold their offices during good behavior, and shall, at stated times, receive for their services, a compensation, which shall not be diminished during their continuance in office.

Section 2.

1. The judicial power shall extend to all cases, in law and equity, arising under this Constitution, the laws of the United States, and treaties made, or which shall be made, under their authority;—to all cases of admiralty and maritime jurisdiction;—to controversies to which the United States shall be a party;[7]—to controversies between two or more States;—between a State and citizens of another State;—between citizens of different States;—between citizens of the same State claiming lands under grants of different States, and between a State, or the citizens thereof, and foreign States, citizens or subjects.
2. In all cases affecting ambassadors, other public ministers and consuls, and those in which a State shall be party, the Supreme Court shall have original

jurisdiction. In all the other cases before mentioned, the Supreme Court shall have appellate jurisdiction, both as to law and fact, with such exceptions, and under such regulations as the Congress shall make.

3. The trial of all crimes, except in cases of impeachment, shall be by jury; and such trial shall be held in the State where the said crimes shall have been committed; but when not committed within any State, the trial shall be such place or places as the congress may by law have directed.

Section 3.

1. Treason against the United States shall consist only in levying war against them, or in adhering to their enemies, giving them aid and comfort. No person shall be convicted of treason unless on the testimony of two witnesses to the same overt act, or on confession in open court.
2. The Congress shall have power to declare the punishment of treason, but no attainder of treason shall work corruption of blood, or forfeiture except during the life of the person attained.

Article IV

Section 1.

Full faith and credit shall be given in each State to the public acts, records, and judicial proceedings of every other State. And the Congress may by general laws prescribe the manner in which such acts, records and proceedings shall be proved, and the effect thereof.

Section 2.

1. The citizens of each State shall be entitled to all privileges and immunities of citizens in the several States.[8]
2. A person charged in any State with treason, felony, or other crime, who shall flee from justice, and be found in another State, shall on demand of the executive authority of the State from which he fled, be delivered up to be removed to the State having jurisdiction of the crime.
3. No person held to service or labor in one State under the laws thereof, escaping into another, shall, in consequence of any law or regulation therein, be discharged from such service or labor, but shall be delivered up on claim of the party to whom such service or labor may be due.[9]

Section 3.

1. New States may be admitted by the Congress into this Union; but no new State shall be formed or erected within the jurisdiction of any other State, nor any State be formed by the junction of two or more States, or parts of States, without the consent of the legislatures of the States concerned as well as of the Congress.

2. The Congress shall have power to dispose of and make all needful rules and regulations respecting the territory or other property belonging to the United States; and nothing in this Constitution shall be so construed as to prejudice any claims of the United States, or of any particular State.

Section 4.

The United States shall guarantee to every State in this Union a republican form of government, and shall protect each of them against invasion; and on application of the legislature, or of the executive (when the legislature cannot be convened) against domestic violence.

Article V

The Congress, whenever two thirds of both Houses shall deem it necessary, shall propose amendments to this Constitution, or, on the application of the legislatures of two thirds of the several States, shall call a convention for proposing amendments, which in either case shall be valid to all intents and purposes, as part of this Constitution, when ratified by the legislatures of three fourths of the several States, or by conventions in three fourths thereof, as the one or the other mode of ratification may be proposed by the Congress; Provided that no amendment which may be made prior to the year one thousand eight hundred and eight shall in any manner affect the first and fourth clauses in the ninth section of the first article; and that no State, without its consent, shall be deprived of its equal suffrage in the Senate.

Article VI

1. All debts contracted and engagements entered into, before the adoption of this Constitution, shall be as valid against the United States under this Constitution, as under the Confederation.[10]
2. This Constitution, and the laws of the United States which shall be made in pursuance thereof; and all treaties made, or which shall be made, under the authority of the United States, shall be the supreme law of the land; and the judges in every State shall be bound thereby, any thing in the Constitution or laws of any State to the contrary notwithstanding.
3. The senators and representatives before mentioned, and the members of the several State legislatures, and all executive and judicial officers, both on the United States and of the several States, shall be bound by oath or affirmation to support this Constitution; but no religious test shall ever be required as a qualification to any office or public trust under the United States.

Article VII

The ratification of the conventions of nine States shall be sufficient for the establishment of this Constitution between the States so ratifying the same.

Done in Convention by the unanimous consent of the States present the seventeenth day of September in the year of our Lord one thousand seven hundred and eighty-seven, and of the independence of the United States of America the twelfth. In witness whereof we have hereunto subscribed our names.

[Names omitted]

Articles in addition to, and amendment of, the Constitution of the United States of America, proposed by Congress, and ratified by the legislatures of the several States, pursuant to the fifth article of the original Constitution.

Amendment I

[First ten amendments ratified December 15, 1791]

Congress shall make no law respecting an establishment of religion, or prohibiting the free exercise thereof; or abridging the freedom of speech, or of the press; or the right of the people peaceably to assemble, and to petition the government for a redress of grievances.

Amendment II

A well regulated militia, being necessary to the security of a free State, the right of the people to keep and bear arms, shall not be infringed.

Amendment III

No soldier shall, in time of peace be quartered in any house, without the consent of the owner, nor in time of war, but in a manner to be prescribed by law.

Amendment IV

The right of the people to be secure in their persons, houses, papers, and effects, against unreasonable searches and seizures, shall not be violated, and no warrants shall issue, but upon probable cause, supported by oath or affirmation, and particularly describing the place to be searched, and the persons or things to be seized.

Amendment V

No person shall be held to answer for a capital or otherwise infamous crime, unless on a presentment or indictment of a grand jury, except in cases arising in the land or naval forces, or in the militia, when in actual service in time of war or public danger; nor shall any person be subject for the same offense to be twice put in jeopardy of life or limb; nor shall be compelled in any criminal case to be a witness against himself, nor be deprived of life, liberty, or property, without due process of law; nor shall private property be taken for public use, without just compensation.

Amendment VI

In all criminal prosecutions, the accused shall enjoy the right to a speedy and public trial, by an impartial jury of the State and district wherein the crime shall have been committed, which district shall have been previously ascertained by law, and to be informed of the nature and cause of the accusation; to be confronted with the witnesses against him; to have compulsory process for obtaining witnesses in his favor, and to have the assistance of counsel for his defense.

Amendment VII

In suits at common law, where the value in controversy shall exceed twenty dollars, the right of trial by jury shall be preserved, and no fact tried by a jury shall be otherwise reexamined in any court of the United States, than according to the rules of the common law.

Amendment VIII

Excessive bail shall not be required, nor excessive fines imposed, nor cruel and unusual punishments inflicted.

Amendment IX

The enumeration in the Constitution of certain rights shall not be construed to deny or disparage others retained by the people.

Amendment X

The powers not delegated to the United States by the Constitution, nor prohibited by it to the States, are reserved to the States respectively, or to the people.

Amendment XI

[January 8, 1798]

The judicial power of the United States shall not be construed to extend to any suit in law or equity, commended or prosecuted against one of the United States by citizens of another State, or by citizens or subjects of any foreign State.

Amendment XII

[September 25, 1804]

The electors shall meet in their respective States, and vote by ballot for President and Vice President, one of whom, at least, shall not be an inhabitant of the same State with themselves; they shall name in their ballots the person voted for as President, and in distinct ballots, the person voted for as Vice President, and they shall make distinct lists of all persons voted for as President and of all persons voted for as Vice President, and of the number of votes for each, which lists they shall sign and certify, and transmit sealed to the seat of the government of the United States, directed to the President of the Senate;—The President of the Senate shall, in the presence of the Senate and House of Representatives, open all the certificates and the votes shall then be counted;—The person having the greatest number of votes for President, shall be the President, if such number be a majority of the whole number of electors appointed; and if no person have such majority, then from the persons having the highest numbers not exceeding three on the list of those voted for as President, the House of Representatives shall choose immediately, by ballot, the President. But in choosing the President, the votes shall be taken by States, the representation from each State having one vote; a quorum for this purpose shall consist of a member or members from two thirds of the States, and a majority of all the States shall be necessary to a choice. And if the House of Representatives shall not choose a President whenever the right of choice shall devolve upon them, before the fourth day of March next following, then the Vice President shall act as President, as in the case of the death or other constitutional disability of the President. The person having the greatest number of votes as Vice President shall be the Vice President, if such number be a majority of the whole number of electors appointed, and if no person have a majority, then from the two highest numbers on the list, the Senate shall choose the Vice President; a quorum for the purpose shall consist of two thirds of the whole number of Senators, and a majority of the whole number shall be necessary to a choice. But no person constitutionally ineligible to the office of President shall be eligible to that of Vice President of the United States.

Amendment XIII

[December 18, 1865]

Section 1.

Neither slavery nor involuntary servitude, except as a punishment for crime whereof the party shall have been duly convicted, shall exist within the United States, or any place subject to their jurisdiction.

Section 2.

Congress shall have power to enforce this article by appropriate legislation.

Amendment XIV

[July 28, 1868]

Section 1.

All persons born or naturalized in the United States, and subject to the jurisdiction thereof, are citizens of the United States and of the State wherein they reside. No State shall make or enforce any law which shall abridge the privileges or immunities of citizens of the United States; nor shall any State deprive any person of life, liberty, or property, without due process of law; nor deny to any person within its jurisdiction the equal protection of the laws.

Section 2.

Representatives shall be apportioned among the several States according to their respective numbers, counting the whole number of persons in each State, excluding Indians not taxed. But when the right to vote at any election for the choice of electors for President and Vice President of the United States, representatives in Congress, the executive and judicial officers of a State, or the members of the legislature thereof, is denied to any of the male inhabitants of such State, being twenty-one years of age, and citizens of the United States, or in any way abridged, except for participating in rebellion, or other crime, the basis of representation there shall be reduced in the proportion which the number of such male citizens shall bear to the whole number of male citizens twenty-one years of age in such State.

Section 3.

No person shall be a senator or representative in Congress, or elector of President and Vice President, or hold any office, civil or military, under the United

States, or under any State, who having previously taken an oath, as a member of Congress, or as an officer of the United States, or as a member of any State legislature, or as an executive or judicial officer of any State, to support the Constitution of the United States, shall have engaged in insurrection or rebellion against the same, or given aid or comfort to the enemies thereof. But Congress may by a vote of two thirds of each House, remove such disability.

Section 4.

The validity of the public debt of the United States, authorized by law, including debts incurred for payment of pensions and bounties for services in suppressing insurrection or rebellion; shall not be questioned. But neither the United States nor any State shall assume or pay any debt or obligation incurred in aid of insurrection or rebellion against the United States, or any claim for the loss or emancipation of any slave; but all such debts, obligations, and claims shall be held illegal and void.

Section 5.

The Congress shall have the power to enforce, by appropriate legislation, the provisions of this article.

Amendment XV

[March 30, 1870]

Section 1.

The right of citizens of the United States to vote shall not be denied or abridged by the United States or by any State on account of race, color, or previous condition of servitude.

Section 2.

The Congress shall have power to enforce this article by appropriate legislation.

Amendment XVI

[February 25, 1913]

The Congress shall have power to lay and collect taxes on incomes, from whatever source derived, without apportionment among the several States, and without regard to any census or enumeration.

Amendment XVII

[May 31, 1913]

The Senate of the United States shall be composed of two senators from each State, elected by the people thereof, for six years; and each senator shall have one vote. The electors in each State shall have the qualifications requisite for electors of the most numerous branch of the State legislature.

When vacancies happen in the representation of any State in the Senate, the executive authority of such State shall issue writs of election to fill such vacancies: *Provided*, That the legislature of any State may empower the executive thereof to make temporary appointments until the people fill the vacancies by election as the legislature may direct.

This amendment shall not be so construed as to affect the election or term of any senator chosen before it becomes valid as part of the Constitution.

Amendment XVIII[11]

[January 29, 1919]

After one year from the ratification of this article, the manufacture, sale, or transportation of intoxicating liquors within, the importation thereof into, or the exportation thereof from the United States and all territory subject to the jurisdiction thereof for beverage purposes is thereby prohibited.

The Congress and the several States shall have concurrent power to enforce this article by appropriate legislation.

This article shall be inoperative unless it shall have been ratified as an amendment to the Constitution by the legislatures of the several States, as provided in the constitution, within seven years from the date of the submission hereof to the States by Congress.

Amendment XIX

[August 26, 1920]

The right of citizens of the United States to vote shall not be denied or abridged by the United States or by any State on account of sex.

Congress shall have the power to enforce this article by appropriate legislation.

Amendment XX

[January 23, 1933]

Section 1.

The terms of the President and Vice President shall end at noon on the 20th day of January and the terms of Senators and Representatives at noon on the 3d day of January, of the years in which such terms would have ended if this article had not been ratified; and the terms of their successors shall then begin.

Section 2.

The Congress shall assemble at least once in every year, and such meeting shall begin at noon on the 3d day of January, unless they shall by law appoint a different day.

Section 3.

If, at the time fixed for the beginning of the term of President, the President-elect shall have died, the Vice President-elect shall become President. If a President shall not have been chosen before the time fixed for the beginning of his term, or if the President-elect shall have failed to qualify, then the Vice President-elect shall act as President until a President shall have qualified; and the Congress may by law provide for the case wherein neither a President-elect nor a Vice President-elect shall have qualified, declaring who shall then act as President, or the manner in which one who is to act shall be selected, and such person shall act accordingly until a President or Vice President shall have qualified.

Section 4.

The Congress may by law provide for the case of the death of any of the persons from whom, the House of Representatives may choose a President whenever the right of choice shall have devolved upon them, and for the case of the death of any of the persons from whom the Senate may choose a Vice President whenever the right of choice shall have devolved upon them.

Section 5.

Sections 1 and 2 shall take effect on the 15th day of October following the ratification of this article.

Section 6.

This article shall be inoperative unless it shall have been ratified as an amendment to the Constitution by the legislatures of three-fourths of the several States within seven years from the date of its submission.

Amendment XXI

[December 5, 1933]

Section 1.

The Eighteenth Article of amendment to the Constitution of the United States is hereby repealed.

Section 2.

The transportation or importation into any State, Territory, or possession of the United States for delivery or use therein of intoxicating liquors in violation of the laws thereof, is hereby prohibited.

Section 3.

This article shall be inoperative unless it shall have been ratified as an amendment to the Constitution by conventions in the several States, as provided in the Constitution, within seven years from the date of the submission thereof to the States by the Congress.

Amendment XXII

[March 1, 1951]

No person shall be elected to the office of the President more than twice, and no person who has held the office of President, or acted as President, for more than two years of a term to which some other person was elected President shall be elected to the office of the President more than once.

But this article shall not apply to any person holding the office of President when this article was proposed by the Congress, and shall not prevent any person who may be holding the office of President, or acting as President during the term within which this article becomes operative from holding the office of President or acting as President during the remainder of such term.

This article shall be inoperative unless it shall have been ratified as an amendment to the Constitution by the legislatures of three-fourths of the several States within seven years from the date of its submission to the States by the Congress.

Amendment XXIII

[March 29, 1961]

Section 1.

The District constituting the seat of Government of the United States shall appoint in such manner as the Congress may direct.

A number of electors of President and Vice President equal to the whole number of Senators and Representatives in Congress to which the District would be entitled if it were a state, but in no event more than the least populous State; they shall be in addition to those appointed by the States, but they shall be considered, for the purposes of the election of President and Vice President, to be electors appointed by a State; and they shall meet in the District and perform such duties as provided by the twelfth article of amendment.

Section 2.

The Congress shall have power to enforce this article by appropriate legislation.

Amendment XXIV

[January 23, 1964]

Section 1.

The right of citizens of the United States to vote in any primary or other election for President or Vice President, for electors for President or Vice President or for Senator or Representative in Congress, shall not be denied or abridged by the United States or any State by reason of failure to pay any poll tax or other tax.

Section 2.

The Congress shall have power to enforce this article by appropriate legislation.

Amendment XXV

[February 10, 1967]

Section 1.

In case of the removal of the President from office or of his death or resignation, the Vice President shall become President.

Section 2.

Whenever there is a vacancy in the office of the Vice president, the President shall nominate a Vice President who shall take office upon confirmation by a majority of both Houses of Congress.

Section 3.

Whenever the President transmits to the President pro tempore of the Senate and the Speaker of the House of Representatives his written declaration that he is unable to discharge the powers and duties of his office, and until he transmits to them a written declaration to the contrary, such powers and duties shall be discharged by the Vice President as Acting President.

Section 4.

Whenever the Vice president and a majority of either the principal officers of the executive departments or of such other body as Congress may by law provide, transmit to the President pro tempore of the Senate and the Speaker of the House of Representatives their written declaration that the President is unable to discharge the powers and duties of his office, the Vice President shall immediately assume the powers and duties of the office as Acting President.

Thereafter, when the President transmits to the President pro tempore of the Senate and the Speaker of the House of Representatives his written declaration that no inability exists, he shall resume the powers and duties of his office unless the Vice President and a majority of either the principal officers of the executive departments or of such other body as Congress may by law provide, transmit within four days to the President pro tempore of the Senate and the Speaker of the House of Representatives their written declaration that the President is unable to discharge the powers and duties of his office. Thereupon Congress shall decide the issue, assembling within forty-eight hours for that purpose if not in session. If the Congress, within twenty-one days after receipt of the latter written declaration, or, if Congress is not in session, within twenty-one days after Congress is required to assemble, determines by two-thirds vote of both Houses that the President is unable to discharge the powers and duties of his office, the Vice President shall continue to discharge the same as Acting President otherwise, the President shall resume the powers and duties of his office.

Amendment XXVI

[June 30, 1971]

Section 1.

The right of citizens of the United States who are eighteen years of age or older to vote shall not be denied or abridged by the United States or by any State on account of age.

Section 2.

The Congress shall have power to enforce this article by appropriate legislation.

Notes

1. See the Sixteenth Amendment.
2. See the Fourteenth Amendment.
3. See the Seventeenth Amendment.
4. See the Seventeenth Amendment.

5. See the Sixteenth Amendment.
6. Superseded by the Twelfth Amendment.
7. See the Eleventh Amendment.
8. See the Fourteenth Amendment, Sec. 1.
9. See the Thirteenth Amendment.
10. See the Fourteenth Amendment, Sec. 4.
11. Repealed by the Twenty-first Amendment.

APPENDIX 3

Federalist 10 (1787–1788) James Madison

November 22, 1787

To the People of the State of New York

Among the numerous advantages promised by a well-constructed union, none deserves to be more accurately developed than its tendency to break and control the violence of faction. The friend of popular governments never finds himself so much alarmed for their character and fate as when he contemplates their propensity to this dangerous vice. He will not fail, therefore, to set a due value on any plan which, without violating the principles to which he is attached, provides a proper cure for it. The instability, injustice, and confusion introduced into the public councils have, in truth, been the mortal diseases under which popular governments have everywhere perished, as they continue to be the favorite and fruitful topics from which the adversaries to liberty derive their most specious declamations. The valuable improvements made by the American constitutions on the popular models, both ancient and modern, cannot certainly be too much admired; but it would be an unwarrantable partiality to contend that they have as effectually obviated the danger on this side, as was wished and expected. Complaints are everywhere heard from our most considerate and virtuous citizens, equally the friends of public and private faith and of public and personal liberty; that our governments are too unstable; that the public good is disregarded in the conflicts of rival parties; and that measures are too often decided not according to the rules of justice and the rights of the minor party, but by the superior force of an interested and overbearing majority. However anxiously we may wish that these complaints had no foundation, the evidence of known facts will not permit us to deny that they are in some degree true. It will be found, indeed, on a candid review of our situation, that some of the distresses under which we labor have been erroneously charged on the operation of our governments; but it will be found, at the same time, that other causes will not alone account for many of our heaviest misfortunes and, particularly, for that prevailing and increasing distrust of public engagements and alarm for private rights which are echoed from one end of the continent to the others. These must chiefly, if not wholly, effects of the unsteadiness and injustice with which a factious spirit has tainted our public administrations.

By a faction I understand a number of citizens, whether amounting to a majority or minority of the whole, who are united and actuated by some common impulse of passion, or of interest, adverse to the rights of other citizens, or to the permanent and aggregate interests of the community.

There are two methods of curing the mischiefs of faction: the one, by removing its causes; the other, by controlling its effects.

There are again two methods of removing the causes of faction: the one, by destroying the liberty which is essential to its existence; the other, by giving to every citizen the same opinions, the same passions, and the same interests.

It could never be more truly said than of the first remedy that it was worse than the disease. Liberty is to faction what air is to fire, an aliment without which it instantly expires. But it could not be a less folly to abolish liberty, which is essential to political life, because it nourishes faction than it would be to wish the annihilation of air, which is essential to animal life, because it imparts to fire its destructive agency.

The second expedient is as impracticable as the first would be unwise. As long as the reason of man continues fallible and he is at liberty to exercise it, different opinions will be formed. As long as the connection subsists between his reason and his self-love, his opinions and his passions will have a reciprocal influence on each other; and the former will be objects to which the latter will attach themselves. The diversity in the faculties of men, from which the rights of property originate, is not less an insuperable obstacle to a uniformity of interests. The protection of these faculties is the first object of government. From the protection of different and unequal faculties of acquiring property, the possession of different degrees and kinds of property immediately results; and from the influence of these on the sentiments and views of the respective proprietors ensues a division of the society into different interests and parties.

The latent causes of faction are thus sown in the nature of man; and we see them everywhere brought into different degrees of activity, according to the different circumstances of civil society. A zeal for different opinions concerning religion, concerning government and many other points, as well of speculation as of practice; an attachment to different leaders ambitiously contending for pre-eminence and power; or to persons of other descriptions whose fortunes have been interesting to the human passions, have, in turn, divided mankind into parties, inflamed them with mutual animosity, and rendered them much more disposed to vex and oppress each other than to co-operate for their common good. So strong is this propensity of mankind to fall into mutual animosities, that where no substantial occasion presents itself, the most frivolous and fanciful distinctions have been sufficient to kindle their unfriendly passions and excite their most violent conflicts. But the most common and durable source of factions has been the various and unequal distribution of property. Those who hold and those who are without property have ever formed distinct interests in society. Those who are creditors, and those who are debtors, fall under a like discrimination. A landed interest, a manufacturing interest, a mercantile interest, a moneyed interest, with many lesser interests, grow up of necessity in civilized nations and divide them into different classes, actuated by different sentiments and views. The regulation of these various and interfering interests forms the principal task of modern legislation and involves the spirit of party and faction in the necessary and ordinary operations of government.

No man is allowed to be a judge in his own cause, because his interest would certainly bias his judgment, and, not improbably, corrupt his integrity. With equal, nay with greater reason, a body of men are unfit to be both judges and parties at the same time; yet what are many of the most important acts of legislation but so many judicial determinations, not indeed concerning the rights of single persons, but concerning the rights of large bodies of citizens? And what are the different classes of legislators but advocates and parties to the causes which they determine? Is a law proposed concerning private debts? It is a question to which the creditors are parties on one side and the debtors on the other. Justice ought to hold the balance between them. Yet the parties are, and must be, themselves the judges; and the most numerous party, or in other words, the most powerful faction must be expected to prevail. Shall domestic manufacturers be encouraged, and in what degree, by restrictions on foreign manufacturers? are questions which would be differently decided by the landed and the manufacturing classes, and probably by neither with a sole regard to justice and the public good. The apportionment of taxes on the various descriptions of property is an act which seems to require the most exact impartiality; yet there is perhaps, no legislative act in which greater opportunity and temptation are given to a predominant party to trample on the rules of justice. Every shilling with which they over-burden the inferior number is a shilling saved to their own pockets.

It is in vain to say that enlightened statesmen will be able to adjust these clashing interests and render them all subservient to the public good. Enlightened statesmen will not always be at the helm. Nor, in many cases, can such an adjustment be made at all without taking into view indirect and remote considerations, which will rarely prevail over the immediate interest which one party may find in disregarding the rights of another or the good of the whole.

The inference to which we are brought is that the *causes* of faction cannot be removed and the relief is only to be sought in the means of controlling its *effects*.

If a faction consists of less than a majority, relief is supplied by the republican principle, which enables the majority to defeat its sinister views by regular vote. It many clog the administration, it may convulse the society; but it will be unable to execute and mask its violence under the forms of the Constitution. When a majority is included in a faction, the form of popular government, on the other hand, enables it to sacrifice to its ruling passion or interest, both the public good and the rights of other citizens. To secure the public good, and private rights against the danger of such a faction, and at the same time to preserve the spirit and the form of popular government, is then the great object to which our inquiries are directed. Let me add that it is the great desideratum by which alone this form of government can be rescued from the opprobrium under which it has so long labored and be recommended to the esteem and adoption of mankind.

By what means is this object attainable? Evidently by one of two only. Either the existence of the same passion or interest in a majority at the same time must be prevented, or the majority, having such coexistent passion or interest, must be rendered, by their number and local situation, unable to concert and carry into effect schemes of oppression. If the impulse and the opportunity be suffered to coincide, we well know that neither moral nor religious motives can be relied on as an adequate control. They are not found to be such on the injustice and violence of individuals, and lose their efficacy in proportion to the number combined together, that is, in proportion as their efficacy becomes needful.

From this view of the subject it may be concluded that a pure democracy, by which I mean a society consisting of a small number of citizens who assemble and administer the government in person, can admit of no cure for the mischiefs of faction. A common passion or interest will, in almost every case, be felt by a majority of the whole; a communication and concert results from the form of government itself; and there is nothing to check the inducements to sacrifice the weaker party or an obnoxious individual. Hence it is that such democracies have ever been spectacles of turbulence and contention; have ever been found incompatible with personal security or the rights of property; and have in general been as short in their lives as they have been violent in their deaths. Theoretic politicians, who have patronized this species of government, have erroneously supposed that by reducing mankind to a perfect equality in their political rights, they would at the same time be perfectly equalized and assimilated in their possessions, their opinions, and their passions.

A republic, by which I mean a government in which the scheme of representation takes place, opens a different prospect and promises the cure for which we are seeking. Let us examine the points in which it varies from pure democracy, and we shall comprehend both the nature of the cure and the efficacy which it must derive from the Union.

The two great points of difference between a democracy and a republic are: first, the delegation of the government, in the latter, to a small number of citizens elected by the rest; secondly, the greater number of citizens and greater sphere of country over which the latter may be extended.

The effect of the first difference is, on the one hand, to refine and enlarge the public views by passing them through the medium of a chosen body of citizens, whose wisdom may best discern the true interest of their country and whose patriotism and love of justice will be least likely to sacrifice it to temporary or partial considerations. Under such a regulation, it may well happen that the public voice, pronounced by the representatives of the people, will be more consonant to the public good than if pronounced by the people themselves, convened for that purpose. On the other hand, the effect may be inverted. Men of factious tempers, of local prejudices, or of sinister designs, may, by intrigue, by corruption, or by other means, first obtain the suffrages, and then betray the interests of the people. The question resulting is, whether small or extensive republics are most favorable to the election of proper guardians of the public weal; and it is clearly decided in favor of the latter by two obvious considerations.

In the first place it is to be remarked that however small the republic may be, the representatives must be raised to a certain number in order to guard against the cabals of a few; and that however large it may be, they must be limited to a certain number in order to guard against the confusion of a multitude. Hence, the number of representatives in the two cases not being in proportion to that of the constituents, and being proportionally greatest in the small republic, it follows that if the proportion of fit characters be not less in the large than in the small republic, the former will present a greater option, and consequently a greater probability of a fit choice.

In the next place, as each representative will be chosen by a greater number of citizens in the large than in the small republic, it will be more difficult for unworthy candidates to practice with success the vicious arts by which elections are too often carried; and the suffrages of the people being more free, will be more likely to center on men who possess the most attractive merit and the most diffusive and established characters.

It must be confessed that in this, as in most other cases, there is a mean, on both sides of which inconveniencies will be found to lie. By enlarging too much the number of electors, you render the representative too little acquainted with all their local circumstances and lesser interests; as by reducing it too much, you render him unduly attached to these, and too little fit to comprehend and pursue great and national objects. The federal Constitution forms a happy combination in this respect; the great and aggregate interests being referred to the national, the local and particular to the State legislatures.

The other point of difference is the greater number of citizens and extent of territory which may be brought within the compass of republican than of democratic government; and it is this circumstance principally which renders factious combinations less to be dreaded in the former than in the latter. The smaller the society, the fewer probably will be the distinct parties and interests composing it; the fewer the distinct parties and interests, the more frequently will a majority be found of the same party; and the smaller the number of individuals composing a majority, and the small the compass within which they are placed, the more easily will they concert and execute their plans of oppression. Extend the sphere and you take in a greater variety of parties and interests; you make it less probably that a majority of the whole will have a common motive to invade the rights of other citizens; or if such a common motive exists, it will be more difficult for all who feel it to discover their own strength and to act in unison with each other. Besides other impediments, it may be remarked that, where there is a consciousness of unjust or dishonorable purposes, communication is always checked by distrust in proportion to the number whose concurrence is necessary.

Hence, it clearly appears that the same advantage which a republic has over a democracy in controlling the effects of faction is enjoyed by a large over a small republic—is enjoyed by the Union over the States composing it. Does this advantage consist in the substitution of representatives whose enlightened views and virtuous sentiments render them superior to local prejudices and to schemes of injustice? It will not be denied that the representation of the Union will be most likely to possess these requisite endowments. Does it consist in the greater security afforded by a greater variety of parties, against the event of any one party being able to outnumber and oppress the rest? In an equal degree does the increased variety of parties comprised within the Union increase this security? Does it, in fine, consist in the greater obstacles opposed to the concert and accomplishment of the secret wishes of an unjust and interested majority? Here again the extent of the Union gives it the most palpable advantage.

The influence of factious leaders may kindle a flame within their particular States but will be unable to spread a general conflagration through the other States. A religious sect may degenerate into a political faction in a part of the Confederacy; but the variety of sects dispersed over the entire face of it must secure the national councils against any danger from that source. A rage for paper money, for an abolition of debts, for an equal division of property, or for any other improper or wicked project, will be less apt to pervade the whole body of the Union than a particular member of it; in the same proportion as such a malady is more likely to taint a particular county or district than an entire State.

In the extent and proper structure of the Union, therefore, we behold a republican remedy for the diseases most incident to republican government. And according to the degree of pleasure and pride we feel in being republicans ought to be our zeal in cherishing the spirit and supporting the character of federalists.

APPENDIX 4

Federalist 51
James Madison

February 6, 1788
To the People of the State of New York

To what expedient, then, shall we finally resort for maintaining in practice the necessary partition of power among the several departments, as laid down in the Constitution? The only answer that can be given is that as all these exterior provisions are found to be inadequate, the defect must be supplied by so contriving the interior structure of the government as that its several constituent parts may, by their mutual relations, be the means of keeping each other in their proper places. Without presuming to undertake a full development of this important idea I will hazard a few general observations which may perhaps place it in a clearer light, and enable us to form a more correct judgment of the principles and structure of the government planned by the convention.

In order to lay a due foundation for that separate and distinct exercise of the different powers of government, which to a certain extent is admitted on all hands to be essential to the preservation of liberty, it is evident that each department should have a will of its own; and consequently should be so constituted that the members of each should have as little agency as possible in the appointment of the members of the others. Were this principle rigorously adhered to, it would require that all the appointments for the supreme executive, legislative, and judiciary magistracies should be drawn from the same fountain of authority, the people, through channels having no communication whatever with one another. Perhaps such a plan of constructing the several departments would be less difficult in practice than it may in contemplation appear. Some difficulties, however, and some additional expense would attend the execution of it. Some deviations, therefore, from the principle must be admitted. In the constitution of the judiciary department in particular, it might be inexpedient to insist rigorously on the principle first, because peculiar qualifications being essential in the members, the primary consideration ought to be to select that mode of choice which best secures these qualifications, second, because the permanent tenure by which the appointments are held in that department must soon destroy all sense of dependence on the authority conferring them.

It is equally evident that the members of each department should be as little dependent as possible on those of the others for the emoluments annexed to their offices. Were the executive magistrate, or the judges, not independent of the legislature in this particular, their independence in every other would be merely nominal.

But the great security against a gradual concentration of the several powers in the same department consists in giving to those who administer each department the necessary constitutional means and personal motives to resist encroachments of the others. The provision for defense must in this, as in all other cases, be made commensurate to the danger of attack. Ambition must be made to counteract ambition. The interest of the man must be connected with the constitutional rights of the place. It may be a reflection on human nature that such devices should be necessary to control the abuses of government. But what is government itself but the greatest of all reflections on human nature? If men were angels, no government would be necessary. If angels were to govern men, neither external nor internal controls on government would be necessary. In framing a government which is to be administered by men over men, the great difficulty lies in this: You must first enable the government to control the governed; and in the next place, oblige it to control itself. A dependence on the people is, no doubt, the primary control on the government; but experience has taught mankind the necessity of auxiliary precautions.

This policy of supplying, by opposite and rival interests, the defect of better motives, might be traced through the whole system of human affairs, private as well as public. We see it particularly displayed in all the subordinate distributions of power, where the constant aim is to divide and arrange the several offices in such a manner as that each may be a check on the other—that the private interest of every individual may be a sentinel over the public rights. These inventions of prudence cannot be less requisite in the distribution of the supreme powers of the State.

But it is not possible to give to each department an equal power of self defense. In republican government, the legislative authority necessarily predominates. The remedy for this inconveniency is to divide the legislature into different branches and to render them, by different modes of election and different principles of action, as little connected with each other as the nature of their common functions and their common dependence on the society will admit. It may even be necessary to guard against dangerous encroachments by still further precautions. As the weight of the legislative authority requires that it should be thus divided, the weakness of the executive may require, on the other hand, that it should be fortified. An absolute negative on the legislature appears, at first view, to be the natural defense with which the executive magistrate should be armed. But perhaps it would be neither altogether safe nor alone sufficient. On ordinary occasions it might not be exerted with the requisite firmness, and on extraordinary occasions it might be perfidiously abused. May not this defect of an absolute negative be supplied by some qualified connection between this weaker department and the weaker branch of the stronger department, by which the latter may be led to support the constitutional rights of the former, without being too much detached from the rights of its own department?

If the principles on which these observations are founded be just, as I persuade myself they are, and they be applied as a criterion to the several State constitutions, and to the federal Constitution, it will be found that if the latter does not perfectly correspond with them, the former are infinitely less able to bear such a test.

There are, moreover, two considerations particularly applicable to the federal system of America, which place that system in a very interesting point of view.

First. In a single republic, all the power surrendered by the people is submitted to the administration of a single government, and usurpations are guarded against by a division of the government into distinct and separate departments. In the compound republic of America, the power surrendered by the people is first divided between two distinct governments, and then the portion allotted to each subdivided among distinct and separate departments. Hence a double security arises to the rights of the people. The different governments will control each other, at the same time that each will be controlled by itself.

Second. It is of great importance in a republic not only to guard the society against the oppression of its rulers, but to guard one part of the society against the injustice of the other part. Different interests necessarily exist in different classes of citizens. If a majority be united by a common interest, the rights of the minority will be insecure. There are but two methods of providing against the evil: the one by creating a will in the community independent of the majority—that is, of the society itself; the other, by comprehending in the society so many separate descriptions of citizens as will render an unjust combination of a majority of the whole very improbable, if not impracticable. The first method prevails in all governments possessing an hereditary or self-appointed authority. This, at best, is but a precarious security; because a power independent of the society may as well espouse the unjust views of the major as the rightful interests of the minor party, and may possibly be turned against both parties. The second method will be exemplified in the federal republic of the United States. Whilst all authority in it will be derived from and dependent on the society, the society itself will be broken into so many parts, interests and classes of citizens, that the rights of individuals, or of the minority, will be in little danger from interested combinations of the majority. In a free government the security for civil rights must be the same as that for religious rights. It consists in the one case in the multiplicity of interests, and in the other in the multiplicity of sects. The degree of security in both cases will depend on the number of interests and sects; and this may be presumed to depend on the extent of country and number of people comprehended under the same government. This view of the subject must particularly recommend a proper federal system to all the sincere and considerate friends of republican government, since it shows that in exact proportion as the territory of the Union may be formed into more circumscribed Confederacies, or States, oppressive combinations of a majority will be facilitated; the best security, under the republican forms, for the rights of every class of citizen, will be diminished; and consequently the stability and independence of some member of the government, the only other security, must be proportionally increased. Justice is the end of government. It is the end of civil society. It ever has been and ever will be pursued until it be obtained, or until liberty be lost in the pursuit. In a society under the forms of which the stronger faction can readily unite and oppress the weaker, anarchy may as truly be said to reign as in a state of nature where the weaker individual is not secured against the violence of the stronger; and as, in the latter state, even the stronger individuals are prompted, by the uncertainly of their condition, to submit to a government which may protect the weak as well as themselves; so, in the former state, will be more powerful factions or parties be gradually induced, by the like motive, to wish for a government which will protect all parties, the weaker as well as the more powerful. It can be little doubted that if the State of Rhode Island was separated from the Confederacy and left to itself, the insecurity of rights under the popular form of government within such narrow limits would be displayed by such reiterated oppressions of factious majorities that some power altogether independent of the people would soon be called for by the voice of the very factions whose misrule had proved the necessity of it. In the extended republic of the United States, and among the great variety of interests, parties, and sects which it embraces, a coalition of a majority of the whole society could seldom take place on any other principles than those of justice and the general good; whilst there being thus less danger to a minor from the will of the major party, there must be less pretext, also, to provide for the security of the former, by introducing into the government a will not dependent on the latter, or, in other words, a will independent of the society itself. It is no less certain than it is important, notwithstanding the contrary opinions which have been entertained, that the larger the society, provided it lie within a practicable sphere, the more duly capable it will be of self-government. And happily for the *republican cause*, the practicable sphere may be carried for a very great extent by a judicious modification and mixture of the *federal principle*.

Publius

GLOSSARY

Alien and Sedition Acts: Collective name given to four acts passed in 1798 by Congress that restricted freedom of speech and the liberty of foreigners living in the U.S.

American exceptionalism: Attitude attributed to Americans suggesting that they are somehow unique or exceptional to other political societies. Allows them make their own set of rules that are not bound by convention or even world experience.

Anarchy: Literally "rule of no one." Mob rule. Chaotic lawlessness due to absence of governmental control.

Anti-Federalists: Opponents of a strong national government who opposed ratification of the Constitution.

Articles of Confederation: Document which set up a loose confederation of states that comprised the first national government of the US from 1781 to 1788.

Aristocracy: Rule by a small group of a society's "best" citizens, the selection of which may be based on birth, wealth, ability, social standing or ecclesiastical position.

Bill of Attainder: Legislative act that authorizes the detention of a person, declares them guilty and imposes a punishment without a judicial trial.

Calling: A Puritan belief that by pursuing worldly careers virtuously, a person was in effect glorifying God.

Calvin, John: Wrote *Institutes of the Christian Religion* in which he advocated purification of the church. His theology was the foundation of the Puritan religion.

Civil Liberty: Also known as moral self-governance; the freedom to do what is right.

Common Law: Body of law developed in England by judicial decisions over time. Based on custom and precedent, it constitutes the basis of the English legal system.

Constitution: A written or otherwise recognized agreement within a polity setting forth principles and laws upon which a government will be structured and guaranteeing certain rights to the people.

Counterpoise: Control mechanism by which the self-interest of one group is balanced against that of another.

Covenant: A formal agreement among the members of a political or religious group which functions as a founding document.

Deists: Those who subscribe to a school of religious thought which emphasizes that a basic morality is inborn in every person which can be accessed by the use of reason and a belief in a Creator who is the rational architect of an ordered universe.

Demagogue: A leader who gains power by appealing to popular passions and prejudices, and by championing the cause of the common people.

Democracy: Governmental system in which ultimate political authority lies with the people, through direct participation.

Despotism: Oppressive rule by a single individual.

Divine Right of Kings: Theory that legitimized the power of kings over the people by holding that kings inherited their authority directly from God.

English Bill of Rights: Incorporated in 1689, gave civil and political rights to the people and political supremacy to Parliament.

Ex Post Facto: Literal translation means "after the fact". When discussed in conjunction with law, refers to making something a crime and holding a person accountable to it after the action has already occurred.

Farewell Address: Given by George Washington in September, 1796. Influenced by Alexander Hamilton, it cautioned against political parties and foreign entanglements.

Federalism: The division of power between the national and state government. Both act directly upon the people, are supreme within their proper spheres of authority, and must consent to constitutional change.

The Federalist: Collection of eight-five essays written by James Madison, Alexander Hamilton and John Jay, published under the name "Publius." It was the deepest and most searching inquiry into republican government ever written.

Federalists: Supporters of a strong national government who favored ratification of the Constitution.

Founding Myth: Imparts a sense of legitimacy to a political society by identifying the founding as sanctioned by the gods, by nature, by truth and justice, or by history.

Good Society: Escapes the human predicament by providing benefits for its citizens.

Great Oughts: Certain rights drafted into the Constitution in rhetorical rather than legal language. Meant to be shining standards of rights that people *ought to have*, and were to be used as rallying points for citizens. In the US Constitution they comprise freedom of expression, freedom of conscience and the right to privacy.

Implied Rights: Rights not specifically identified in the Constitution that have been inferred from stated rights.

Jay, John: Prominent New York statesman and first Chief Justice of the Supreme Court. His participation in writing the essays in *The Federalist* was limited because of ill health.

Jay's Treaty: Negotiated with Britain in 1794. The US made major concessions to avert a war over the seizure of US ships. The terms were extremely unfavorable to US interests.

Judicial Review: Power of the court to declare laws or acts of the legislative or executive branches unconstitutional. The Constitution is supreme law, acts contrary to it are unconstitutional, and thus null and void.

Kentucky and Virginia Resolutions: Response sponsored by Thomas Jefferson and James Madison to the Alien and Sedition Acts. In 1798, they argued that the states created the federal union and retained the right to declare acts of the federal government null and void if not deemed beneficial or benevolent.

Lawgiver: One who divines the laws of society and sets them forth, endowing the law with a special sanctity.

Laws of Nature: Refers to the notion that there are laws that govern the entire universe giving it order and purpose. These laws are presumed to be accessible to all human beings, endowing them with certain unalienable rights. (See Natural Law.)

Libel: The publication of knowingly false or malicious statements that damage someone's reputation.

Liberalism: Philosophy emphasizing individual freedom from several forms of restraint, including government.

Loyal opposition: Philosophical stance citizens take when they oppose something the government has done or is doing, yet remain loyal to the form of government itself.

Marbury v. Madison: Supreme Court decision of 1803 that created the precedent of "judicial review."

Mercantilism: British economic system that sought the self-sufficiency of Great Britain through favorable balances of trade with colonies and other imperialist policies. (See study aid for chapter 4.)

Monarchy: Rule by one, usually a king or queen.

Nationalists: Not a recognized group, but a label applied by historians to supporters of a strong central government.

Natural Law: The notion that there are laws that govern the entire universe giving it order and purpose. Natural Laws are the moral equivalent to physical laws like gravity.

Natural Liberty: Freedom to do whatever one desires.

Nature: The transcendent entity that gives order to the universe.

New Jersey/Paterson Plan: The small states' rebuttal to the Virginia Plan in which states would retain sovereignty and representation would be equal for each state regardless of population.

Oligarchy: Oppressive rule by an elite few.

Particular Founding: Correct only at a given time, in a given place and for a given people. (See study aid for chapter 1.)

Patria: The concept of a fatherland that evokes patriotic feelings and that stands above a particular government or political organization of society.

Patronage: A system whereby one person in society is dependent upon some higher person—his patron—for many of the opportunities of life. Operates on the basis of favors and informal understandings.

Polity: A group of people who organize themselves into a political society.

Popular Government: Based on the natural rights concept that ultimate power rests with the people who can create, alter, or abolish government. Differs from a democracy in that popular government avoids direct contact with the people, opting instead for representative institutions operating as filters between the people and the government.

Predestination: Doctrine that God, in consequence of his foreknowledge of all events, infallibly guides those who are destined for salvation or damnation.

Puritanism: A religious movement and culture based on Calvinistic doctrine and characterized by simplicity and austere living. This was done in an effort to "purify" the Church of England. (See study aid for chapter 2.)

Representation: A system where selected leaders represent a larger group of citizens.

Republic: A government in which supreme power resides in a body of citizens entitled to vote, and which is exercised by elected representatives responsible to the citizens and who govern according to law.

Republicanism: A political philosophy derived from ideas associated with classical antiquity which recognized the ultimate political power of society as being vested in the "people" and which strove to balance the inherent interests within society in order to produce a stable and effective government.

Rule-mindedness: A characteristic need in humans to create order by making and obeying rules.

Rule of Law: In the private sphere of human activities, the government may not act except in the enforcement of a known, general rule. For the rule of law to exist, five elements must be present: Generality, Prospectivity, Publicity, Consent and Due Process.

Scholasticism: A mode of inquiry that employs human reason to reach conclusions about theological questions.

Sedition: Actions or criticisms that incite rebellion or discontent against established government.

Seditious libel: Political doctrine which held that the government had the right to punish those who spoke against it with malice.

Separation of Powers: The structural design in government that places rule-making and rule-enforcing powers in separate hands. The officials of each branch are selected by different procedures, have different terms of office, and are independent of one another. Designed to prevent tyranny.

Sovereignty: The power to make final or ultimate decisions, especially in regards to government.

State of nature: Hypothetical existence in which all humans lived in perfect freedom without society or government. All had the same rights granted by Nature and therefore enjoyed a basic equality.

Statesmanship: A quality necessary to some degree for republican leaders. Characterized by virtue (areté) as exhibited by putting public interests above private, showing enormous personal integrity, and exercising wise, practical, far-sighted leadership.

Stretch-points: Portions of the Constitution written ambiguously so as to leave them open to various interpretations about how far the powers of the government would extend and into what areas.

Structure: An overall organizational system composed of structural controls, using the predominance of self-interest to channel human nature toward a desirable outcome.

Timocracy: Government in which love of honor is the ruling principle.

Tocqueville, Alexis de: French statesman and author. His two-volume work, *Democracy in America* (1835), makes detailed observations about American society and politics.

Tribalism: One aspect of human nature. Reflects human desire to be with their own kind. Creates a sense of community and common purpose.

Tyranny: Rule of will. The use of the power of government to further the ends of a person or group.

Universal Founding: Correct for all people at all times and in all situations.

Utopia: A futuristic place of ideal perfection (especially in laws, government, and social conditions).

Virginia Plan/Resolves: Plan backed by the nationalists and large states at the Constitutional Convention, which proposed creating one large, unified republic in which representation would be decided by population.

Winthrop, John: Leader of the first Puritan colony in America. His vision of the Puritan colony as a city upon a hill is fundamental to the American character.

Zoon Politikon: Man as a political animal that is naturally interested in government and society.

PHOTOGRAPH CREDITS

Introduction

page 2. *A View of New Orleans Taken from the Plantation of Marigny*, by Boqueto de Woiserie, courtesy of the Chicago Historical Society

Chapter 1

page 8. "The Liberty Bell," courtesy of the Philadelphia Convention and Tourist Bureau.

page 13. "Cambodia: Skulls of Khmers," by Brian Eades, *The Observer*, 1980, courtesy of Hulton-Getty Archives.

page 19. "Theodore Roosevelt," from The Verdict, courtesy of the Harvard College Library, Cambridge, Massachusetts.

page 26. *Colonel and Mrs. James A. Whiteside, Son Charles and Servants*, by James Cameron, courtesy of the Hunter Museum of Art, Chattanooga, Tennessee.

Chapter 2

page 36. *Country Life, Vermont*, courtesy of the Vermont Agency of Development.

page 39. "World's Highest Standard of Living," by Margaret Bourke-White, reprinted from *Life* Magazine, 1937, courtesy of Time, Inc.

page 41. "Scopes Trial," courtesy of Associated Press/Wide World Photos.

page 43. Copperplate engravings from *The Book of Trades*, courtesy of the American Antequarian Society, Worcestor, Massachusetts.

page 44. "Christ and the Rich Young Ruler," by Neinrich Hofmann, courtesy of C. Morrison Conroy Co., Inc., Charlotte, North Carolina.

page 48 *American Progress (Manifest Destiny)*, by John Gast, courtesy of The Granger Collection, New York.

Chapter 3

page 58 *Freedom of Speech*, by Norman Rockwell. Printed by permission of the Norman Rockwell Family Trust. Copyright © 1943 the Norman Rockwell Family Trust.

page 61. *The Course of Empire: Consummation of Empire*, by Thomas Cole, courtesy of the New York Historical Society.

page 67. *Broadway, North from Cortlandt Street*, by J.J. Fogerty, courtesy of the New York Historical Society.

page 69. "Nazi Victims," courtesy of the Library of Congress.

page 76. Courtesy of PhotoDisc, Inc.

Chapter 4

page 88. *Common Sense*, by Thomas Paine, courtesy of the Library of Congress.
page 91. "Skull & Crossbones Stamp," reprinted from the *Pennsylvania Journal and Weekly Advertiser*, October 31, 1765 , courtesy of The Granger Collection, New York.
page 93. "Join or Die," courtesy of The Granger Collection, New York.
page 98. *Lexington*, engraved from a painting by Alonz Chappel, courtesy of the Library of Congress.
page 107. *George Washington*, courtesy of the Library of Congress.
page 111. *Washington Ascending into Heaven*, courtesy of Winterthur Museum, Winterthur, Delaware.

Chapter 5

page 120. Courtesy of the National Archives.
page 125. *Taking of the Bastille*, courtesy of Hulton-Getty.
page 129. *John Adams*, courtesy of the White House Historical Association.
page 133. "Hitler Reviewing Hitler Youth at a Rally," courtesy of the National Archives.
page 134. "Civil Rights Demonstration," by Bill Hudson, courtesy of Associated Press/Wide World Photos.

Chapter 6

page 152. *Against All Odds*, by Karin Draper, reprinted by permission of the artist.
page 157. *James Madison*, courtesy of The Granger Collection, New York.
page 161. *Independence Hall*, courtesy of the Library of Congress.
page 167. *George Washington Presiding at the Convention*, courtesy of the Free Library of Philadelphia.
page 169. *The Artist in His Museum*, by Charles Willson Peale, courtesy of the Pennsylvania Academy of Fine Arts, Philadelphia.

Chapter 7

page 178. *The Ideas of the Constitutional Convention*, by Samuel Paul Baird, reprinted by permission of the artist.
page 187. *Second Street, North from Market Street and Christ Church, Philadelphia*, by William Birch, courtesy of Independence National Historical Park, Philadelphia.
page 190. "Slave Auction Handbill," courtesy of the American Antiquarian Society, Worcestor, Massachusetts.
page 197. "Remember, gentlemen . . . Constitution," by Peter Steiner, reprinted from *The New Yorker*, 1982, courtesy of CartoonBank.com.

Chapter 8

page 206. *Dawn*, by Katy Neale, reprinted by permission of the artist.
page 210. "Suffragettes," courtesy of the Library of Congress.
page 231. *Benjamin Franklin*, by Charles Wilson Peale, courtesy of the American Philosophical Society, Philadelphia.

Chapter 9

page 240. *The American Institution*, by Aimee Morgan, reprinted by permission of the artist.
page 251. Photograph by United Press International, courtesy of Corbis.
page 256. "Am I not a woman and a sister?" courtesy of the Library of Congress.
page 258. *The Cornell Farm*, by Edward Hicks, courtesy of the National Gallery of Art, Washington, D.C.
page 260. "The Inclusive Republic," courtesy of Associated Press/Wide World Photos.
page 263. "Confidence in the Constitution," courtesy of The Granger Collection, New York.

Chapter 10

page 270. *Commemoration of the Constitution*, by Amos Doolittle, courtesy of the John Carter Brown Library, Brown University, Providence, Rhode Island.
page 273. "A Home for Everybody," courtesy of the Library of Congress.
page 276. Symbol for The Philadelphia Society for Promoting Agriculture, courtesy of the Library of Congress.
page 281. *The Federalist Papers*, courtesy of The Granger Collection, New York.
page 282. *Washington's Triumphal Entry into New York City, 1783*, courtesy of the Abby Aldrich Rockefeller Folk Art Center, Williamsburg, Virginia.
page 286. *Philadelphia Election Day*, by John L. Krimmel, courtesy of the Winterthur Museum, Winterthur, Delaware.

Chapter 11

page 294. *We the People*, by Merrilee Allred, reprinted by permission of the artist.
page 296. "Tiananmen Square," by Jeff Widener, courtesy of Associated Press/Wide World Photos.
page 306. Daniel Schorr, couresy of Corbis-Bettman.
page 310. *Bill of Rights*, courtesy of The Granger Collection, New York.

Chapter 12

page 320. *Jefferson, Hamilton & Washington*, by Constantino Brumidi, courtesy of Hulton-Getty.

page 330. *Execution of King Louis XVI on January 21, 1793*, courtesy of Erich Lessing/Art Resource, New York.

page 333. *Monticello*, courtesy of the Library of Congress.

page 335. *George Washington*, by Gilbert Stuart, courtesy of the White House Historical Association.

page 339. "Congressional Pugilists," courtesy of the Collection of the New York Historical Society.

page 341. "Republicans Turn Out . . . ," courtesy of the Collection of the New York Historical Society.

page 344. *USS Constitution*, courtesy of The Granger Collection, New York.

Conclusion

page 354. *Constitution*, by Nancy Layton, reprinted by permission of the artist.

page 357. "Calipatria, California," by Dorthea Lange, courtesy of the Library of Congress.

page 361. "Presidential Campaign, 1840," courtesy of The Granger Collection.

page 363. "Civil War, Gettysburg, Third Day," courtesy of the Library of Congress.

page 367. *Tontine Coffee House*, by Francis Guy, courtesy of the Collection of the New York Historical Society.

page 369. "The Star-Spangled Banner," courtesy of the Smithsonian Institution.

INDEX

Note: Page numbers followed by the letter *f* indicate figures.